The Evolution of Literature

Legacies of Darwin in European Cultures

152

Internationale Forschungen zur
Allgemeinen und
Vergleichenden Literaturwissenschaft

Begründet von Alberto Martino und in Verbindung mit

herausgegeben von

Norbert Bachleitner
(Universität Wien)

Redaktion: Paul Ferstl und Rudolf Pölzer

Anschrift der Redaktion:
Institut für Vergleichende Literaturwissenschaft, Sensengasse 3A , A-1090 Wien

The Evolution of Literature

Legacies of Darwin in European Cultures

Edited by
Nicholas Saul and Simon J. James

Rodopi

Amsterdam - New York, NY 2011

Cover Image: Lucas Swann

Cover design: Pier Post

Le papier sur lequel le présent ouvrage est imprimé remplit les prescriptions de "ISO 9706:1994, Information et documentation - Papier pour documents - Prescriptions pour la permanence".

The paper on which this book is printed meets the requirements of " ISO 9706:1994, Information and documentation - Paper for documents - Requirements for permanence".

Die Reihe "Internationale Forschungen zur Allgemeinen und Vergleichenden Literaturwissenschaft" wird ab dem Jahr 2005 gemeinsam von Editions Rodopi, Amsterdam – New York und dem Weidler Buchverlag, Berlin herausgegeben. Die Veröffentlichungen in deutscher Sprache erscheinen im Weidler Buchverlag, alle anderen bei Editions Rodopi.

From 2005 onward, the series "Internationale Forschungen zur Allgemeinen und Vergleichenden Literaturwissenschaft" will appear as a joint publication by Editions Rodopi, Amsterdam – New York and Weidler Buchverlag, Berlin. The German editions will be published by Weidler Buchverlag, all other publications by Editions Rodopi.

ISBN: 978-90-420-3397-9
E-Book ISBN: 978-94-012-0684-6

Printed in The Netherlands

Contents

Acknowledgements

This book emerges from an interdisciplinary comparative literature conference *The Evolution of Literature: Legacies of Darwin in European Cultures*, which was held at St Mary's College, Durham University, from 4-6 April 2008, and designed to celebrate both the two hundredth birthday of Charles Darwin and the one hundred and fiftieth birthday of his most important work, the *Origin of Species*. It contains not only the papers delivered at that conference but also some of the most distinguished contributions to a public lecture series at Durham, *The Legacies of Charles Darwin*, which preceded the conference in Epiphany and Easter Terms 2007. Both of these events were most generously sponsored by the interdisciplinary Institute of Advanced Study at Durham University. To the Institute, and especially to its directors Ash Amin, Michael O'Neill and Susan Smith, we would like to express our immense gratitude – not only for their financial support, but also for their imagination and encouragement. The list of others to whom the organisers and editors owe their thanks is long. Through its Conference Grant scheme the British Academy provided further indispensable financial support at just the right moment in the project's gestation, and so made possible the attendance of a number of distinguished guest speakers. The Faculty of Arts and Humanities and the School of Modern Languages and Cultures at the University of Durham made generous contribution to the finances. People, as well as institutions, helped to bring our plans to fruition. Heather A. Fenwick, Administrator of Modern Languages, ran the show behind the scenes. The Department of English Studies funded a research assistant, Allison Weeter, to organise the conference in situ, and the University's Christopherson-Knott Fund financed a further research assistantship for Tony Patterson to copy-edit the volume. Christine Shuttleworth magically produced an index in time. John McCarthy and colleagues at Rodopi provided copious (and invincible) advice during the editing and publication phase. To all of them the editors are indebted. Any faults in the work remain our own.

Nicholas Saul
Simon J. James

Durham University, November 2010

Introduction: The Evolution of Literature

Nicholas Saul and Simon J. James

'Finally, it has not escaped our notice that the more we learn about the human genome, the more there is to explore.' As Ian McEwan noted at the time, the conclusion of the Human Genome Sequencing Consortium's report in *Nature* contained a self-conscious literary allusion, a homage to the arch-understatement of Crick and Watson's epoch-making paper on the DNA double helix.[1] The words 'it has not escaped our notice' are a literary device designed to locate the authors in their predecessors' tradition of creativity: both joining and distancing, acknowledging debt but also establishing originality, and proclaiming a new paradigm with a decently veiled euphoria. Such a moment might be seen to demonstrate an instance of the informed literary sensibility sometimes, if wrongly, felt by humanities specialists to be lacking in scholars of the natural sciences.

Since even before C.P. Snow's *The Two Cultures* (1959, one hundred years after *The Origin of Species* and one hundred and fifty after Darwin's own birth), the notion of 'the divide' has been a recurring topos in would-be interdisciplinary literary scholarship.[2] One recent literary figuration is the grimly hilarious confrontation between the paradigms of natural science and of radical constructivism in McEwan's *Solar* (2010).[3] If scientists from Goethe and Georg Büchner to Primo Levi, at least, have built elegant bridges across the chasm that apparently divides the sciences from literature, then, how best might writers and critics approach the sciences from our own side?

Darwin himself provided no comprehensive theory of aesthetics – at best only a suggestion – and his own literary taste dismayed his peers.[4] Across Europe even in Darwin's own time, however, writers of fiction from

---

1 J. D. Watson, F. H. Crick, 'Genetic Implications of the Structure of Deoxyribonucleic Acid', *Nature* 171 (1953), 964-67; Ian McEwan, 'Literature, Science, and Human Nature', in *The Literary Animal. Evolution and the Nature of Narrative*, ed. by Jonathan Gottschall and David Sloan Wilson (Evanston, IL: Northwestern University Press, 2005), pp.5-19 (p.19).

2 C. P Snow, *The Two Cultures and The Two Cultures: A Second Look*, ed. by Stefan Collini (Cambridge: Cambridge University Press, 1993).

3 Ian McEwan, *Solar* (London: Jonathan Cape, 2010), pp.130-37.

4 See Janet Browne, *Charles Darwin: A Biography*, 2 vols (London: Jonathan Cape, 1995-2002), II: *The Power of Place* (2002), pp.68-71. There is a fascinating study of Darwin's aesthetic taste by Jonathan Smith, *Charles Darwin and Visual Culture* (Cambridge: Cambridge University Press, 2006); see too Horst Bredekamp, *Darwins Korallen: Frühe Evolutionsmodelle und die Tradition der Naturgeschichte* (Berlin: Wagenbach, 2006).

Thomas Hardy to Émile Zola, Jules Verne to Wilhelm Raabe, Edward Bellamy to Lafcadio Hearne, engaged earnestly with the emergent and newly dominant idea of Darwinian evolution. Pioneering thinkers such as Henri Bergson and Wilhelm Bölsche, Max Nordau and Herbert Spencer, Leslie Stephen and H. G. Wells attempted to apply the new evolutionary model to the history, theory and practice of literature. Today, Darwinian paradigms inform and inspire the work of creative writers such as McEwan, A. S. Byatt, Dietmar Dath and Michel Houellebecq.

The engagement of such minds is a recognition of the extraordinary and charismatic epistemic authority which Darwin's theory has possessed since its ascent to becoming the canonical, foundational theory of the life sciences. In the form of evolutionary psychology, it now also increasingly seeks to dominate the social sciences.[5] As with, say, the comparable theories of Freud or Copernicus, most readers believe they know the theory even with no direct knowledge of its founding texts. Darwinism has attained an illusory immediacy before readers' consciousnesses, has itself become orthodoxy to the extent that even most humanities scholars would not call it into question. At the same time, however, this science of human (and other) origins has failed to dominate academic research within the humanities, where the Kantian tradition still rules aesthetics and the Durkheim school cultural theory.[6] Yet many feel with Daniel Dennett that if Darwin's theory is indeed true and universal, then it must, in some manner still to be adequately defined, have its impact upon our understanding of one of the fundamental human behaviours, the production and consumption of art in general and literature in particular.[7]

Writers in one tradition such as Joseph Carroll, Steven Pinker, Karl Eibl and Brian Boyd, and writers in another from Gillian Beer and Daniel Pick to Paul Weindling, have sought to bridge the gap, yet no theory unifying literary behaviour with Darwinian notions has yet crystallised.[8] Both Carroll

---

5 See on this Louise Barrett, Robin Dunbar and John Lycett, *Human Evolutionary Psychology* (Princeton, NJ: Princeton University Press, 2002).

6 Clifford Geertz's stress on variousness is an outstanding exception here. See *The Interpretation of Cultures: Selected Essays* (London: Fontana, 1993), pp.33-83, especially pp.48ff.

7 Daniel Dennett, *Darwin's Dangerous Idea: Evolution and the Meanings of Life* (Harmondsworth: Penguin, 1995), p.521.

8 See Gillian Beer, *Darwin's Plots: Evolutionary Narrative in Darwin, George Eliot and Nineteenth-Century Fiction* (London: Routledge, 1983); also the same author's *Open Fields: Science in Cultural Encounter* (Oxford: Clarendon Press, 1996); Daniel Pick, *Faces of Degeneration: A European Disorder, c. 1848-1918* (Cambridge: Cambridge University Press, 1989); also Mike Hawkins, *Social Darwinism in European and American Thought 1860-1945: Nature as Model and Nature as Threat* (Cambridge: Cambridge University

and Eibl note rightly that the humanities tend to insist on the autonomy of the text from social and cultural processes, and so maintain their independence from naturalist scientific discourse.[9] However, rather than seek an account that does justice both to accepted scientific theory and the cognitive richness of aesthetic experience, Carroll often makes the literary text subservient to a dominant discourse of a version of evolutionary psychology. While Darwin's findings now unarguably have more enduring empirical validity than the findings of Freud or Marx, Carroll's – admittedly ingenious, erudite, resourceful – practice thus tends, like the vulgar Marxist of Freudian literary criticisms of old, to reduce the cognitive product of the text to an allegory of the theory that has been applied to interpret it.[10] The literary text, when murdered to dissect in this light, is reduced to only a dependent function – usually involving behavioural training in the mating game. Such a mode of criticism might acknowledge universally, and without saving irony, that all single men in possession of whatever fortune must always be in want of a wife, or that what Darwin teaches us both in literature and in life is that all that female humans are looking for is a mate, and males for a fight. (The Victorian sexual politics of Darwin's own writing are by no means an irrelevance here.)

Like all theoretical methodologies, the literary Darwinism practiced by those who follow Carroll tends to privilege certain varieties of text in which the kinds of data in which it is interested are more visible. Hence plot receives more attention than literary language; readings of texts that foreground sexual competition or mating are more commonly discussed than less primal and more evolved forms of mimesis; depictions of the natural world over modern, urban experience. Canonicity is admitted in authors who fit into normative notions of human nature, and withheld from those who deviate from it: Homeric narratives of competition and the New-

Press, 1997); Paul Weindling, *Health, Race and German Politics between National Unification and Nazism, 1870-1945* (Cambridge: Cambridge University Press, 1989); see also Peter Sprengel, *Darwin in der Poesie: Spuren der Evolutionslehre in der deutschsprachigen Literatur des 19. und 20. Jahrhunderts* (Würzburg: Königshausen & Neumann, 1998); and Werner Michler, *Darwinismus und Literatur: Naturwissenschaftliche und literarische Intelligenz in Österreich 1859-1914* (Vienna, Cologne, Weimar: Böhlau, 1999). See also Joseph Carroll, *Literary Evolution, Human Nature, and Literature* (New York, London: Routledge, 2004); Steven Pinker, *How the Mind Works* (New York: Norton 1997); Karl Eibl, *Animal poeta: Bausteine der biologischen Kultur- und Literaturtheorie* (Paderborn: Mentis, 2004); Brian Boyd, *On the Origin of Stories: Evolution, Cognition and Fiction* (Cambridge: MA, Belknap Press of Harvard University, 2009).

9 Carroll, *Literary Darwinism*, p.215.

10 Cf. Joseph Carroll, *Evolution and Literary Theory* (Columbia: University of Missouri Press), pp.40, 223, 304; 'Human Nature and Literary Meaning: A Theoretical Model Illustrated with a Crtique of *Pride and Prejudice*', *Literary Darwinism*, pp.187-216.

Comedy-to-Austen-and-beyond story of successful courtship overwhelmingly predominating.[11] This literary criticism risks becoming only an observation of the mechanistic working-through of epigenetic rules, or a recursive explanation of characters' motives according to Darwinian folk psychology.[12] Darwin's texts, rarely mediated through his later nineteenth-century or modern interpreters, are taken as gospel, as Kantian *a priori*: T. H. Huxley's insistence on the separateness or even dissonance of whole spheres of human living, such as ethics, from the cosmic process is ruled out.[13] Since Darwin's theories have now been so widely assimilated, this literary Darwinism can never itself match the epistemological shock of Darwin's announcement of humanity's kinship with the animal kingdom. It does attempt, however, to reverse the recent history of developments in literary critical study, exposing them as no more than Emperor's new clothes on a more scientifically verifiable naked body. Carroll's work avowedly claims to overturn the effects of the so-called 'theory revolutions,' reeling away from the vertiginous possibilities of social constructivism's ontological blank slate of being, towards asserting a more comforting set of stable, empirical certainties that occupy the same place as the educated, normative 'common sense' that underpinned the New Criticism or Matthew Arnold's cultural theory.[14]

Eibl's more subtle work frankly accepts the fact of the autonomy of culture and literature, but also seeks to mediate between natural and cultural domains through a reconstructive dialectic based in the origin and function of natural language. Thanks to Eibl's recognition of language's primary

---

11 *Evolution and Literary Theory*, pp.254, 427-28. Compare Frank Kelleter, 'Controversy: A Tale of Two Natures: Worried Reflections on the Study of Literature and Culture in an Age of Neuroscience and Neo-Darwinism', *Journal of Literary Theory* 1.1, 153-89.

12 See John A. Johnson, Joseph Carroll, Jonathan Gotschall and Daniel Kruger, 'Hierarchy in the Library: Egalitarian Dynamics in Victorian Novels', *Evolutionary Psychology* 2008 (6[4]), 715-38, which assessed the ability of 519 University academics specialising in Victorian fiction to distinguish between 'good guys and bad guys', concluding that 'novels apparently enable readers to participate vicariously in an egalitarian social dynamic like that found in hunter-gatherer societies' (718, 715). Cf. *Evolution and Literary Theory*, pp.252-53.

13 On Darwin as Gospel, see 'Introduction', to Charles Darwin, *On the Origin of Species by Means of Natural Selection* (Peterborough, ON: Broadview Press, 2003), pp.9-66 (p.9), *Evolution and Literary Theory*, p.235; on Huxley, *Evolution and Literary Theory*, pp.110-11, 365-66.

14 The claim that Carroll's are 'a set of conceptually neutral categories for the comparison of world views' is made in 'Organism, Environment, and Literary Representation', *Interdisciplinary Studies in Literature and Environment* 9.2 (2002). See also Carroll, 'The Use of Arnold in a Darwinian World', *Literary Darwinism*, pp.3-14, *Evolution and Literary Theory*, p.468.

function of language' as identification, of a word making a particular perceived thing special so that attention dwells on it, Eibl can give a plausible, Darwin-derived account of the characteristic vicariousness, the disinterested pleasure and cognitive-experimental quality of aesthetic experience.[15] However, Eibl's ultimate explanation of art as an imitative pleasure technology providing advantageous life-extending de-stressing only retreads Pinker's 'evolutionary treat' or cheesecake for the mind theory of aesthetics, an approach which (in Pinker's case intentionally) collapses any qualitative distinctions between aesthetic productions.[16]

Such readings, according to Francis Steen, omit:

> vast tracts of aesthetic experience [...] from the frivolous to the sublime. What evolutionary aesthetics has so far failed to provide is a credible framework for understanding the surprising range of aesthetics. Just as significantly, the implicitly underlying assumption that aesthetic pleasure is comparable to the pleasures of sex or food in driving adaptive behaviour [...] is clearly false: the subjective phenomenology of aesthetic enjoyment differs qualitatively from desire. In contrast to hunger and lust, the experience of beauty is its own reward; unlike these it does not find its fulfilment and satiation in possession. To the extent that this is so, we must look for an explanation that honors beauty itself without seeing it as a proxy for something else.[17]

Evolutionary theory is itself not immune from the evolutionary paradigm or from any future paradigm shift of the kind adumbrated by Thomas Kuhn.[18] As Alistair Brown's essay in this volume shows, even the Darwinian paradigm is itself not conceptually neutral: the 'universal solvent' might itself be further dissolved. Biology is itself dependent on physics and mathematics, but as further epistemological frames of reference are revealed, the object of enquiry, the literary text itself, risks receding further into more distant irrelevance, each aesthetic artefact no more than an epiphenomenon, the experience of art only qualia. A biological understanding of art might thus possess no more of an a priori claim to knowledge than, say, a

---

15 Eibl, pp.12f.

16 Pinker, pp.310-19.

17 Francis Steen, 'A Cognitive Account of Aesthetics', in *The Artful Mind: Cognitive Science and the Riddle of Human Creativity*, ed. by Mark Turner (Oxford: Oxford University Press, 2006), pp.57-71 (p.58).

18 Thomas Kuhn, *The Structure of Scientific Revolutions*, 2nd edn (Chicago: Chicago University Press, 1970). Cf. Carroll, *Evolution and Literary Theory*, pp.449-65; 'Kuhn like Derrida has failed to register the major change in conceptual order initiated by Darwin, and as a result he has consigned himself to an intellectual status of the second-rate' (p.457).

philosophical understanding of sport, a neurological interpretation of law, or a chemical explanation of table manners.

While primate behaviour and *Hamlet* may share certain universals, it is surely the respects in which they differ that establish both as worthy of our attention; eliding the qualitative differences between them as data makes both for poor science and poor literary criticism. Interdisciplinary research gains less from a simplifying cross-disciplinary translation, and more from a mutual respect for the complexity of other fields of enquiry, if seeking to achieve convergent, rather than divergent, trends of debate.[19] Evolutionary psychology can too often insist on universality: if all human nature is the same, then, essentially so all literature becomes the same (if science can be frequently misunderstood on the artistic side of the divide, so too might art be on science's part). It is striking to what divergent evaluations and functionalisations Darwinist notions can be subjected, how far wholly contrary arguments look for support to Darwin's authority; indeed how difficult it sometimes is to discover the authentically Darwin in what is claimed to be Darwinist. Equally striking is the extent to which this set of contributors sees the approach of what may now be called traditional literary Darwinism to be problematic – in terms of assessing literary form, quality, hermeneutic content and experience. Perhaps most striking is the extent to which contributors to this volume, despite their reservations, still sustain their theoretical approaches to literature with Darwinian energies, from anthropological holism, to the reconstruction of strategies of reader-response, and the coming of a truly evolutionary digital literature. If Darwinism possesses a vocabulary to speak of literariness, it is surely that of complexity and variety. *The Evolution of Literature* seeks, like the best kind of scientific enquiry, to recognise the significance of difference over sameness.[20] Literary texts are, as Carroll acknowledges, ultimately a product or consequence of human DNA, but while art is, like culture, made up, contingent, the human experience of literature cannot be wholly reducible to code, and is capable of achieving a discrete individuality.[21] The essays in this collection make a case in Darwinian terms for the special complexity, the recognised distinctiveness, of the aesthetic subjects discussed in a way that

19 See The Third Culture: Literature and Science, ed. by Elinor Shaffer (Berlin: de Gruyter, 1998); also John A. McCarthy, Remapping Reality: Chaos and Creativity in Science and Literature: Goethe–Nietzsche-Grass (Amsterdam, New York: Rodopi, 2006); The Arts and Sciences of Criticism, ed. by Patricia Waugh and David Fuller (Oxford: Oxford University Press, 1999), Mark Turner, The Literary Mind: The Origins of Thought and Language (Oxford: Oxford University Press, 1997). Cf. Carroll, Literary Darwninism, pp.ixf., 110, and Eibl, Animal poeta, pp.9f.

20 Mark Turner, 'The Art of Compression', *The Artful Mind*, pp.93-113.

21 *Evolution and Literary Theory*, p.461.

does justice to both the variousness and the distinctiveness both of human cognitive capacity and of the literary.[22]

This volume proceeds from a public lecture series and conference held at Durham University (under the auspices of its interdisciplinary Institute for Advanced Study), which looked to address these and other issues in a fresh and differentiated way. *The Evolution of Literature* asks why the Darwinian model has not yet widely taken hold in the practice of literary analysis, and looks towards the formation of an authentically Darwinist, evolutionary aesthetic. It examines both evolution *in* literature, and the evolution *of* literature. It looks at literary representations of Darwinism both historically and synchronically, at how a theory of literature might be derived from evolutionary theory, and indeed how evolution as a process might be regarded as itself aesthetic. It complements these theoretical and historical dimensions of enquiry with the comparative dimension, examining how literary and philosophical discourse have refracted the Darwinian idea since 1859 in the major European literatures of Britain, France and Germany. Among the questions asked against this background are: What have been the representations of Darwinian evolutionary theory in literature since the late nineteenth century? What are the leading paradigms in theory and in literature for renovating the evolutionary model? What were, and are, the differences in British, French, German paradigms of literary Darwinian reception? How, if at all, did Darwinian modes of thought hybridise across national borders? What is the future of the Darwinian mode?

Our volume collects some nineteen essays. Nine emerge from English literature, four from French, four from German; one addresses the history of Darwinian ideas in the nineteenth century, another the evolution of language itself. Three of them outline bases for further enquiry. Joseph Emonds considers the steps by which language's discrete properties of phonology and syntax might have evolved in the first place, and makes a claim for the singular importance of primate vision in this process. David Knight deploys a mighty panorama of the differentiated reception of key issues in Darwin's theory in late-Victorian and Edwardian life and thought, while Paul Weindling clarifies the relation of authentic Darwinism with the aberrations of Social Darwinism and eugenics practised in its name.

The essays that follow examine the reception of Darwinism as an impulse to contemporary literary creativity in nineteenth-century national cultures. Here, generations and degeneration are a strong transnational theme. David Amigoni shows how in England writers and scientists when

---

22 For a fascinating interrogation of this issue, see Jon Adams, *Interference Patterns: Literary study, Scientific Knowledge, and Disciplinary Autonomy* (Lewisburg, PA: Bucknell University Press, 2007). Cf. *Evolution and Literary Theory*, p.105

celebrating Darwin adapted notions of filiation, literal and metaphorical, even to criticise and subvert his legacy into a form of post-Lamarckism; Samuel Butler in particular acts as a next-generation poetic developer of evolutionary thought. Anna Barton shows that Alfred Tennyson and Robert Chambers responded to the challenge of paternalist Darwinism in terms of creating a foundation for meaning for the following generations. Simon J. James reads the apocalyptic outcome of Wells's *The Time Machine* (1895) as a dual warning of the consequences of degeneration, presenting the alternatives of literature as potentially a civilising instance of human evolution, or of human evolutionary degeneration itself hastening the end of literature. Christopher Lloyd traces how the French tradition from Hugo and Maupassant to Mirbeau views the repeated emergences of the bestial ape from the human, and how the post-evolutionary tradition comes to naturalise this aspect of the monstrous, taking a decreasingly ethical or optimistic line on the extent to which humanity could or should be saved from the monster within. David Baguley challenges the received idea of the chain of generative influence between Zola and Darwin by noting just how little Darwin Zola actually encountered, but also shows how Zola's later work engages with the great themes of the age, from descent to determination to degeneration. Within the German reception of Darwin, Nietzsche also knew Darwin less than is commonly supposed, concludes John McCarthy, but he travelled by a more independent path to similar conclusions about aesthetic self-generation and the creativity of the evolutionary process. Nicholas Saul notes that the late nineteenth-century reception of Darwinian evolutionary theory proper in literature passed through two distinct phases. Wilhelm Jensen's *Legacy of Blood* (1869), for instance, articulates a polemic against the cognitive evolutionary value of the Bible and a plea for literature's adaptive cognitive power to release the individual's creative potential against inheritance. In the late 1880s, theorists such as Wilhelm Bölsche and Max Nordau radically developed Spencerian ideas to make a claim for sexual selection as the origin of all aesthetic beauty and the cognitive power of literature as a powerful weapon in the human struggle for cultural existence. David Midgley also charts the influence of a selectively imagined 'Darwin' on successive generations. Midgley shows how in Germany, thanks to Kantorowicz's translation, Bergson's meta-Darwinian *élan vital* became (the somewhat less gainly) *Lebensschwungkraft*, and in this shape made its way through the writings of Musil and Döblin to infiltrate post-1900 German philosophical Darwinism.

Modern writers still profit from and reflect upon the generative power radiated by Darwinian thought. Patricia Waugh shows how Darwin's holistic anthropology locates the self emphatically in the body and deprivileges received Cartesian rationalist epistemology in favour of sensual and emotional response, as dramatised in the minute reconstructions of consciousness of McEwan's *Saturday*. Virginia Woolf's and James Joyce's characters too experience consciousness of self less through processing information than by adaptive emotional and intelligent interaction in their environment. Louise Lyle and Douglas Morrey testify to the vitality of Darwinism in modern French literature, a tradition still permeated with Lamarckism and generating wildly opposed interpretations of the human condition. In the context of Darwin's more mixed French reception, Lyle reveals how Vercors and Romain Gary use evolutionary concepts to problematise the notion of humanity and defend a more complex notion of humanity against racism in postwar France. If Lyle emphasises the emancipatory energies latent in Darwinism, Morrey reveals the bleaker legacy in Michel Houellebecq's recent dystopian, posthuman novels, which characterise human existence as defined deterministically by the struggle for life and sexual selection, so appealing regressively to a tendentious, even nineteenth-century reading of Darwin. If Houellebecq ultimately suggests that humanity can nevertheless overcome this determinism and will its own replacement by something better, then the foundation of his analysis, so Morrey argues, is more a desire for the religion which science replaced.

The closing essays focus on the future of literary Darwinism, suggesting, like Saul, that a more informed sense of the history of evolutionary thinking will benefit its future. Wendy Wheeler demonstrates that even post-New Synthesis Darwinism relies on an unexamined notion of the sign; the reading of signs is not just a supplement to existence but constitutes living itself. Thinking along Peirceian lines about how humans as a biological species perceive the semiosphere can do justice to the magnificent richness and diversity of both the natural and aesthetic spheres, and even return us to a sense of enchantment in the world. John Holmes's discussions of the flaws of Grant Allen's evolutionary aesthetic – that a more advanced literary taste is the product of reaching a more advanced stage of biological evolution – is also the weakness of Edward O. Wilson's notion of consilience, where the best art is that which expresses human nature most accurately. Today's theorists of literary evolution can learn from John Addington Symonds, who applies the principles of evolutionary theory to evolution itself and urges readers to examine both texts and their own judgements upon them as the products of evolutionary processes. Jon Adams is similarly sceptical about previous applications of evolutionary theory to art. His shrewd analysis of the controversy between Steven Pinker and Joseph Carroll turns on Carroll's desire to rescue high literary Darwinism by devising a means for it, against Pinker's leveller's perspective, to assess literary quality; the difficulty of measuring a text's adherence to a preset scale of scientific verisimilitude leads to a disturbing tendency towards normalising moral judgements.

The theoretical perspectives of both Katja Mellmann and Alastair Brown discuss the extent to which the recognition of the essentially aesthetic and cognitive quality of literature within Darwinian aesthetic paradigms can release critics from sterile sociobiology. Mellmann notes how little received Darwinian theory has to offer in describing the phenomenon of literariness, but then proposes that an adaptationist heuristics of evolutionary psychology and structural literary theory can reveal correlations of literary structure and affective response that can be used to answer authentically literary questions. Brown also demonstrates a genuinely Darwinian analysis of literature's structural elements. While recognising the evident differences of the narratives of computer games and novels, such analysis can uncover striking similarities between the algorithm-based evolutionary digital game *The Sims* and the way in which literature builds complex worlds from a set of ground rules, even suggesting a parallel between the game and the *Bildungsroman*, and pointing to the ways in which the new forms of a future, evolutionary digi-literature might function.

So it is that from this chapter in literary evolution there slowly emerges a kind of order, from the tangled web in the hedgerow to a view of some open fields beyond. On the basis of our essays, Darwinism might still have a secure future in literary studies. Historically, the range of Darwinism's functions as generator of ideas and form in literary creation is still unfolding across three major European literatures. Evolutionary theory is itself, of course, not immune from further evolution and as humans continue to develop, so does its behaviours such as the writing of literature, and of literary criticism.

The Law of Higgledy-piggledy: Charles Darwin's Inheritance, his Legacy and the Moral Order of Nature

David Knight

Abstract: The philosopher-physicist John Herschel condemned Darwin's theory as 'the law of higgledy-piggledy'. Paley, Wordsworth, and Davy saw a moral order in nature, the handiwork of a benevolent Deity; others saw tooth and claw. Could Darwin's open-ended and statistically-based mechanism of natural selection accompany the idea that Nature had moral authority or displayed Design? Asa Gray thought so; Henry Drummond found hope in progress; but for Huxley morality meant defying the struggle for existence, putting (with Tennyson) the ape and the tiger behind us. Debate continues: but surely, as Coleridge suggested, we get back from nature what we put in.

On 10 December 1859, back in Down after his spell on a water-cure in Ilkley and anxiously awaiting reactions to the *Origin of Species,* Charles Darwin wrote to Charles Lyell:

> I have heard by round-about channel that Herschel says my book 'is the law of higgledy-piggledy'. – What this exactly means I do not know, but it is evidently very contemptuous. – If true this is a great blow & discouragement.[1]

Sir John Herschel was an immensely-respected natural philosopher, the son of Sir William, the eminent astronomer who discovered Uranus.[2] At Cambridge he had been one of the gang of four who had jerked the prestigious mathematical course out of barren Newton-worship and into the nineteenth century with their translation of a French textbook in 1815. He refused to specialise, and after losing the election for the Presidency of the Royal Society to the Duke of Sussex in 1830 (something he afterwards saw as a great blessing), he went to South Africa to observe the southern stars and nebulae. There he entertained Darwin and Robert FitzRoy when in 1836 *HMS Beagle* called at the Cape on her voyage home. On his return, Herschel found himself a hero, becoming a sage or pundit, a seer photographed in old age with flying white hair by Margaret Cameron.[3] Before embarking, Darwin had read and delighted in Herschel's *Preliminary Discourse* (1830), a classic work in philosophy of science, where Herschel, expounding what he saw as Newtonian method, emphasised the need for an

---

1 Charles Darwin, *The Correspondence of Charles Darwin,* ed. by F. Burkhardt et al., 17-vols (Cambridge: Cambridge University Press, 1991), VII, p.425; and see also IX, 1994, pp.135-36. Adrian Desmond and James Moore, *Darwin* (London: Penguin, 1991).

2 *John Herschel, 1792-1871*, ed. by Desmond King-Hele (London: Royal Society, 1992).

3 Steven Ruskin, *John Herschel's Cape Voyage: Private Science, Public Imagination, and the Ambitions of Empire* (Aldershot: Ashgate, 2004).

explanation to be a 'vera causa' rather than a hypothetical mechanism.[4] Evolution was an old hypothesis, proposed by Charles Darwin's grandfather Erasmus Darwin and by Jean-Baptiste Lamarck at the turn of the century, and in the anonymous and sensational *Vestiges* in 1844.[5] Darwin believed that in natural selection he had found a *vera causa* for it, making it scientific. In the *Origin,* development was a process full of sex and violence, but definitely law-driven; it was not mere chance, 'a tale told by an idiot, full of sound and fury, signifying nothing'.[6] That was why he was so upset that Herschel, to whom he had had an advance copy sent, was not impressed.

Herschel was a devout man. In a poem on 'Man the Interpreter of Nature,' a theme from Francis Bacon, whose philosophy he much admired, he saw the world as a great design only brought to fulfilment with the creation of mankind:

> Yet what availed, alas! These glorious forms of Creation,
> Forms of transcendent might – Beauty with Majesty joined,
> None to behold, and none to enjoy, and none to interpret?
> Say! Was the WORK wrought out! Say, was the GLORY complete?
> What could reflect, though dimly and faint, the INEFFABLE PURPOSE
> Which from chaotic powers, Order and Harmony drew?
> What but the reasoning spirit, the thought and the faith and the feeling?
> What but the grateful sense, conscious of love and design?
> Man sprang forth at the final behest. His intelligent worship
> Filled up the void that was left. Nature at length had a soul.[7]

Newton had asked 'whence arises all that Order and Beauty which we see in the World?,' and science for Herschel, as for Michael Faraday and other contemporaries, was a form of worship, reading God's book of nature; wherein a moral order could be expected.[8]

Admiring, even deifying, nature was not new: a sixteenth-century proverb goes 'Nature is no butcher'.[9] In the classic text of Romanticism

---

4 John F. W. Herschel, *Preliminary Discourse on the Study of Natural Philosophy*, ed. by Michael Partridge (New York: Johnston, 1966).

5 James A. Secord, *Victorian Sensation: The Extraordinary Publication, Reception, and Secret Authorship of Vestiges of the Natural History of Creation* (Chicago: Chicago University Press, 2000).

6 William Shakespeare, *Macbeth,* V. 5. 26-28.

7 John F. W. Herschel, *Essays from the Edinburgh and Quarterly Reviews, with Addresses and Other Pieces* (London: Longman, 1857), p.737; David Knight, *Public Understanding of Science: A History of Communicating Scientific Ideas* (London: Routledge, 2006), pp.44-61.

8 Isaac Newton, *Opticks,* 4th edn, (New York: Dover, n.d., [1730]).

9 *The Oxford Dictionary of English Proverbs*, ed. by W. G. Smith and P. P. Wilson (Oxford: Oxford University Press, 1970), p.310.

Lyrical Ballads (1798), William Wordsworth published these lines critical of closet experts:

> One impulse from the vernal wood
> May teach you more of man;
> Of moral evil and of good,
> Than all the sages can. [10]

Nature should be our guide, rather than fallible human moralists. In Wordsworth's next stanza, we find the line, 'we murder to dissect': which indeed happened literally, with Burke and Hare, selling corpses to the medical school in enlightened Edinburgh where Darwin studied for two years before going to Cambridge. The second edition of *Lyrical Ballads* was seen through the press in London by Humphry Davy, who later wrote:

> Oh, most magnificent and noble nature!
> Have I not worshipped thee with such a love
> As never mortal man before displayed?
> Adored thee in thy majesty of visible creation,
> And searched into thy hidden and mysterious ways
> As Poet, as philosopher, as Sage? [11]

Pantheistic delight in Nature, whose works were far superior to those of Man (her ape) went down well with the audiences Davy attracted, in competition with other entertainments, in Regency London.[12] Anna Laetitia Barbauld in her poem *1811* sought to recapture this enthusiasm – imagining a later tourist seeking to:

> Call up sages whose capacious mind
> Left in its course a track of light behind;
> Point where mute crowds on Davy's lips reposed,
> And Nature's coyest secrets were disclosed.[13]

This kind of Romantic pantheism was prominent among Victorian natural philosophers, and when the journal *Nature* was founded in 1869 it

---

10 Samuel Taylor Coleridge and William Wordsworth, *Lyrical Ballads,* ed. by Michael Schmidt (London: Penguin, 1999), p.101.

11 David Knight, *Humphry Davy: Science and Power,* 2nd edn (Cambridge: Cambridge University Press, 1998), p.vii.

12 David Knight, *Science and Spirituality: The Volatile Connection* (London: Routledge, 2004), pp.82-91.

13 *Romantic Women Poets, 1770-1838: An* Anthology, ed. by Andrew Ashfield (Manchester: Manchester University Press, 1995), p.25.

had on its mast-head Wordsworth's lines 'To the solid ground of Nature | Trusts the mind that builds for aye'.[14]

Davy wrote of his distaste for the evolutionary speculations of *'materialists,'* and of his delight in leaving the dissecting room when a 'walk into the green fields or woods by the banks of rivers brought back my feelings from nature to God … I saw *love* as the creative principle in the material world'.[15] But there was another, grimmer, way of looking at things. David Hume in his posthumously-published dialogue on natural religion (1779) had suggested that 'the whole presents nothing but the idea of a blind Nature, impregnated by a great vivifying Principle, and pouring forth from her Lap, without Discernment or parental Care, her maim'd and abortive Children'.[16] Curious artifices of nature everywhere brought pain and torture. Instead of a wise mother-goddess, Nature might be a cruel stepmother. The haunting phrase 'the struggle for existence' was coined a decade before Darwin was born by Thomas Malthus, writing of our savage ancestors:

> Restless from present distress; flushed with the hope of fairer prospects; and animated with the spirit of hardy enterprize, these daring adventurers were likely to become formidable adversaries to all who opposed them … And when they fell in with any tribes like their own, the contest was a struggle for existence. [17]

Malthus's book was to give Darwin the clue he needed when seeking his *vera causa* after his return from the *Beagle's* voyage; but the dynamic vision of constant struggle seemed hardly benign, or moral. In Britain, it bred pessimism about the possibility of alleviating poverty; elsewhere, it was invoked to explain 'manifest destiny' in North America, Australia and Siberia as European colonists displaced aborigines.

For the polymath Erasmus Darwin, struggle and predation were indeed facts about the world; and his later life was spent in years of revolutions and world war.[18] His evolutionary poem, *The Temple of Nature,* contains the gloomy lines:

> Air, earth, and ocean, to astonish'd day
> One scene of blood, one mighty tomb display!

---

14 *Nature* 1 (1869), 1.

15 Humphry Davy, *Consolations in Travel: Or the Last Days of a Philosopher* (London: Murray, 1830), pp.219-20.

16 David Hume, *On Religion*, ed. by A. Wayne Colver and J. V. Price (Oxford: Oxford University Press, 1976), p.241.

17 T. R. Malthus, *An Essay on the Principle of Population* (London: Johnson, 1798), pp.47-48.

18 *The Genius of Erasmus Darwin*, ed. by Christopher Upham Murray Smith and Robert Arnott (Aldershot: Ashgate, 2005).

From Hunger's arm the shafts of Death are hurl'd,
And one great Slaughter-house the warring world![19]

Yet this Darwin was an optimist, seeing overall evolutionary progress everywhere, from the origin of the Solar System up to contemporary society. Darwin's poetry was with other left-wing writings ruthlessly parodied by William Pitt, George Canning and Hookham Frere in their *Anti-Jacobin*, surely one of the very few pieces of government propaganda to be funny. He had written the *Loves of the Plants,* popularising Linnaeus; they presented geometry as the loves of the triangles – and also, mocking Robert Southey, used the words 'Higgledy, piggledy' as a refrain in a 'jacobinical' poem exalting anarchy.[20]

William Paley's *Natural Theology* was published in 1802, the year Erasmus Darwin died, shortly before *The Temple of Nature* was published. Paley too was aware of predation, and knew well that wolves and tigers are manifestly well-equipped to prey upon weaker creatures; but to him predation might be seen as euthanasia: better to be eaten by a crocodile than by a cancer. He was a utilitarian, and believed that 'The world was made with a benevolent design. Nor is the design abortive. It is a happy world after all.'[21] Watching shrimps sporting in the surf, he was sure they were having a good time. Primarily, however, the book is about design. Paley's long argument, of a legal rather than logical character, a rope rather than a chain of reasoning, making a case beyond reasonable doubt, was undoubtedly powerful. It has impressed Richard Dawkins; it impressed the young Darwin, who happened to live in Paley's old rooms at Christ's College, Cambridge, and the *Origin* is a long cumulative argument like Paley's.[22] If one or two of the examples seem unpersuasive, the rest of the strands may still hold: and readers may go some of the way, and then find themselves on a slippery slope towards the conclusion, as both Paley and Darwin hoped they would.

Not everyone liked Paley's approach. To the evangelical Thomas Gisborne, the world as we see it is not a perfect, or a happy place: since the Fall, it is cursed, a Botany Bay or gulag, a place of punishment, pain and death for sinners. Redeeming faith is not achieved by balancing probabilities or utilitarian arguments, but is the gift of grace to the meek-hearted; and

---

19 Erasmus Darwin, *The Temple of Nature; Or, the Origin of Society: A Poem with Philosophical Notes* (London: Johnson, 1803), p.134.

20 *The Anti-Jacobin,* 11 December 1797, and 16 April 1798; *Poetry of the Anti-Jacobin* (London: Wright, 1799), pp.20, 108-29, 134-41.

21 William Paley, *Natural Theology*, ed. by Matthew D. Eddy and David Knight (Oxford: Oxford University Press, 2006), p.238.

22 Richard Dawkins, *The Blind Watchmaker,* new edn (London: Penguin, 2006), p.4.

doubt is a sin.[23] There was a moral order in the creation, but humans spoilt it, and only traces remain. To Coleridge also the attempt to build up 'evidences' for religion was mistaken, when one should instead make people feel the need for it. But he also disliked both Paley's utilitarian philosophy, and his mechanical world-view: in *Natural Theology,* the world and its denizens are all compared to watches, going like clockwork. The world of Coleridge, Davy and other Romantics was dynamic. We human beings endure like waterfalls through the flux of our constituent particles, and apparent stability is the result of forces in balance. Further, Coleridge came to believe, in Neo-platonic or Kantian vein, that we project upon nature our feelings and expectations. In his 'Dejection' ode he wrote:

> O Lady! We receive but what we give,
> And in our life alone does Nature live:
> Ours is her wedding garment, ours her shroud![24]

It was an inanimate cold world, where happiness was elusive and nature amoral.

On 5 July 1844 Darwin completed an account of his evolutionary views, which he circulated to a small group of friends; and requested that if he should die, his widow should publish it. In October of that same year, Robert Chambers brought out *Vestiges,* setting out a cosmic evolutionary scheme (as Erasmus Darwin had done), and creating an enormous furore that alarmed Darwin. The book, anonymously-written by a well-read publisher of educational and improving works, was very readable: but its hypothetical character, and its many slips and blunders in scientific accuracy, annoyed natural philosophers and naturalists (including Darwin and Thomas Huxley); while its 'atheism' or materialism outraged clergy. With hindsight, the men of science missed the wood for the trees; while the perspective was (like Erasmus Darwin's) not atheistic but Deistic, God being a remote First Cause who had launched a system that would govern itself by laws He had imposed. The result of the many hostile reviews was that the book sold very well, and evolution became a hot topic, while guessing the author became a great game. Chambers was concerned about reconciling the benevolence of God with the evidence of the bloodthirstiness of nature:

> The system here unfolded does not imply the most perfect love or regard on the part of the Deity towards his creatures… It may be that, while we are committed

---

23 Thomas Gisborne, *The Testimony of Natural Theology to Christianity* (London: Cadell and Davies, 1818).

24 Samuel Taylor Coleridge, *Selected Poems,* ed. by Richard Holmes (London: HarperCollins, 1996), p.180.

> to take our chance in a natural system of undeviating operation, and are left with apparent ruthlessness to endure the consequences of every collision into which we knowingly or unknowingly come with each law of the system, there is a system of Mercy and Grace behind the screen of nature.[25]

This is no great consolation to the widow, the orphan or the sufferer: we are indeed launched into Victorian doubt.

The great spokesman for doubt was Alfred Tennyson, like Herschel photographed as a sage by Margaret Cameron. He read both *Vestiges* and Charles Lyell's *Principles of Geology* (1830-33) – an 'attempt to explain the former changes of the Earth's surface by reference to causes now in operation'.[26] This involved vast periods, a dark backward and abysm of time, and slow changes; and as he wrote his great poem *In Memoriam* (1850) this new science was a leaden echo, indicating an indifferent or hostile Nature:

> [Man] who trusted God was love indeed
> And love creation's final law –
> Tho' Nature, red in tooth and claw
> With ravine, shriek'd against his creed –

Tennyson's poetry particularly delighted Huxley, who would quote it in lectures and who ensured that the Royal Society was represented at his funeral. An important part of Darwin's inheritance was the perception of human insignificance in a vast, sublime and indifferent scheme of things: that was not something he invented.

Malthus's 'struggle for existence' had also entered deeply into British consciousness, even though his dire predictions had not come about in an era of population explosion. Samuel Smiles published *Self-Help* in 1859, and getting on in a competitive world was a deeply ingrained objective among Darwin's contemporaries. It was perhaps a particularly masculine trait: Dinah Craik, a novelist from a Nonconformist background, author of *John Halifax, Gentleman* (1857), wrote sardonically in 1858 'That first "law of nature", self-preservation, is – doubtless, for wise purposes – imprinted pretty strongly on the mind of the male sex'.[27] Darwin, though living a sheltered and well-heeled existence in Down, did not have to look far for ideas that ruthless competition was the key to understanding society.

---

25 [Robert Chambers], *Vestiges of the Natural History of Creation*, ed. by James A. Secord (Chicago: Chicago University Press, 1994), pp.383-84.

26 Charles Lyell, *Principles of Geology: Being An Attempt to Explain the Former Changes of the Earth's Surface, by Reference to Causes Now in Operation*, ed. by Martin J.S. Rudwick (Lehre: Cramer, 1970).

27 *The Oxford Dictionary of English Proverbs*, 3rd edn, ed. by Frank Percy Wilson and Joanna Wilson (Oxford: Oxford University Press, 1970), p.712.

Darwin himself had always contemplated nature with love rather than horror, even when as a young man he delighted in shooting. Indeed, after taking us through the struggle for existence and its consequent natural selection leading to development over millions of years, he ended the *Origin of Species* on a Wordsworthian note:

> It is interesting to contemplate an entangled bank, clothed with many plants of many kinds, with birds singing on the bushes, with various insects flitting about, and with worms crawling through the damp earth, and to reflect that these elaborately-constructed forms, so different from each other, and dependent on each other in so complex a manner, have all been produced by laws acting around us… Thus, from the war of nature, from famine and death, the most exalted object which we are capable of conceiving, namely, the production of the higher animals, directly follows. There is a grandeur in this view of life, with its several powers, having been originally breathed into a few forms or into one; and that whilst this planet has gone cycling on according to the fixed law of gravity, from so simple a beginning endless forms most beautiful and most wonderful have been, and are being, evolved.[28]

The last word in the book is thus 'evolved,' but the word 'evolution' was not used in the first edition of the *Origin* because to Darwin it smacked of design and purpose – it was a word used of the growth of the foetus, for example. He used 'development' instead, to indicate that he perceived not progress up some scale of nature (he disliked terms like 'higher' and 'lower' applied to organisms), but opportunistic filling of niches. The barnacles that had needed to be scraped off *HMS Beagle* at intervals became the object of his close study during the years he was collecting evidence for his theory. Not long before his voyage, John Vaughan Thomson, a military surgeon stationed in Cork, had demonstrated the life cycle of the barnacle: though it behaved like a mollusc, its young were like tiny shrimps, and its relatives were not the limpets but the crabs and lobsters.[29] In middle life they settle down with a vengeance, gluing themselves to a rock (or ship); and in some species Darwin described in his great (and very dull) monograph, the male has become 'parasitic,' degenerating into a stomach and sex organs living within the shell of the female.[30] Social mobility in Victorian England implied going down as well as up: barnacles had gone down in the world, but had found a niche and flourished mightily as a result – there are millions of them

---

28 Charles Darwin, *The Origin of Species* (London: Murray, 1859), pp.489-90.

29 John V. Thomson, *Zoological Researches and Illustrations, 1828-1834*, ed. by AlwyneWheeler (London: Society for Bibliography of Natural History, 1968).

30 *Correspondence*, IV, p.128; Charles Darwin, *A Monograph of the Sub-class Cirripedia*, vol. I (London: Ray Society, 1851), pp.189-203.

everywhere. Whether one could with Paley say that they were, like shrimps, happy, is not clear.

Darwin greatly admired Alexander von Humboldt, especially for the vivid evocations of tropical forests and mountains in Humboldt's 'personal narrative' of his journeys in Latin America.[31] Darwin had sought to model his own account of his own voyage upon Humboldt's writings, and the very handsome large volumes of the *Zoology* of the voyage that Darwin edited on his return (1838-43), getting experts to describe his collections, are reminiscent of the magnificent tomes Humboldt had published in Paris when he had got back (Darwin was always good at getting people to help).[32] The structure of his book Darwin took from Paley, setting out the theory of natural selection, and then supporting it with a series of probabilistic chapters (including one candidly discussing difficulties) constituting the strands of a cumulative argument, that twisted together will support its weight. From the *Concordance* we can see that 'Nature' appears some 240, 'probable/y' about 160, 'possible/y about 80, and 'perhaps' some 70 times in the book.[33] 'Evolution' and Herbert Spencer's phrase 'survival of the fittest' found their way into later editions of the *Origin.* Although 'God' appears once, and 'creator' seven times in the first edition, Darwin (a Church of England ordinand when he embarked on *HMS Beagle)* was by 1859 a theist or perhaps an agnostic, and could not believe that the Earth's history represented the working-out of any overall plan by a benevolent deity.

That was not how it seemed to everyone among his converts. The Harvard botanist Asa Gray, long a correspondent, using Darwin's theory to account for the relationships of the North American flora to those of Asia and Europe, believed that the theory showed how God had acted; and in 1860 wrote an essay, 'Natural Selection not inconsistent with Natural Theology,' to demonstrate it. Darwin, though not convinced, got this published in Britain at his own expense. Gray further saw that Darwin had 'brought back teleology to natural history … usefulness and purpose come to the front again as working principles of the first order'.[34] Huxley's

---

31 Alexander von Humboldt, *Personal Narrative of Travels to the Equinoctial Regions of America, During the Years 1799-1804,* trans. by Thomasina Ross (London: Bohn, 1852). Darwin used an earlier edition.

32 Charles Darwin, ed., *The Zoology of the Voyage of HMS Beagle*, ed. by Charles Fleming (Wellington, NZ: Nova Pacifica, 1980); Janet Browne, *Charles Darwin: Voyaging* (London: Cape, 1995), pp.104, 226, 365, 409, 486.

33 Paul H. Barratt, Donald J. Weinshank and Timothy T. Gottleber, *A Concordance to Darwin's Origin of Species, First Edition* (Ithaca, NY: Cornell University Press, 1981).

34 Asa Gray, *Darwiniana: Essays and Reviews Pertaining to Darwinism*, ed. by A.Hunter Dupree (Cambridge, MA: Harvard University Press), 1963, pp.72-145 (294).

exclamation on reading the *Origin* (which he reviewed for *The Times* on Boxing Day 1859) 'How extremely stupid not to have thought of that' is well-known; but he also noted much later that while Darwin had dealt the death-blow to Paleyan teleology: 'There is a wider teleology which is not touched by the doctrine of Evolution, but is actually based upon the fundamental proposition of Evolution.'[35] The Darwinian can safely use the language of purpose and design, knowing that if challenged these can be explained in terms of adaptation and survival value.

Few were much disturbed by the thought that crabs and barnacles had a common ancestor, but for humans and orangutans or other apes it was a not-unfamiliar but highly contentious matter. Huxley argued that to a visitor from Saturn (the seventh heaven) it would be apparent that just as horses, donkeys and zebras belong to one natural group, so do we and the apes: and his famous and much-reproduced procession of skeletons is not an evolutionary but a taxonomic illustration.[36] He had no trouble with simian ancestry; but Alfred Russel Wallace, the natural-history 'collector' whose letter from Malaysia to his hero Darwin had precipitated the writing of the *Origin,* and who can be seen as the co-discoverer of natural selection, believed that there was a radical discontinuity. In an essay in his book, modestly entitled *Darwinism*, he wrote:

> Those who accept the spiritual nature of man, as not in any way inconsistent with the theory of evolution, but as dependent on those fundamental laws and causes which furnish the very materials for evolution to work with [...] will be relieved from the crushing mental burthen imposed upon those who – maintaining that we, in common with the rest of nature, are but products of the blind eternal forces of the universe [...] have to contemplate a not very distant future in which all this glorious earth shall be as if it had never existed; [...] that the slow growths of our race struggling towards a higher life, all the agony of martyrs, all the groans of victims, all the evil and misery and undeserved suffering of the ages, all the struggles for freedom, all the efforts towards justice, all the aspirations for virtue and the wellbeing of humanity, shall absolutely vanish, and, "like the baseless fabric of a vision, leave not a wrack behind".[37]

Wallace was a great travel-writer, a socialist who believed in the nationalisation of the land, and a spiritualist, who delighted in the progress

---

35 *The Life and Letters of Charles Darwin*, ed. by Francis Darwin (London; Murray, 1887), pp.197, 201.

36 T. H. Huxley, *Man's Place in Nature,* (1863), in *The Major Prose of Thomas Henry Huxley*, ed. by Alan P. Barr (Athens: Georgia University Press, 1997), pp.20-153. Adrian Desmond, *Huxley: The Devil's Disciple* (London: Joseph, 1994-97).

37 Alfred Russel Wallace, *Darwinism,* 2nd edn (London: Macmillan, 1890), pp.476-77; his (mis)quotation is from William Shakespeare, *The Tempest* IV. 1. 151.

made during the wonderful nineteenth century.[38] Like Huxley (but unlike Darwin) he was a poor man, who had known something of the struggle for existence at first hand. In foreseeing global cooling and the heat-death of the world, he was drawing upon the work of William Thomson, Lord Kelvin, in thermodynamics; a theme taken up in 1895 by H. G. Wells in his scientific romance, *The Time Machine* (1895), where evolutionary divergence and degeneration is also prominent.[39]

Wallace believed that the spiritual, mathematical and other abilities of humankind that would have been no use to our primitive ancestors, showed that a gulf separated us from other animals. Huxley had no time either for Wallace's spiritualism or for Gray's belief in God; but he was concerned for morality. 'Atheism' had been a matter of being indifferent to the Ten Commandments and the prospect of a Last Judgement: when Marianne Evans lost her faith, and went to London, she moved in with a married man – thus demonstrating to respectable contemporaries the connection between religious doubt and immorality. Writing novels as George Eliot, it was only when her partner died and she married John Walter Cross that she could be accepted in respectable society. Edward Collins picked up the link between unbelief and immorality in a poem aimed at Huxley and his circle, 'The Positivists':

> Life and the universe show Spontaneity;
> Down with ridiculous notions of Deity!
> Churches and creeds are all lost in the mists;
> Truth must be sought with the Positivists!
>
> Wise are their teachers beyond all comparison,
> Comte, Huxley, Tyndall, Mill, Morley and Harrison;
> Who will adventure to enter the lists,
> With such a squadron of Positivists?
>
> Social arrangements are awful miscarriages;
> Cause of all crime is our system of marriages;
> Poets with sonnets, and lovers with trysts,
> Kindle the ire of the Positivists.
>
> Husbands and wives should be all one community,
> Exquisite freedom with absolute unity;

---

[38] Martin Fichman, *An Elusive Victorian: The Evolution of Alfred Russel Wallace* (Chicago: Chicago University Press, 2003).

[39] William Thomson Kelvin (Lord Kelvin), *Popular Lectures and Addresses,* 3 vols (London: Macmillan, 1888), I, pp.356-429, III, 1894, pp.10-131; H. G. Wells, *The Time Machine*, ed. by Patrick Parrinder (London: Penguin, 2005).

Wedding rings worse are than manacled wrists, -
Such is the creed of the Positivists.

There was an APE in the days that were earlier;
Centuries passed, and his hair became curlier;
Centuries more gave a thumb to his wrist, -
Then he was a MAN – and a Positivist.

If you are pious, (mild form of insanity,)
Bow down and worship the mass of humanity,
Other religions are buried in mists;
We're our own gods, say the Positivists.[40]

In fact, Huxley (who coined the term 'agnostic') was contemptuous of Comte's Positive Philosophy (which did not readily allow for theories such as evolution, not directly testable) and of his Religion of Humanity; and he lived an exemplary moral life, not believing that if we are akin to apes we have the right to monkey about.[41]

In 1894, then a grand old man near the end of his life, Huxley was invited to deliver the Romanes Lecture in Oxford, and took as his theme 'Evolution and Ethics'. Picking up from Malthus (and echoing Tennyson), he saw social justice and progress in defying rather than following nature:

> For his successful progress, throughout the savage state, man has been largely indebted to those qualities which he shares with the ape and the tiger; his exceptional physical organization; his cunning, his sociability, his curiosity and his imitativeness; his ruthless and ferocious destructiveness when his anger is roused by opposition … Whatever differences of opinion may exist among experts there is general consensus that the ape and tiger methods of the struggle for existence are not reconcilable with sound ethical principles […] Social progress means the checking of the cosmic process at every step and the substitution for it of another, which may be called the ethical process.[42]

He saw just as clearly as conservative critics the threat of 'Social Darwinism,' ideas from political economy increased in potency by a passage through biology, to undermine social relations and our duty to love our

---

40 *Everyman's Book of Victorian Verse*, ed. by John Richard Watson (London: Dent, 1982), pp.240-41.

41 Paul White, *Thomas Huxley: Making the 'Man of Science'* (Cambridge: Cambridge University Press, 2003); *Thomas Henry Huxley's Place in Science and Letters*, ed. by Alan P. Barr (Athens: Georgia University Press, 1997).

42 T. H. Huxley, 'Romanes Lecture', in *Major Prose*, pp.309-44, (pp.311-2, 327); Alfred Tennyson, *In Memoriam*, p.133 (section 118).

neighbour; and like Darwin saw that evolution was not just a benign progressive force.[43]

The biologist and populariser Grant Allen also perceived the grim side of evolution.[44] In a 'ballade,' the refrains of the verses go:

> For the fittest will always survive
> While the weakliest go to the wall. [...]
>
> So the handsomest managed to wive,
> While the ugliest went to the wall. [...]
>
> And the strongest continued to thrive,
> While the weakliest went to the wall.
>
> Prince, in our civilized hive,
> Now money's the measure of all;
> And the wealthy in coaches can drive
> While the needier go to the wall. [45]

Amidst the anxieties about 'degeneration' and the growth of a savage underclass, some evolutionists such as Wallace, Huxley and Allen saw the need for democracy and meritocracy rather than eugenics.[46]

But others rejoiced to see science on the side of getting on, and self help. Henry Drummond, whose Lowell Lectures in Boston were delivered in the same year as Huxley's, and given the title *The Ascent of Man* in an elegant allusion to Darwin's *Descent of Man* (1871), perceived it that way:

> Evolution has ushered new hope into the world. The supreme message of this age is that all Nature is on the side of the man who tries to rise. Evolution, development, progress are not only on her programme, they are her programme. For all things are rising, all worlds, all planets, all stars and suns.[47]

There were after all many reactions to Darwin, and in Britain the mainstream interpretation, as 'evolution' became a crucial feature of the

---

43 Mike Hawkins, *Social Darwinism in European and American Thought, 1860-1945: Nature as Model and Nature as Threat* (Cambridge: Cambridge University Press, 1997).

44 Jonathan Smith, 'Grant Allen, Physiological Aesthetics, and the Dissemination of Darwin's Theory', in *Science Serialized: Representation of the Sciences in Nineteenth-Century Periodicals*, ed. by George Cantor and Sally Shuttleworth (Cambridge, MA: MIT Press, 2004), pp.285-305.

45 *Everyman's Book of Victorian* Verse, ed. by J. R. Watson, p.249.

46 Daniel Pick, *Faces of Degeneration: A European Disorder, c1848-c1918* (Cambridge: Cambridge University Press, 1989).

47 Henry Drummond, *The Ascent of Man* (London: Hodder, 1894), p.435.

explanation of almost anything, was more benign than Darwin's.[48] With Darwin safely enshrined in Westminster Abbey, evolution and progress, as in Drummond and in the spirit of Lamarck and Erasmus Darwin, became synonymous; and in that form, development was accepted by almost all. Thus Frederick Temple, subsequently Archbishop of Canterbury, declared in the Bampton Lectures in Oxford in 1884 that 'to the many partial designs which Paley's Natural Theology points out, and which still remain what they were, the doctrine of Evolution adds the design of a perpetual progress.'[49] Temple was not too concerned with apparently needless pain and waste, which were like imperfections in a half-completed picture, or the larva of a beautiful insect, or a saint nevertheless changing from glory to glory.

The last word can go to Arthur Balfour, who would become Prime Minister and President of the British Association for the Advancement of Science, be responsible for the Declaration that led ultimately to the founding of Israel, and towards the end of his life be de facto Britain's first Minister for Science. In 1895, already a very prominent Conservative politician, he published *Foundations of Belief*, directed particularly against agnostic naturalism: Huxley was invited to review it for the *Nineteenth Century*, but fell ill, and never completed the answer to his satisfaction before he died. Balfour was sure that all knowledge rested upon beliefs, and that while science was the right guide to follow in its sphere, it was hopeless beyond it:

> If on the naturalistic hypothesis the sentiments associated with beauty seem like a poor jest played on us by Nature for no apparent purpose, those that gather round morality are, so to speak, a deliberate fraud perpetrated for a well-defined end [...] competitive advantage in the struggle for existence.[50]

His interests were in music, art and ethics, and here Darwinism seemed to let him down: these things could not be reduced to something to do with natural selection. In agreeable and readable prose, he urged his readers not to be too bothered about consistency in their world-views, but to be reasonable.

When it came to contemplating moral lessons from nature, Darwin's legacy was thus not so very different from his inheritance: mixed. Darwin was the heir, like all men of science of his generation, of both the

---

48 *Disseminating Darwinism: The Role of Place, Race, Religion, and Gender*, ed. by Ronald L. Numbers and John Stenhouse (Cambridge: Cambridge University Press, 1999).

49 Frederick Temple, *The Relations Between Religion and Science* (London: Macmillan, 1885), pp.116-17.

50 Arthur James Balfour, *The Foundations of Belief: Being Notes Introductory to the Study of Theology* (London: Longman, 1895), p.79.

Enlightenment of Hume, and the Romantic Movement of Wordsworth. The word 'scientist' was coined at a BAAS meeting in Cambridge while he was away on *HMS Beagle*, by William Whewell, by analogy with 'artist'; but it did not catch on until about 1882, the year of Darwin's death. 'Natural Philosopher' (and Darwin was 'Philos' to his shipmates) implied a broad view, where implications were taken seriously; and was an activity for gentlemen (and a few ladies). But the Victorian age was one of specialisation, of expertise, of qualifications and new professions, and maybe, by the end, of 'two cultures'. Science had been popularised and practised, and widely understood, in terms of natural theology. Huxley and his associates succeeded in their aim of displacing clergy as custodians of right thinking, making science highly important culturally. Henry Sidgwick, Balfour's tutor, was the first layman to hold the chair of moral philosophy at Cambridge. There is no doubt that it was much easier for Wordsworth or Davy to find moral order and authority in Nature rather than from sages than it became a hundred years later: but Darwin should not get all the blame for that, and surely Coleridge was right, that we get back from Nature what we put in.

'Our Racial Friends': Disease, Poverty and Social Darwinism, 1860-1940

Paul Weindling

Abstract: While most commentators on Social Darwinism explore the implications of natural selection, this chapter explores sexual selection.

I

While Darwinists disseminated scientific insights into evolution and heredity, underlying these were far-reaching changes in *Weltanschauung* and society. Biologists in the period 1860 to 1920 pioneered a sophisticated understanding of reproductive biology, and cellular processes of development. German biologists' studies of chromosomes laid the foundations of genetics. The expansion of biology went with a wider set of values and interventions at a socio-cultural level. Although Darwinism is generally associated with individualist ideas of a struggle for survival, it was rare to find this in pure unadulterated form. Far more prevalent were organic philosophies of social cohesion. Biologists used ideas of correlation and integration in order to stabilise social structures during the upheavals of modernisation. They saw the impact of urban expansion, new industries and mass migration in the context of physiological repercussions. The quantity and physiological quality of the burgeoning industrial population represented a particular concern.

Biological solutions to social problems of poverty and disease have been interpreted as a soulless and inhumane agenda for eradication of the unfit, and racial war. Although the eugenic quest for maintaining the biological equilibrium was firmly postulated by Francis Galton in the 1880s, eugenic organisations and institutions coincided with the twentieth-century rise of the welfare state. As a consequence, eugenics is often interpreted as institutionalised Social Darwinism. The claim is that a logic of extermination culminated in selective measures to weed out the mentally and socially 'unfit'. Selective welfare and sterilisation appear as preludes for the killing of the mentally ill, and the unleashing of the Holocaust. This essay will argue that this presumed progression misunderstands evolutionary theory as debated at the time.

Direct links have been made between Darwin and Hitler. Hannsjoachim Koch in his overview *Der Sozialdarwinismus* of 1973 spoke of the thread connecting Darwin to Auschwitz.[1] The social historian Karl Heinz Roth coined the chilling phrase, 'the Final Solution to the Social Problem' to

---

1 Hannsjoachim Koch, *Der Sozialdarwinismus: Seine Genese und sein Einfluß auf das imperialistische Denken* (Munich: Beck, 1973).

condemn how social policy was permeated by an exterminatory eugenics.[2] The implication is that Darwinism's shattering of Christian values had a brutalising effect. Richard Weikart has detected an ethic of destruction linking Darwin to Hitler's racial values, and through these to the Holocaust. One must however note the terms 'struggle' and 'selection' had a variety of ideological roots, and seeing a branching tree in their lineage going back to Darwin is highly questionable. The conventional historiography is of an exterminatory synthesis of Darwin's natural selection and Herbert Spencer's principle of 'survival of the fittest'.[3] These moral imperatives were incorporated in the new eugenic societies, founded in the decade prior to WW1, and projected into the programmes of the ultra-right in the interwar period.

Evolutionary theories show a rich theoretical variety, and cannot be reduced to a single principle – that of natural selection. Darwin's own ideas were a complex amalgam of natural selection and other mechanisms of evolution, so that natural selection was a significant but not the exclusive engine of evolution. Sexual selection has attracted much attention in Darwin studies, but less so as regards the socio-cultural impact of Darwinism.[4] By examining the emergence of an evolutionary perspective on reproduction, we see that Darwin's theory of sexual selection had a profound, and hitherto underestimated effect on social values. His theory reflected how increasingly, marriage and the reproduction of the feckless poor lay at the heart of Victorian solutions to the problem of poverty.

Victorian evolutionary social theory was pervaded by ideas of development and growth. The issues of German *Darwinismus/Sozialdarwinismus* and of eugenics were informed by the late Victorian debates. The commentaries on the situation in Britain – often couched as a critique of Anglo-Saxon individualism and laisser-faire – fed into 'racial hygiene' (as it was known in Germany), and thereby impacted on the emergent welfare state. Understanding these issues in Britain helps clarify the more extreme Weimar German and Nazi responses to Darwinism. We need therefore to clarify how the issues of selection and fitness were addressed, whether these were of natural or sexual selection.

2 Götz Aly and Karl Heinz Roth, *Die restlose Erfassung: Volkszahlen, Identifizieren, Aussondern im Nationalsozialismus* (Berlin: Rotbuch, 1984).

3 Richard Weikart, *From Darwin to Hitler: Evolutionary Ethics, Eugenics, and Racism in Germany* (New York. Palgrave, 2004), also Andre Pichot, *The Pure Society. From Darwin to Hitler* (London: Verso, 2009).

4 Helena Cronin, *The Ant and the Peacock* (Cambridge: Cambridge University Press, 1993), e.g. pp.234-45.

Finally, the issue inevitably arises whether German views on breeding out the unfit were a prelude to genocide.

II

Victorian science was public and participatory with a popular interest in natural history and a wide readership of periodicals. This was an era before the arrival of the professional 'biologist'. At this transitory stage there were full time naturalists like Darwin and collectors like Alfred Wallace, but full time professors of zoology or botany were few. 'Biology' and 'biologist' were new terms, making it clear that the same natural laws applied to plants, animals and – by extension – the human species. Darwin's ideas on natural selection and evolution were profoundly shaped by Victorian political economy.[5] The evolutionary terminology of 'accumulation,' 'production' [of new varieties, PW], 'division of labour,' and 'class' all suggest a fluid interaction of natural history with political economy. Feminist historians have linked evolutionary writings on 'domestication' and courtship to the representation of women in Victorian fiction.[6]

Biological ideas permeated the Victorian debate on poverty. The social historian Gareth Stedman Jones – presciently, for this was at a point in the 1970s when social history was about to make an academic breakthrough – depicted the social typology of the labourer in terms of a biological discourse on poverty. He observed that '[b]iologism provided a framework for a comprehensive theory of hereditary urban degeneration,' and that casual employment was a form of the degeneracy.[7] These insights into poverty and Darwinism have still to be explored. The grime and dirt of the poor was a sign that they were degenerating into African savages; Africa was 'a pauper lunatic asylum without keepers'.[8] In Reverend Charles Kingsley's *Water Babies*, published in 1863, the 'ugly, black, ragged' Tom, the chimney-sweep's boy, was 'a little black ape'. He plunged atavistically into the stream and metamorphosed into a water baby with gills: this was a redemptive evolution. Kingsley depicted the biological baptism in terms of a spiritual Lamarckism, a cleansing reversion. Sooty Tom was lucky to escape the

---

5 Adrian Desmond and James Moore, *Darwin* (London: Michael Joseph, 1991).

6 Eveleen Richards, 'Darwin and the Descent of Woman', in *The Wider Domain of Evolutionary Thought*, ed. by D. R. Oldroyd and I. Langham (Dordrecht: Reidel,1983), pp.57-111. Richards, 'Huxley and Woman's Place in Science: The "Woman Question" and the Control of Victorian Anthropology', in *History, Humanity and Evolution*, ed. by James R. Moore (Cambridge: Cambridge University Press, 1989), pp.253-84.

7 Gareth Stedman Jones, *Outcast London. A Study in the Relationships between Classes in Victorian Society* (Oxford: Oxford University Press, 1971), p.130. Note the references to Darwin and Darwinism, pp.5, 8, 130, 286-7, 313, 329.

8 See Stedman Jones too on 'the sympathy and intelligence', in *Outcast London*, p.vi.

clutches of dissection by Professor Owen and Professor Huxley.[9] One might read this allusion as the birth of a chillingly modern literary trope – the scientist experimenting on children.

Yet Tom's redemption is better seen in terms of education and moral progress as evolutionary characteristics of a higher sensibility. It was no mere act of compassion but accorded with the laws of evolution and – by extension – social progress. Education could be seen in physiological terms as stimulus and response on the basis of 'natural reactions'. Such an evolutionary Darwinism illuminates the Victorian discourse on poverty. For the social fabric was held together by instincts, common sympathies and behaviour patterns, which could be explained as according with the laws of biology.

Herbert Spencer made extensive use of Lamarckism and embryology in his influential synthetic philosophy. Spencer's law of organic correlation held that the greater the division of labour, the greater the compensating integration must be. This principle exercised a profound influence on Victorian biology (especially the embryology of the nervous system) and neurology. Spencer himself adopted an ethic of redemptive philanthropic altruism rather than exterminatory survival of the fittest as in a higher civilised society.[10]

Darwin for his part argued against the idea that the races of men were distinct species, and consequently denied their 'clear distinctive characters'.[11] He stressed that Mind transcended any physical differences.[12] In the *Descent of Man*, he found merit in how 'civilised men' check nature's efforts to eliminate the weak: 'We build asylums for the imbecile, the maimed, and the sick; we institute poor laws; and our medical men exert their utmost skill to save the life of everyone to the last moment [...]. Thus the weak members of civilised societies propagate their kind.'[13] The historian Richard Soloway sees Darwin as giving way to weakness, in outlining 'the instinct of sympathy,' and that if we would sacrifice sympathy to hard reason, humans would lose 'the noblest part of our nature'. Darwin comforted himself with the observation that 'the weaker and inferior members of society do not

9 Gay Weber, 'Science and Society in Nineteenth Century Anthropology', *History of Science*, 12 (1974) 260-83.

10 Herbert Spencer, 'Progress: Its Law and Cause', *Westminster Review*, 67 (1857), 445-57, 451, 454-56, 464-65. Mark Francis, *Herbert Spencer and the Invention of Modern Life* (Stocksfield: Acumen, 2007), pp.190-92.

11 *Descent*, pp.174-76.

12 *Descent*, p.178.

13 Richard Soloway, 'Demography and Degeneration', p.53, citing Darwin, *Descent of Man*, pp.501-02. The extract occurs on pp.133-34.

marry so freely as the sound.'[14] But this is not mere weakness. For in higher animals – and by extrapolation human societies – natural selection is supplanted by sexual selection. Darwin's prime concern is sexual selection and marriage. We see this in embryo in Darwin's bachelor calculations of 1838 whether he should endure the celibacy of a Cambridge professor or 'Marry, marry, marry'.[15] Herbert Spencer at one time drew up a numerical calculation on marriage and emigration but he decided that he would compromise his independence by marrying. He wrote in 1856 to his friend Potter, the father of the celebrated Beatrice [later, Webb]:

> Moral and intellectual beauties do not by themselves attract me; and owing to the stupidity of our educational system it is rare to find them united in a good physique. Moreover, there is the pecuniary difficulty. Literature, and especially philosophical literature pays badly. If I married I should have to kill myself to get a living.[16]

For Darwin, animal mating – and human marriage – were central to sexual selection. Ornament and courting replace predatory natural selection as we advance up the evolutionary scale. Darwin was interested in marriage rates of the poor, and infant mortality. He cited observations of Francis Galton that the poor marry early and the responsible marry late – thereby having fewer children:[17]

> The advancement of the welfare of mankind is a most intricate problem: all ought to refrain from marriage who cannot avoid abject poverty for their children; for poverty is not only a great evil, but tends to its own increase by leading to recklessness in marriage. On the other hand, as Mr. Galton has remarked, if the prudent avoid marriage, whilst the reckless marry, the inferior members tend to supplant the better members of society. (p.618)

Darwin was impressed by observations by William Rathbone Greg, a scion of Manchester merchants associated with Quarry Bank Mill at Styal. Greg in 1868 wrote 'On the Failure of 'Natural Selection' in the Case of Man'.[18] Greg saw a racial dichotomy between the profligate Irishman, who 'multiplies like rabbits,' and the 'frugal, foreseeing, self-respecting, ambitious

---

14 *Descent*, p.134.

15 Darwin, 'This is the Question Marry Not Marry' [Memorandum on marriage], (7.1838) CUL-DAR210.8.2 *Darwin Online* <http://darwin-online.org.uk> accessed 16 October 2009.

16 Herbert Spencer, *Autobiography*, 2 vols (London: Willams and Norgate: 1904), I, p.478.

17 Darwin, *Descent*, p.133.

18 *Fraser's Magazine* (September 1868), 353-62.

Scot,' who marries late (p.361). Greg admits 'a republic is *conceivable* in which paupers should be forbidden to propagate.' But the social tendency is counter-selective, not least because 'we absolutely refuse to let the poor, the incapable and the diseased die.' But no people would ever endure a selectively based set of laws because of the 'interference and control' involved. The key Victorian rituals of marriage and philanthropy became transformed into biological processes.

Darwin finds sound scientific reasons for assisting needy families rather than accepting the 'hard reason' of Galton and Greg. During the Irish famine Darwin condemned the Corn Laws, avoided potatoes and distributed penny bread tickets.[19] Darwin's actual behaviour towards the poor blended philanthropy with the self-help gospel of Samuel Smiles – whose book *Self-Help* was coincidentally published in 1859 (when *Origin of Species* appeared). Darwin was treasurer of two charitable clubs in the Kent village of Down: one was the Down Coal and Clothing Club. Its aim was 'to encourage saving Habits in the Poor by adding the Honorary Subscription to whatever they themselves pay in weekly'. The other club was a Sick or Friendly Club which he enrolled with the Registrar of Friendly Societies. His co-sponsor was his neighbour, Sir John Lubbock, Baron Avebury, who published on worker ants, and lectured the club members. During the farm slump of 1877 Darwin had to persuade the village labourers that it was improvident to dissolve the society and share the proceeds: only the surplus should be distributed. Interestingly, among Darwin's papers are the Societies' subscription books. Darwin also sponsored a fund for distressed scientists. Darwin thus espoused an ethic of altruism.

Darwin commended Lubbock for showing that all civilised nations are descended from the barbarian, and that the 'savages are independently able to raise themselves a few steps on the scale of civilisation' (*Descent*, p. 144). Darwin goes on to turn religion on its head by seeing it as a characteristic of higher evolution, by observing how: 'Indeed, belief in a God, hating sin and loving righteousness, is a mark of evolutionary progress. Indeed, it is Man, who has risen' (p.145).

We must conclude that Darwin was no social Darwinist in the conventionally understood sense. Scientifically, Darwin blended natural selection with Lamarckian ideas of adaptation to the environment; the transmissibility of acquired characters to the next generation; and the instinctive ability of animals to learn.

The crucial point concerning Darwin's views on human evolution was his observation of how in higher organisms sexual selection replaced natural

---

19 Desmond and Moore, *Darwin*, pp.333-34.

selection. The *Descent of Man* elaborated the evolution of a range of physical characteristics of animal colours and ornamentation to explain sexual differences and mating behaviour. Darwin saw human love and social sympathy as rooted in family ties and evolution. The desire for praise for moral actions was similarly an evolutionary characteristic (pp.610-11). Individual reason and the notion of an omniscient Deity were part of the evolutionary process.

The co-discoverer of natural selection, Alfred Russel Wallace gave more consistent support for the universality of Natural Selection than Darwin. But unlike Darwin he was unable to discern a naturalistic basis for morality, social action and what he called the realm of the spirit. Wallace's spiritualism meant that mind – and thus human action – could not be explained by selection. Wallace was a convinced socialist, and supported Chartist land reform as President of the Land Nationalisation Society.

Herbert Spencer appears to be a prime candidate for promulgating an exterminatory Social Darwinism. His principle of 'survival of the fittest' (dating from his *Social Statics*,1851) was taken over by Darwin in the *Origin* in the 1869 edition. Spencer's hostility to the state is legendary, and in the 1850s he attacked the inhumanity of the New Poor Law as the result of the superabundance of government. Again, the higher instinct of altruism comes into play. He concluded *The Principles of Biology*, first published in 1867, with the prediction that population increase resulted not only in individuation but also in higher emotional and intellectual life. This in turn restricted numbers to an average of two children per pair.[20]

III

While Darwin and Spencer had a liberal faith in moral and social progress, Galton represents a profound break. Where Darwin saw constant accumulation of variation, Galton's law of ancestral inheritance saw an immutable burden of the genetic past. His contribution to *Macmillan's Magazine* of December 1875, 'Hereditary Talent and Character,' marked a turning point: by rejecting notions of moral improvement in favour of inheritance. He rejected Darwin's Lamarckian notion of the circulation of hereditary particles. He argued that drunkards might have imbecile children and outbreaks of disease were likely to be caused by latent physiological characteristics. Galton called the conglomerate of hereditary particles, the

20 Herbert Spencer, *Principles of Biology*, 2 vols (London: Williams and Norgate, 1864-67), ch. 13: 'Human Population in the Future', II, pp.494-508.

'stirp,' and compared it to a nation; moreover, he compared 'the germs that achieve development to its foremost men'.[21]

Politically, the new Galtonian paradigm represented a meritocracy within a conservative social structure. Galton adopted the Gaussian bell curve structure to portray the distribution of social traits. He derived this from the statistician Quetelet's analysis of the distribution of chest size of Scottish soldiers. The elite had to be identified by competitive examinations: it was a recipe for professional elitism. One might see this as a Disraelian emphasis on inherited tradition as a foil to Gladstonian moral progress.

Galton rejected aristocratic inheritance of wealth and title. Similarly, poverty and crime could be seen not as moral lapse but as the outcome of decadent and mentally defective stock. His ideas on Hereditary Genius could be used to discourage the perpetuation of an underclass of less desirable qualities. Here lie the germs of Galton's eugenics, which he steadily outlined during the 1870s, and he took steps to encourage public interest in collecting pedigrees.

Galton's philanthropy was characteristically different to Darwin's coal club: Galton used his wealth to found a Eugenics Records Office, an institute and a chair at University College London. This was modern style philanthropy to determine the cause of social problems. The eugenic utopia Kantsaywhere owed its existence to Mr Neverwas leaving his property to the ruling council with the stipulation that no income should be spent on the support of the 'naturally feeble'.

How were the feeble to be recognised? One of Galton's technical inventions was the composite photograph – we can see how this was the basis for a biological typology of human types. Another was finger printing. Here he combined forces with the former surveyor Herbert Spencer in 1877 with the idea that social types have distinctive inherited features.

Galton's answers were marriage certificates and competitive examinations. Galton demanded marriage certificates for mental, physical and hereditary fitness issued by a state authority.

IV

In 1880s Britain the fear arose that medical advances and welfare were damaging the nation's racial stock by keeping degenerates alive. Infectious diseases were 'our racial friends'. Stamping them out only served to perpetuate Poor Types.[22] A German academic – Alexander Tille – went so

---

21 Francis Galton, 'A Theory of Heredity', *The Contemporary Review* (December 1877), pp 80-95, 86.

22 John Berry Haycraft, *Darwinism and Race Progress*, 2nd edn (London: Sonnenschein, 1895, 1900), pp.54, 57.

far as to claim that the high infant mortality among the poor in London's East End meant this was a 'Nationalheilanstalt' or 'national sanatorium'.

The today neglected Social Darwinist John Berry Haycraft shows that by the 1890s a younger generation was applying 'the biological standpoint' to social problems. Haycraft represented a new provincial biology and professionalisation, as well as internationalisation. Born in 1857, Haycraft studied at Edinburgh University, where he took degrees in medicine and in 1878 in public health, and – significantly – at Leipzig.[23] After spells as interim Professor of Physiology at Edinburgh and Mason and Queen's College Birmingham he was in March 1893 appointed the first Professor of Physiology at the new Welsh School of Medicine at the College of South Wales and Monmouthshire, located in the rapidly expanding coal metropolis of Cardiff.[24]

Haycraft first developed his views in the context of the Edinburgh Health Society. Founded in 1881 under the patronage of the Earl of Rosebery, it saw university academics and physicians engaged in giving lectures, issuing penny tracts and publications and making representations to the authorities on matters like cremation and overcrowding.[25] The Society published Health Lectures for the People. Lecturers preached self-improvement – in the guise of dietetics and personal hygiene. In 1891 Haycraft lectured to the Edinburgh Health Society on the 'Importance of Ideals of Health and Beauty towards Race Progress'.[26] He married an artist, suffragette and scion of an Irish dynasty, Lilie Stacpoole.[27]

The Royal College of Physicians in London invited Haycraft to deliver three Milroy Lectures. These prestigious lectures were held in March 1894, and appeared in *The Lancet*, providing wide dissemination of his ideas to the medical profession.[28] He expanded the lectures into his book *Darwinism and*

---

23 *The Times*, 23 April 1878.

24 *The Times*, 2 January 1923.

25 *Annual Meeting*, 13 February 1886, in *Health Lectures For the People*, VI (Edinburgh: MacNiven & Wallace 1886), pp.131-52.

26 *Health Lectures For the People* ser. 2, XI (1891), pp.17-32.

27 'Obituary of John Berry Haycraft, 1888-1969, born in Edinburgh 10 August 1888', in *Lives of the Fellows of the Royal College of Surgeons of England 1965-1973* (London: Pitman Medical, 1981), pp.154-55; compare *British Medical Journal* 20 (1969), I, 582. Deborah Cherry, *Painting Women: Victorian Women Artists* (London: Routledge, 1993), p.94, p.103. Haycraft was married three times.

28 J. B. Haycraft, Lecture I: 'The Milroy Lecture on Darwinism and Race Progress. Delivered before Members of the Royal College of Physicians at the Examination Hall, Victoria Embankment, Feb. 15th, 1894', in *The Lancet*, 17 February 1894, no. 3677, pp.383-88. Lecture II: 'Disease and Natural Selection, February 20th 1894', in *The Lancet*, 24 February 1894, no. 3678, pp.453-7; Lecture III: 'Crime and Incapacity', *The Lancet*, pp.520-23.

Race Progress. Haycraft took up Alfred Marshall's diagnosis of 1885 that 'charity and sanitary regulations are keeping alive, in our large towns, thousands of such persons who would have died fifty years ago.'[29] His views impressed Alexander Tille, a lecturer in German in Glasgow from 1890 until 1900, when his pro-Boer sympathies aroused student animosities. Tille's *Von Darwin bis Nietzsche* was published in 1895, and it had a formative influence on the radical feminist Helene Stoecker. She fell in love with Tille and enrolled in 1898 at the University of Glasgow. Tille adopted Haycraft's ideas of high child mortality as strengthening the race: as indicated above, the East End of London, with its high mortality rates (due, he thought, not least to alcohol and sexual excess), was for Tille an exterminatory yet redemptive 'Nationalheilanstalt'.[30]

Haycraft began by asking whether modern philanthropic effort was not damaging racial health? Deterioration was inevitable, unless 'selective agencies are maintained'. A key issue was that of poverty and the poor laws, highlighting 'our unfortunate use of the word 'Poor'.[31] Haycraft's approach was historical. He looked back to Tudor and Stuart discrimination between the deserving and undeserving poor, which he equated with the biologically degenerate. He criticised modern poor laws for treating the worthy and respectable men, women and children, people with excellent physique and mental capacity, in the same way as the unworthy. These are destitute either through deficiencies of education, or their special surroundings. The 'lazy and vicious' criminal class were lumped together with the worthy and respectable in the same rooms and wards of the poorhouse. He condemned 'the present disgraceful method of treating those without means of subsistence'. He cited a case in a North Country poorhouse: a soldier's widow, whose husband and three sons (all soldiers) had been killed on active service, was left without relatives. She supported herself and lived soberly until old age, when feebleness and commencing gangrene compelled her to seek the poorhouse: 'I saw her in the next bed to a drunken prostitute. The one woman had given of her body to the country's defence, the other had given of her body to its ruin, and yet the country treated them both alike because they were alike in want of bread' (p.99). It was an argument for an alternative measure of biological worth to the economic.

Haycraft prescribed reform of the poorhouse: it should become more like a hospital for chronic disease patients for the worthy, but the criminals

---

29 Cited in Stedman Jones, *Outcast London*, p.287. Compare Arnold White's *The Problems of a Great City* (London: Remington, 1887).

30 Wilfried Schungel, *Alexander Tille (1866-1912): Leben und Ideen eines Sozialdarwinisten* (Husum: Matthiesen, 1980), pp.47-48.

31 Haycraft, *Darwinism and Race Progress*, p.97.

should be strictly segregated to prevent their reproduction. The corollary was that professions should be open to talent of any social provenance rather than only the economically privileged of the English universities and public schools. He also prescribed marriage reform – the selection of parents as race-producers.

The third class of poor were destitute from 'innate incapacity'. Vagrants were 'the direct descendants of our more savage ancestry'. (Haycraft's evocation of Neolithic ancestors is echoed by Thomas Hardy's presentation of the Durbeyvilles and the Mayor of Casterbridge as primitive types). Such types had survived by the charity of monasteries and the Poor Law over generations. Haycraft feared their reproducing, and demanded segregation. He favoured a law for the Public Custody of Children to curb the rights of feckless fathers.

We now come to the exterminatory part of Haycraft: he argued that medical advances were increasing the numbers of the unfit, by diminishing diseases, which would otherwise carry off the weaker members of society. Phthis and scrofula were 'our racial friends' (p.54). Disease had a selective function. Yet he drew from this conclusions in line with Darwinian sexual selection: that the key issue was marriage: 'One wants marriage not with a view to the amassing of acres but for the perpetuating of fine physique' (p.523). His focus was on the positive issue of arousing 'family pride' rather than on the negative. Whereas Darwin stressed constant variation as a progressive force, Haycraft used the hereditary theories of Galton and the German biologist August Weismann to argue that sexual conjugation was inherently conservative.[32]

Haycraft represented a new breed of professional biologist, prescribing solutions to medical, human and social problems. He ultimately took the view that: 'The Masses must be Taught the Main Facts of Heredity and Evolution.'[33] The medical expert should prescribe the laws of social behaviour.

V

T.H. Huxley agreed that 'that the organization of society upon a new and purely scientific basis is not only practicable, but is the only political object much worth fighting for'.[34] The science was one of organic social cohesion. But Huxley recognised that natural selection was inappropriate as a basis for social action, and he was concerned about the prospect of a selective state. He used the metaphor of the selective gardener in 1892 when he

---

32 Haycraft, 'The Role of Sex', *Natural Science*, 7 (1895), 193-200, 245-50, 342-44.

33 'The Role of Sex', p.166.

34 Huxley, *Essays Selected from Lay Sermons* (New York: Appleton, 1871), p.94.

contemplated the destructive effects of Darwinian ideology. He raised the issue of whether a biological authority has the right to grade, exclude and eradicate supposedly surplus populations. He recognised that a new situation arises when the administrator limits the effects of natural selection. Thus 'hygienic precautions would check, or remove, the natural cause of disease'.[35] This was something not so unrealistic in the wake of the recent bacterial discoveries, and of the new serum therapy and vaccines for rabies and diphtheria.

Huxley raised the spectre of 'evolutionary regimentation,' as an abhorrent prospect. He saw population increase as destabilising. The administrator, like the gardener, could be tempted by eradicating or excluding the superfluous among the population. The excess included 'the hopelessly diseased'.[36] Huxley had prophetic insight into the exterminatory implications of natural and artificial selection for the chronic sick, but he insisted that ethics immunised humanity against the gardening state, protecting against any inhumane course of action.

VI

Turning to Germany, there was intense interest in the British discussions. Ernst Haeckel is customarily seen as the leading prophet of German Darwinism. He is associated with ideas of applying natural selection to human society. Yet he was no orthodox Darwinist. On the one hand he was a Lamarckian; on the other Darwin observes that Haeckel prioritised sexual selection when he explained 'the different races of men' in a branching tree from the primitive primal type. He observed how the ancient Spartans killed the weak, sick and deformed newborn. This is a text cited by historians of Nazi 'euthanasia' (the term was a euphemism for the killing of racial and social undesirables) to show Haeckel as a forerunner of the Holocaust implemented on the basis of an evolutionary understanding of racial biology. But – to go back to Haeckel's original words – he clearly saw the practice as an antithesis of modern civilised society rather than a model to follow.[37]

Texts by Huxley, Spencer, Haycraft, and Tille had a strong impact in Germany. The German biologist August Weismann drew on British Social Darwinists. He rejected the Lamarckism of Herbert Spencer, but he introduced works based on strict Darwinian natural selection by the political commentator Benjamin Kidd, and Haycraft. Kidd visited Weismann in 1890

---

35 T. H. Huxley, 'Prolegomena,' in *Evolution and Ethics* (London: Macmillan, 1893), p.44.

36 'Prolegomen', p.45.

37 Ernst Haeckel, *The History of Creation*, 2 vols (New York: Appleton, 1896), vol. 1, 152-53.

to prepare a major review of his theories.[38] Kidd used Weismann's principle of immutability of the *Keimplasma* ('germ plasm') to refute socialism.[39] Haycraft's work was translated into German as *Naturliche Auslese und Rassenverbesserung*. Along with the treatise by Alfred Ploetz, *Grundlinien der Rassen-Hygiene* Haycraft made Weismann's determinist concept of germ plasm central to Social Darwinism. This displaced the environmentalist belief in progress of Ernst Haeckel and Herbert Spencer, whose ideas of social altruism were underpinned by Lamarckism.

Ernst Haeckel read Spencer's theory of organic integration as necessitating a strong state in an evolutionary advanced society. Ploetz argued that a humane eugenics consists of reproductive controls. Schallmayer produced the scheme of the physician as state official. People should have an annual medical examination, and only the fit should be allowed to reproduce. Control on marriage was an answer to so-called 'racial poisons' as TB, VD and alcoholism. Custodial institutions would prevent the defective from reproducing. Schallmayer articulated a clear model of eugenic medicine. He advocated a shift from the physician taking care of the sick individual to becoming a state official. Inhabitants should be subjected to annual screening and issued with health passports. Sanitary inspection on borders and immigration controls were the corollary. He saw the possibility of a United States of Europe running collective institutions.[40]

The social hygienist Alfred Grotjahn's *Fortpflanzungshygiene* ('reproductive ethics') placed controls on reproduction at the centre of Weimar social hygiene policy. The concern was to weed out supposedly 'inferior' persons who could cause the spread of psychological and physical disabilities. Within the framework of a democratic state, professional experts sought to identify the so-called *Minderwertige* – 'those of lesser value'. On the other hand, those of high eugenic worth were to be encouraged to have large families, and there were special incentives for the *Kinderreiche* ('child rich'). Eugenics should be seen as part of the *Menschenökonomie* ('human economy') designed to promote a fit and strong population. Ethnic minorities as Jews and Roma were not the targets of Weimar eugenic welfare schemes, although

---

38 Weismann papers, Freiburg University, Kopierbuch 1, Bl 739, 784, 803; D.P. Crook, *Benjamin Kidd* (Cambridge: Cambridge University Press, 1984); Paul J.Weindling, *Health, Race and German Politics between National Unification and Nazism, 1870-1945* (Cambridge: Cambridge Unversity Press, 1989).

39 Benjamin Kidd, *Soziale Evolution*, mit einem Vorwort von Professor August Weismann (Jena: Gustav Fischer, 1895).

40 Wilhelm Schallmayer, *Die drohende physische Entartung der Culturvölker und die Verstaatlichung des ärztlichen Standes* (Berlin, Neuwied: Heuser, 1895); *Vererbung und Auslese. Grundriss der Gesellschaftsbiologie und der Lehre vom Rassedienst* (Jena: Fischer, 1918).

psychopaths, vagrants, the mentally ill and disabled were. *Erbkarteien* ('hereditary health registers') were compiled in prisons and welfare offices. Population was the 'organic capital of the state'. These concepts allowed for a biologisation of public health and social welfare Grotjahn saw the registration of pregnancies as a way of preventing high rates of abortion. On the other hand where pregnancies were deemed likely to result in defects, there was support for eugenic abortion. The clinics distributed many thousands of pamphlets on eugenics.[41]

We can thus see Weimar welfare – and the same would apply to welfare in the 'Red Vienna' of the 1920s – as a development of Darwin's sexual selection. The idea of a struggle for survival in society was attenuated by the welfare state, as well as sickness insurance for the family.

VII

Nazism represented a cataclysmic social upheaval, and a new set of rationales for violence and social action. A shift from sexual selection to natural selection reinforced the Nazi ideology of a racial struggle. The Nazis attempted to revive the primitive vigour of the German race. Rebirth and national rejuvenation would come through recovering this primitive vigour. Here was a type of biogenetic recapitulation à la Haeckel but with the values transposed away from biology to politics and culture. Darwin's natural selection now became a racial slogan for a war against the weak as a racial burden. This marked a profound contrast to Weimar welfare, where eugenics as underpinned by sexual selection was the overarching paradigm.

Historians have overlooked the rationales for this shift, preferring instead to draw lines of direct continuity without examining the biological rationales. The objects of attention shifted from a sociobiological view of problem groups in society to racial categories, especially Aryan and Jew. There were continuities, as when the Nazis seized hereditary databanks to identify the supposedly 'a-social,' hereditary criminals and the sexually perverse. Compulsory sterilisation might be seen as a rapidly introduced transitional measure, which was not explicitly racial according to the original formulation; but was imposed within a racial context with the idea of strengthening the German race. Attitudes shifted from the population and social welfare to the race. We see this with the Nuremberg laws for segregating Jews, as providing a context for marriage laws imposed at the same time.

Nazi values of race purity and fixed types sit uneasily with Darwinism: first, race was not a fixed value with Darwin. Blood purity was a mystic

41 Weindling, *Health, Race and German Politics*, pp.426-27.

notion. The Nuremberg Laws were concerned to segregate Jews from contact, especially sexual contact, with Germans. The Blood Protection Law forbade all sexual relations between Germans and non Germans. This effectively forbade marriages and sexual relations between Germans, Jews and Nonwhites alike.

VIII: Some Conclusions

Few of the writers and thinkers discussed here, it is now evident, were sole advocates of natural selection in society. A broadened view of their understanding of Social Darwinism as based on both natural *and* sexual selection thus illuminates ideas of welfare on the basis of biological value. Spencer shows that Victorian discourse on poverty emphasised solidarity as the virtue of a civilised society.

In addressing the problem of Nazi eugenics we need to move away from ill-defined myths of racial utopias, which neither explain what the Nazi racial planners intended, nor the means by which they could be achieved. Science should not eclipse how myth and symbol – Nordic, Greek, Aryan – had a powerful appeal, but how they fused with racial science and population theory in ways that were highly specific. Similarly, images of animal societies had a powerful role. Using these natural symbols had immense significance in the forging of biology as an agency of racial policy. They underpinned schemes of racial rejuvenation, which were based on the idea of recovering the primitive vigour of the Aryan, Germanic, or Teutonic race. The converse was destruction of races and pathogenic human types who posed a threat to racial regeneration: among these were the physically disabled, the mentally ill, homosexuals, a-socials, hereditary criminals, and alleged racial undesirables as Jews and Roma. Biology was indeed central to Nazi Germany and the genesis, planning and implementation of the Holocaust. But retrospectively ascribing the guilt of genocide to Darwin deviates from Darwin's own, well-documented evolutionary notions of civilisation, altruism, and the similarity of mentality whatever the physiological differences among humankind.

From Primate to Human in Two Easy Steps

Joseph Emonds[1]

Abstract: Several properties unique to human language arguably derive from 'Duality': two superimposed combinatorial systems, phonology and syntax. However, the discreteness in both systems is generally overlooked. Syntax, a human language 'trademark,' is strangely based on exactly those discrete categories plausibly present in primate vision, possibly primate cognition's only discrete categories. A first evolutionary step projected these discrete categories out of the 'here and now' into a computational screen. A second step dissociated the phonology atoms from meaning, leading to large lexicons and referential displacement.

I. Basic Hallmarks of Human Language

Since generative grammarians typically claim that humans have an innate species-specific propensity for natural language, the question of its *evolution* is immediately posed. Curiously, syntacticians have not spoken out on this issue. Authors of the few studies that relate generative syntax to evolution (most since 1990) are *not* primarily syntacticians, such as Pinker and Bickerton.[2]

Absence of reflection on this issue is not due to any dearth of research on primate communicative abilities. In fact, extensive research in this domain tends overall to confirm a qualitative break between non-human primate communication and human language. Using some non-trivial linguistic properties, this essay will try to elucidate the nature of this 'evolutionary leap,' and suggest two specific steps which must have required evolutionary (genetic) changes beyond adaptation of pre-existing mental tools. Needless to say, in any partly speculative exercise, parsimony plays a central role, so I will assume that properties of communication systems are derivative whenever this is broadly plausible.

My characterisation of 'two primary evolutionary steps' rests not only on some almost obvious natural language properties, but also on a relatively

---

1 This essay somewhat shortens 'From Primate to Human in Two Easy Steps', in *Grammar and Beyond: Essays in Honour of Lars Hellan*, ed. by M. Vulchanova and T. Åfarli (Oslo: Novus, 2005), and is reprinted here by kind permission of editors and publisher.

2 Steven Pinker, *The Language Instinct: How the Mind Creates Language* (New York: William Morrows, 1994), and Derek Bickerton, *Language and Human Behavior* (Seattle: University of Washington Press, 1995). Also: Renato Sabbatini, 'Language and Evolution', Part 5 of 'The Evolution of Intelligence', in *Brain and Mind Magazine* (2001).

developed perspective on the structure generally called the 'lexicon' of its speakers. The argument here elablorates a minority position 'suggested by the seventeenth-century French philosopher Étienne Condillac…[that] has not found favour among linguists or anthropologists…'.[3] I will develop my ideas step by step without any preview of 'where we are going,' and beg the reader's indulgence for this exposition.

To situate the leap(s) into language, perhaps the best starting point is to review some properties of human language laid out in two essays from the 1960s, (i) the somewhat popularised 'Origins of Speech' by Charles Hockett and (ii) an early generative counterpart, 'The Past,' by Noam Chomsky.[4]

As an American structuralist (writing at the dawn of the generative age), Hockett's goal was to enumerate and examine possible evolutionary aspects of natural language. His study lists about a dozen properties of 'communication systems,' and suggests that natural language constitutes a cumulative grouping of properties, none of which are pinpointed as the *differentia specifica* of human language capacity. This perspective is understandable in an era when 'anti-mentalism' and 'behaviourism' were key approaches to human psychology, and Hockett doubtless meant to establish continuity rather than discrepancy between the communicative abilities of humans and their primate cousins.

Chomsky's essay, in contrast, sets out to demonstrate three irreducibly human properties of natural language. After reviewing Hockett's contribution, I will propose an amalgam that tries to extract the essence of all these properties taken together. Precisely because Hockett's study is prejudiced against rather than toward the nativist perspective, it can provide a neutral point of departure for developing some generative-based ideas about an evolutionary linguistic leap. In spite of his tendency to blur the line between primate and human abilities, five of his properties are truly hallmarks of human language.

A. Duality of Patterning. This is Hockett's term for the pervasive fact that natural language consists of two *different superimposed systems*, phonology and syntax. The complex nature of this property merits a somewhat formal description. In any non-trivial communicative system, the smallest elements *in the system* combine by rule to give bigger units.

---

3 Michael Corballis, 'The Gestural Origins of Language', in *American Scientist*, 87 (1999), 138-45 (140).

4 Charles Hockett, 'The Origins of Speech', in *Scientific American*, 'The Origins of Speech', *Scientific American*, 203 (1960), 88-96 ; Noam Chomsky, *Language and Mind.* (New York: Harcourt Brace Jovanovich, 1968).

(1) Bee dancing: Body movements combine to describe food sources to other bees.
Primate signals: Hand claps and/or calls give messages and perhaps also have combinatorial properties.
Human language: Minimal sound units ('phonemes') such as 'w,' 'i,' 't,' etc. combine into morphemes.

Though the man on the street calls the bigger units of human language words, the minimal units of phoneme combination rather correspond to morphemes, because a single word like *interactive* contains three such 'units bigger than phonemes': *inter, act,* and *–ive*. The phonemes thus combine by rules or principles (= the system of phonology) into morphemes. The resulting morphemes are then 'relabelled' with 'meaningful' categories and *combine by yet different rules and principles* into well-formed words and sentences (= the system of syntax). Natural language combines these two systems and hence exhibits 'Dual(ity of) Patterning'.

A fundamental asymmetry in Dual Patterning is that the smaller combinatorial units, the phonemes (or perhaps syllables), typically have no 'meaning' in the system. Only the outputs of phonology, the morphemes, are related to the perceptual and remembered world of the speaker-hearer. The categories used in the larger system are thus 'meaningful' (ANIMATE, PATH, NEGATIVE, PLURAL, COMPARATIVE, etc.).

(2) Duality of Patterning (tentative). In *only human language*, the labelled outputs (morphemes) of combining minimal units, i.e. phonemes, again combine by different principles to make yet bigger units (maximal words, phrases and sentences).

B. Large Lexicon. Hockett discusses 'Traditional Transmission' of a lexicon, focusing on its being learned item-by-item from other speakers. Saussure's well-known 'arbitrariness of the sign' guarantees that at least phonological pairings with concepts in Large Lexicons are not genetic. [5] (A large percentage of the items themselves, *less* their phonological pairings, *may still be genetic.*) In evolutionary terms, a Large Lexicon is an 'acquired trait': most (or all) of it is new in every generation.

A recent study on types of vocabulary, mental storage and provenance of words concludes that modern adults who read moderately store about 20,000 lexical entries in their mother tongue.[6] This estimate of lexical size is

5 Ferdinand de Saussure, *Course in General Linguistics*, trans. by Wade Baskin (New York: McGraw-Hill, 1916).

6 Jean Aitchison, *Words in the Mind* (London: Blackwell, 2003).

not an esoteric extrapolation. Anybody can figure out roughly how many 'words' their vocabulary contains. One simply calculates how many entries one knows on a typical page of a fairly complete dictionary of one's native language. Compound words (*night shift, trigger happy*) and morphologically derived words (*interactive, disinfect*) with their own meanings of course count as separate entries. This number multiplied by the number of pages will almost certainly exceed 15,000 entries. In contrast, animal sign systems number at most a few hundred items, even according to fervid proponents.

As we collect separate properties of natural language, the evolutionary question arises: how many separate mutations are needed to account for them? It is hard to imagine the Large Lexicon as we know it without Dual Patterning. That is, Duality by its nature makes a Large Lexicon possible, and moreover such Lexicons seem recognisably similar to those actually occurring. Suppose simple two syllable words (morphemes) of the form CVCV can be formed choosing first and third consonants from among 30 consonant clusters, and second and fourth vowels from among 5 vowels (e.g., *batri, sleko, trima, ronu, zuye,* etc.). This already yields 30 x 5 x 30 x 5 = 22,500 words, the order of magnitude of a human lexicon – even with no compounds or long words. Of course, occurring languages have many accidental gaps in lexicons, but even if half the 'possible basic morphemes' are missing, compounds, complex morphological formations and long words easily make up for them. Thus, even a simple system with Duality yields lexicons approximating the size of those in natural languages.

While it is unclear how Dual Patterning in itself enhances survival, having words for more things and actions is certainly advantageous. A species with a potential for a large lexicon acquires an evolutionary edge by developing one. In teleological terms, since humans 'needed' Large Lexicons they 'needed' their prerequisite, Duality. Evolutionary parlance would say rather that Large Lexicons are the competitive advantage bestowed on a species capable of Duality of Patterning, while the latter is the actual evolutionary mutation.

C. Displacement. Humans talk about the past and possible futures in detail. They use language to plan, lie, joke, be ironic, mislead, make propaganda, construct imaginary worlds (fantasy, outer space) and discuss distant places. Animals, as far as humans can tell, don't communicate about such absent things. This property Hockett calls Displacement. Interestingly, he takes 'bee dancing,' a system of body movements by which some bees indicate to other bees the location and quality of food sources, as an instance of displacement. But on reflection many researchers judge that such sources are strictly in the bees' 'here and now'; their dancing cannot lie, recount past food gathering, elaborate stories, express unfulfilled wishes, propose new techniques, or invent variations (let's take two hours off; let's eat first, etc.). In contrast, human language is not so limited, unless we beg the question by defining the human 'here and now' as qualitatively different from that of other species.

However, one aspect of human communication is, like bee dancing, limited to a narrowly defined 'here and now,' namely our gestures: beckoning, greeting, rejecting, expressing the heat, showing hunger or boredom, etc. Human gestures either refer to here and now, or language must set the scene for their use, such as placards for mime troupe performances. To my mind, this limitation strongly suggests that the gestural system evolutionarily predates the initial quality that allows for Displacement, namely Dual Patterning. Section V returns to this issue.

We should ask the same question about Displacement as about Large Lexicons. Is it a separate change in the transformation of primates to humans, or might it be a dependent factor? Logically, in a kind of 'prehistoric science fiction,' some species might possess Dual Patterning and potential Large Lexicons, but could only talk about an animal-like 'here and now'. They might well die out. Why?

Displacement allows naming increasingly many absent things, including details about food sources, strategies for avoiding distant dangers, innovative shelter building, etc. The possibility of such discussions, using combinatorial syntax, must bestow enormous evolutionary advantages. And clearly Displacement feeds back into naming ever more absent things. Thus, a Large Lexicon and Displacement plausibly develop in tandem, once Dual Patterning is available.

Though we are speculating somewhat, it seems implausible that species capable of naming and discussing distant, future and desirable things would just not do so. For investigating the human linguistic leap, I therefore take Displacement as a secondary effect of Dual Patterning and the Large Lexicon.

D. Discreteness. All *systems* that humans *invent*, as opposed to inherit, tend to accord a central place to discrete elements, i.e. elements that can be counted. We can count in language (the phonemes in a morpheme or morphemes in a sentence), music (tones on a scale or notes in a passage), writing systems (the graphemes), kinship systems, arithmetic, card games and board games, political categories and divisions, etc. In contrast, countability has no role in animal creations (anthill tunnels, bird nests, beaver dams, bee swarms, etc.). Above the atomic level (i.e. a level unobservable without instruments and human theoretical constructs), no digit above four or five plays a significant role in animal structure or creations; e.g., eight-legged insects doubtless have four legs on each of two sides.

Discreteness is the first natural language property encountered here that seems solidly independent of Dual Patterning. Two factors even suggest that Discreteness is *prior* to Duality. First as just mentioned, essentially any specifically human mental creation centrally involves Discreteness, implying that natural language Duality itself is a special case of a more pervasive mental characteristic. Second, no aspects of Dual Patterning systems fall outside Discreteness. That is, *systems of phonemes and morphemes* introduced under Duality of Patterning consist of *countable combinations of discrete items.* In contrast, loudness of sound, absolute pitch of voice, and physical length of sentences, perceived as varying along continua, are characteristically *not* parts of phonological or syntactic systems.

One can even envision that Discreteness is the foremost natural language property, and that other human systems (kinship, arithmetic, writing systems, music, scientific models, games) are all developments from discrete linguistic systems. Chomsky has expressed the view that at least arithmetic ability depends on natural language capacity, which I find convincing. However, I argue later that mental Discreteness has a pre-human source and that its use in language is an extension, not a fully independent primitive ability.

In any case, among Hockett's properties of natural language at least two are fully independent: Discreteness and Duality of Patterning.

E. Productivity. This is a fifth communication system property Hockett attributes to natural language. Humans produce and understand well-formed sentences and phrases in unlimited numbers. This property is also the first of three discussed in Chomsky's essay: natural language is innovative.[7] The recursive mechanisms of generative grammar have essentially explicated the formal property of innovative productivity, which incidentally seems more characteristic of syntax than phonology. Keeping to a parsimonious treatment of linguistic evolution, we can subsume innovative productivity under a proper definition of Duality. That is, syntax by definition includes a *recursive* system for combining outputs of phonology (morphemes).

> (3) Recursive Duality of Patterning. In only human language, meaningful combinations (= morphemes) of discrete phonemes

[7] Chomsky, *Language and Mind*, ch. 1.

again combine by recursive principles of discrete syntax to make yet bigger units.

F. Freedom from Stimulus Control. Elaborating on Descartes's contributions, Chomsky contrasts the largely automatic communicative impulses of animals with the human ability to use language more freely. On one level, this is not a property of human language but rather a lack of one, i.e. no stimulus/response mapping need be charted. But of course, the implication is not that human language use is random. Rather, 'we can distinguish normal use of language from the ravings of a maniac or the output of a computer with a random element.' Instead of stimulus control, human language use has an even more difficult property to characterise: appropriateness.

G. Appropriateness. Chomsky emphasises that the innovative sentences humans constantly construct and understand are neither random nor deterministic, but typically 'appropriate and coherent' in situations. In contrast, animal communication appears 'stimulus-bound' and 'deterministic'.

Despite Chomsky's convincing Cartesian argument for stimulus-free and situationally appropriate natural language, this may not constitute an independent evolutionary step. Compared to primates, humans control a complex system for meaningful calculations about absent concepts (a Large Lexicon, Displacement and the syntactic part of Duality) far superior to any possessed by other primates. Now, some admittedly mysterious 'survival use law' may dictate that species use survival tools in accord with their potential, at least to a great extent. Precisely because the natural language tool allows discussion of plans, complex alternatives, etc., such a 'survival law' may dictate that it be used effectively, i.e. non-deterministically and appropriately.

In this regard, the (apparently non-communicative) navigation systems of migratory birds also make complex adaptations to novel situations. That is, endowed with complex and modifiable systems, birds somehow must use them 'appropriately'. Perhaps then *any* animal systems, including of communication (rudimentary or human), develop novel uses to enhance survival. So I remain unsure that non-deterministic appropriate use is a specifically linguistic property; Displacement and Appropriateness may just be usage sides of coins whose formal sides are Large Lexicons and Recursive Duality.

II. Why does Human Language Combine so many Species-Specific Properties?

Hockett's and Chomsky's studies brought out seven specifically human language properties. *In an evolving species, which of these might automatically lead to which other(s)?* Could perhaps just one or two crucial developments or mutation(s) lead to human language? (3) above suggests one minimal evolutionary scenario combining three logically independent properties: Duality, Productivity, and Discreteness. However, this formulation is deceptively compact.[8] It glosses over the internal complexity of each subsystem and so may merit decomposition into several developments.

I therefore *decompose* Duality (3) into three separate properties. The different factors are perhaps best analyzed not as Hockett originally did, but rather as follows:

> (4) Lexicon Formation. An inventory of c. 10^2 meaningless discrete elements sequentially combine by principles of vocal compatibility ('phonology') to form stored ('lexical') items.
>
> (5) Lexical Labelling (tentative). Lexical items are associated with ('syntactic') labels drawn from the concepts in an organism's perception and memory.
>
> (6) Computational Syntax. Labelled discrete lexical items combine by syntactic principles into sequences usable at interfaces of understanding ('LF') and transmission ('PF').[9]

I will not further investigate the phonological side of the Lexicon (4) or Computational Syntax (6). This essay pursues rather the following question concerning (5), which an evolutionary perspective can tentatively answer.

> (7) Which among many possible conceptual labels does syntax use, and why are they based on discrete perceptions?

---

8 Additionally I have argued, based on parsimony, that Large Lexicons, Displacement and Appropriateness are all adaptive 'consequences' of Duality; this may eventually prove too simplistic. Recall also, Chomsky's Freedom from Stimulus Control is actually the lack of a property.

9 Chomskyan models propose two outputs of a sentence's syntactic derivation: Logical Form (LF) transferred to the interface with understanding/ use, and Phonological Form (PF) transferred to the auditory/ speech interface.

Unravelling the enigma of Discreteness may also eventually help us answer (8):

> (8) How does discrete syntax underlie other human cognitive inventions (kinship systems, arithmetic, scientific models, writing systems, music, games)?

What follows focuses on (5) and question (7), developing a scenario with two evolutionary steps: one involving Discreteness (section IV) and then a Duality of Patterning mutation (section V). Whichever relation between the two turns out eventually to hold, we need first to review recent lexical research that indicates *the nature of the lexical labels* in syntax (7). This research reveals an internal split in lexicons that plausibly reflects evolution quite directly.

III. A Crucial Division within the Large Lexicon

One way to discuss labels in human mental lexicons is to say that its items contrast in terms of 'binary features': any two items differ by the ± values of at least one feature. Much interesting work on lexicons is still couched in this (fairly primitive) formalism. As said earlier, each speaker controls some 20,000 entries. Consequently human lexicons must contain *many features.* To fix ideas, we can minimally reflect on a few sample features, choosing some that significantly differ from one another, to get an idea of possible variety. We start with features of nouns (N), some of which also occur on the related category PRONOUN:

(i) ±Animate. This important grammatical feature distinguishes e.g. *cat, crowd, us* and *who* from *house, courage, it* and *what.* It also helps determine what kinds of nouns appear in diverse contexts:

> (9) The storm scared the {cat/ crowd/ two of us/ *house/ *courage}.

±Animate is clearly a binary feature. We judge things as animate or not, though sometimes we judge wrongly.

(ii) ±Count, i.e. Count vs. Mass Nouns. This feature is crucial in many languages for determining whether numerals and certain quantifiers (*many, few, every, each*) can modify a noun.[10]

> (10) She presented {four/ few} {new factors/ concerts/ *evidence(s)/music(s)}.
> She didn't present much {*new factor(s)/ *concert(s)/ evidence/music}.

Similar to animateness, a noun is either mass or count, or possibly both. But there is no continuum; mass vs. count is a discrete dichotomy.

(iii) +Temporal, ±Interior, i.e. names of days vs. periods of time. English treats a day and certain temporal stretches as 'surfaces' which one acts 'on,' akin to two dimensional spatial surfaces. Other temporal stretches are analogous to 'volumes' which one acts 'in':

> (11) Let's visit {on/ *in} {Friday/ Easter/ Homecoming Weekend/ our anniversary/ 7-6- 2010/ my lunch break}.
> Let's visit {in/ *on} {June/ Lent/ Homecoming Week/ summer term / 6- 2010 / the next hour / the daytime/ third period}.

This distinction clearly affects grammatical choice of prepositions: 'in' volumes, 'on' surfaces, and 'at' points.

(iv) ±Artefact, i.e., products vs. natural kinds. Various linguistic contexts require one or the other, but not both. The objects in (12a) are possible product names while those in (12b) can only be natural kinds.

> (12) a. They {constructed/ planned} a {park/ space for kids/ pond/ cooking area/ new artificial heart}.
> b. ??They {constructed/ planned} a(n) {cat/ cloud/ sea/ earthquake/healthy heart}.

Again, some items such as *lake* can be in both categories, depending on context.

(v) ±Personal quality. Many nouns for 'personal qualities' have intrinsically positive or negative connotations. Thus, *love, courage, strength, respect, generosity* are 'good,' while *hate, cowardice, weakness, fear, selfishness* are 'bad'.

---

[10] Some nouns (*bread, argument*) can be mass *or* count and appear with both types of modifiers.

We could enumerate long lists of such features: names of things that 'flow' vs. names of things that don't flow. But already these first examples allow some conclusions that elucidate the nature of lexical labels. Although all features help determine which combinations of lexical items 'make sense,' *only some features contribute decisively to syntactic well-formedness* (grammar). Violating expected combinations for non-grammatical features such as 'artefact' or 'personal quality' leads to metaphors, secondary interpretations, etc., but not straightforwardly to ill-formedness. Thus, examples (12b) are possible in constructing theatre sets or devising story plots. Similarly, sentences like *His love was stupid* and *Some old-fashioned selfishness might help* are syntactically perfect. Among the few features just discussed, ±Animate and ±Count are thus *grammatical features,* while many more features (+Artefact, ±Personal Quality, etc.) are not. As a notation for this distinction, I will use upper case F for grammatical features (e.g. ±ANIMATE, ±COUNT) and lower case f for purely semantic features (±Artefact, ±Personal Quality, etc.). The system of syntax properly speaking is then by definition the theory of how upper case F features combine, but is independent of lower case f features.

(vi) Quantifier features. We have been concentrating on features of Nouns. But what of features on grammatical categories like Determiner (DET), which includes numerals and quantifiers? For example, how is *each* different from *every* or *all*? *Each* and *all* but not *every* can 'float' (13a), while *every*, unlike other English quantifiers, cannot appear separated from a noun (13b):

(13) a. The girls will {each/ all/ *every} leave.
b. Those irresponsible students have now{each/all/ *every} left.

The generalisation is that *every* can't appear 'alone'. But for plurality, it is *all* that differs from the other two:

(14) {Each/ every} girl(*s) {is/ *are} late.
All (those) girls {*is/ are} late.

Two features are needed to semantically differentiate three 'universal quantifiers,' and both are also *used in grammar*, i.e. for number agreement and to allow/ forbid separation from Nouns. A range of similar arguments suggests that *every feature that semantically differentiates any pair of DETs is used elsewhere in syntax*, i.e. all are grammatical F. Unlike many groups of Nouns that differ only by purely semantic features f, any two DETs have distinct syntax.

Noun of course is a category with huge lexical membership. Another large category V also has some F used in syntax, but some purely semantic features f as well.

(vii) ±STATE, Activity vs. State. Verbal activities such as *learn* and *buy* differ grammatically from states such as *know* and *own*. Only activities freely allow continuous tenses and imperatives:

(15) a. Mary was {learning / *knowing} calculus at home.
We are {buying/ *owning} bicycles next week.
b. Please {learn / *know} that address immediately.
Don't {buy/ *own} umbrellas today.

Because of these grammatical distinctions, ±STATE must be a grammatical F.

(viii) ±Design Loss. One also easily finds purely semantic verbal features. Direct object 'products' of some 'damage verbs' can still be used for their design, while for others they can't. Some feature ±f' must differentiate these two classes:

(16) a. John's bicycle was {damaged/ harmed/ messed up}, but he uses it still.
b. ??John's bicycle was {ruined/ destroyed/ wrecked}, but he uses it still.

This verbal feature f' means their objects 'lose their intended use'. It is a quite implausible candidate for the syntactic system. Among other things, certain infelicitous uses lead to special meanings rather than ill-formedness. Thus, examples (16b) are acceptable followed by *as a theatre prop*. By comparing features like ±STATE and ±Design Loss, we see that Vs as well as Ns occur with both syntactic F and purely semantic f.

My previous work investigates a range of features on different categories and concludes that *only the lexical categories N, V, A and P* have purely semantic features ±f as well as exhibit grammatical features such as ±ANIMATE, ±COUNT, ±STATE.[11] Items in all other syntactic categories such as DET, AUX, etc. are characterised *with grammatical features alone*. The human syntactic computation, consisting of combinatorial rules for the

---

[11] Joseph Emonds, *Lexicon and Grammar: The English Syntacticon* (Berlin: de Gruyter, 2000), chs. 3, 4.

grammatical features F, therefore *expresses limited but not trivial semantic content* provided by concepts such as ANIMATE, COUNT, PLURAL, STATE, etc. However, syntax *does not include finer grained concepts* such as Artefact, Personal Quality, Flows, Design Loss, etc. Rather, these latter concepts are 'tacked onto' the basic syntactic system only through association with items in four large lexical {'open class') categories labelled N, V, A and P. *These four categories are the sole syntactic 'carriers' of purely semantic features.*

This division into two types of features and types of categories is central to a lexical theory elaborated by some other authors and myself since 1985: a given language consists of a small syntactic lexicon (probably less than 500 members), which I call the *Syntacticon* and Ouhalla the *Grammatical Lexicon.*[12] *All members of categories other than N, V, A and P are in this subcomponent.* They never carry purely semantic features but do have privileges of non-canonical positioning not available for open class items. The latter classes (N, V, A, P) form Ouhalla's *Mental Lexicon* or my *Dictionary,* and encompass 20,000 plus entries. Each such item (i) has at least one purely semantic feature, (ii) appears only in canonical positions for their syntactic category, and (iii) fulfils contextual insertion conditions 'phase-initially,' i.e. prior to syntactic derivations.

The structure of this bifurcated lexical theory lends striking support to the following evolutionary account.

IV. An Evolutionary Scenario for Syntax

If we compare the limited semantic content of syntactic features F to the richer system of purely semantic f, a curious 'anti-correlation' emerges:

> (17) Almost all concepts of human syntax F are among those plausibly associated with cognition of highly developed non-human primates, precisely those who *lack* syntax.

A review of rough meanings of familiar syntactic features confirms their likely presence in advanced primate cognition. Non-human primates recognise living beings (±ANIMATE) and their own species (±HUMAN also functions in syntax), and doubtless distinguish aggregates from discrete sets (±COUNT) and activities from states (±STATE). They differentiate universal (*all, every*) from existential quantification (*some, many*) and reason

---

12 Jamal Ouhalla, *Functional Categories and Parametric Variation* (London: Routledge & Kegan Paul, 1991) and Joseph Emonds, *A Unified Theory of Syntactic Categories* (Berlin: de Gruyter, 1985).

using low integers with distinct grammatical properties.[13] Primates almost certainly conceptualise basic locations (*up/ down, in/ out*), recognise movement, and have some sense of 'excessive': *too hot, too heavy, too far.*

The opposite seems to hold for many specifically human semantic features; they are not in syntax. We assume primates don't clearly distinguish constructed artefacts from natural kinds or know when products are 'usable according to design.' Primates probably don't conceptualise 'temporal periods' longer than a day or classify their peers according to virtues and vices (*respectful, generous, brave*). Strangely then, *a combinatorial syntax of 'non-human semantic features' is precisely what seems to constitute the differentia specifica of humans.* Without evolution, this conclusion would be a paradox. However, an evolutionary approach can make sense of correlation (17).

For ages, our primate ancestors had no syntactic system at all (no F). Moreover, these ancestors certainly *had fewer semantic features {f'}* than today's humans {f}, that is, {f'} C {f}. Primate semantic f' might include concepts of 'flow' and 'fast,' but not of Artefact, Personal Quality, Design Loss, etc. But then a linguistically endowed mutant appeared, passing on this gene to progeny. Eventually, some mutant descendents (a mother-child or sibling grouping) were able to find some *subset {F} of primate concepts {f'}* to construct both (i) a primitive lexicon and (ii) a computational syntax. Their set of syntactic features {F}, chosen from {f'}, included ANIMATE, COUNT, 1 vs. 2 vs. 3, PLURAL, STATE, MOTION, LOCATION, TEMPORAL etc. Precisely because these 'ancestors' had no access to subsequent human culture, the initial building blocks of their syntax (the humanizing innovation) *necessarily drew on semantic concepts present in their pre-human primate culture.*

In this way, an evolutionary perspective on language genesis provides a natural account of (17). Human computational syntax is based on cognitive features controlled by our primate ancestors, and interestingly is still essentially limited to such features. The many human conceptual features introduced into human culture since the invention of syntax are simply not part of syntax, but 'tacked on' (to only the categories N, A, V and P) outside of it.

This evolutionary explanation of (17) leads directly to another challenging topic. Almost certainly the first primate linguistic mutants also had concepts f' of brightness, length, danger, distance, speed, etc. which were not and still are not part of syntactic systems {F}. So we ask:

---

13 Marc Hauser, 'What do Animals Think about Numbers?', *American Scientist*, 88 (2000), 144-51.

(18) How were the first speakers' set of syntactic features {F} related to the totality of their primate concepts {f'}?

The answer depends on what ties together concepts expressed by say the *current* set of grammatical F, sampled in (19).

(19) Noun features: Count vs. Mass; Plurality; ±Animate; ±Human (recognition of our own species); ±Feminine
Determiner features: Demonstratives, based on speaker proximity, e.g. Spanish *este, ese, aquel;* Universal vs. Existential Quantifiers ; features for each of 2, 3, 4 (*once, twice, thrice; first, second, third,* not formed regularly)
Preposition features: Path (*into*) vs. Place (*within*); Temporal vs. Spatial
Verb features: Activity vs. State; Perfect vs. Incomplete Aspect; Real vs. Irrealis 'Mood' (English Tense vs. Modal)

Upon reflection, almost all – perhaps all – of these current syntactic F are likely candidates for primate cognition f'. But many other primate cognitive categories are *not used* in any syntactic system: brightness, loudness, speed perception, non-speaker centred motion, awareness of emotion (sadness, fear, anger, arousal, illness), and essentially all categories of feeling, taste and smell. What key can demarcate early human syntax {F} from so many primate concepts {f'}?

Most features in (19) seem like concepts understood first and foremost as *visible properties.* Not only do we privilege vision in interpreting reality, vision itself divides our reality into discretely categorised entities. Moreover, *vision seems to be the only human (or primate) sense that reports things discretely.* It therefore makes sense to attribute to *our primate vision* the abilities to instantly: count (small) numbers of entities; classify things as ±Animate, ±Human, ±Female and as moving or not; and divide territory into three discrete fields (in Japanese, of the self *kono*, the other *sono*, and neither *ano).* For primates as well as humans, vision is plausibly what divides activities from states and paths from places. Generalizing, the basis of grammatical features F is apparently *the subset of visual features conceptualised discretely.*

Conversely, visual concepts that for both us and primates are continuous never translate into grammatical features. Moreover, *no other sensual mode utilises instantaneous discreteness.* Here are simple, relevant 'home experiments'. It's very difficult to tell by feel alone how many toothpicks or matches are pressing into a single area on your arm or back, or to count discrete sounds heard simultaneously (in a band or moderately noisy street). Counting simultaneous tastes or smells is almost inconceivable. As often observed, linguistic terms for smell and taste are impoverished even in our rich cultural feature inventory; most are metaphors from nouns (*salty*, *oily*, *rotten egg smell*, etc.). Thus, *early humans apparently used no concepts from smell, taste, feel or hearing domains for assembling a syntactic feature set.*

In general, psychology has just *failed to recognise* that discreteness in primate perception, including humans, is *entirely based on vision*, actually static vision. We and primates judge (some of) what we see in discrete terms, though of course not speed, brightness, distance, feeling, sound, taste or smell. Once we recognise that we and primates share discrete visual categories, the following answer to (18) is a natural source for Discreteness:

> (20) Lexical Labelling; replaces (5). Lexical items are associated with ('syntactic') labels drawn from the mutant primates' *discrete visual concepts* (their only conceptually discrete system).

In current syntactic models, grammatical features F in the Syntacticon best reflect this labelling system.

By associating discrete visual categories with labels both understandable and transmittable (without visual stimuli in the immediate here and now), primate mutants were on the road to language. They of course still had to combine these items via Computational Syntax (6). These initial lexical labellings and computational syntax didn't necessarily exhibit Dual Patterning or true Displacement; the primitive lexicon was probably limited to a few hundred items, like sign inventories of today's most competent primates and like a human 'Grammatical Lexicon' or 'Syntacticon'. Their cardinalities are all within the order of 10^2 rather than 10^3.

The novelty in primate mutant syntax was *dissociation of discrete visual concepts from their real world exemplars*, and the computational ability to combine these dissociated labels (categories) in novel and recursive ways:

> (21) Syntax Mutation. The first humans, dissociating discrete visual categories from stimuli, became able to *transmit them as symbols* and *compute their syntactic combinations.*

Thus, rubbing the stomach to signal hunger, one parent could say to another: *we need four things; you get two now and I will get two when you get back*. Or a mother about to give birth could say, *go outside and bring back a woman and other stuff* (signing for water). Such communications clearly have survival value and far exceed abilities observed in primate studies. In such examples, the Syntax Mutation itself accounts for the beginnings of Displacement, which a Large Lexicon later expands.

The diagram (22) schematises a central hypothesis of this essay (21). The smaller oval = {f'} represents primate cognitive features. The larger oval = {f} represents human cognitive and cultural features, many developed since mutation. The grey area represents *primate discrete vision*, the origin of syntax.

(22)

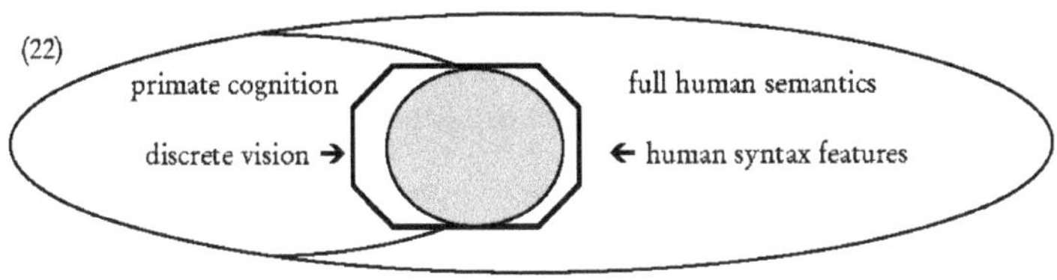

The octagon = *human syntax* = the set {F} = the discrete vision subset of {f'} + any features F' of human syntax developed after mutation (21).

Syntax is thus fundamentally a system of displaced and simplified mental vision, whose elements are morphemes. Moreover this 'impoverished' vision is limited to two dimensions, formally represented in linguistics by 'precede' and some second notion like 'dominate' or 'govern'. Unlike vision, non-discrete elements are filtered out, and simple morphemes lack internal structure accessible to other morphemes. Hence language represents the external world very imperfectly.

Now two final questions about specifically human material outside the smaller oval in (22).

(23) What about semantic f developed only in humans?

The Syntax Mutation Conjecture (21) says nothing about the myriad purely human cognitive and cultural f. These concepts today overflowing our Large Lexicons were simply not available for the Syntax Mutation. *Thus syntactic features are a kind of 'still photo' of immediately pre-human primate culture in the 'evolutionary movie'.* From their absence in syntactic systems, human language predates artefacts, high counting and music. But from their presence in syntax, quantification, plurality, and a sense of complete vs. incomplete actions must predate language.

(24) Are some F' in current syntax not from the set {f'} of pre-human primates?

There doubtless exist some 'post-primate' syntactic F'. For instance, primates plausibly lack Jespersen's mood distinction 'Realis vs. Irrealis' central to many auxiliary systems (English TENSE vs. MODAL). Many linguists say completed vs. uncompleted actions, so-called aspect, is more widespread and acquired earlier than distinctions among (realis) tenses or among future and other (irrealis) modals. Primates might thus have aspect (being able to *see* actions as completed or uncompleted), whereas only humans have developed mood and tense (one doesn't *see* the past, much less the future).[14]

V. Duality again: Large Lexicons

Section II decomposed natural language's Dual Patterning into three components: Lexicon Formation, Lexical Labelling and Computational Syntax. Section IV proposed an evolutionary source of Lexical Labelling, reformulating it as in (20). These labelled items furnish a bare bones grammatical skeleton for Computational Syntax (6).

However, I have not yet linked either (20) or (6) to the essence of Duality, Lexicon Formation (4).

(4) Lexicon Formation. An inventory of c. 10^2 meaningless discrete elements sequentially combine by principles of vocal compatibility ('phonology') to form stored ('lexical') items.

This formulation incorporates a widespread assumption that atomic units (morphemes) in the earliest Computational Syntax were phonological. However, the essence of the Syntax Mutation (21) was *transforming visual categories into symbols*. Almost by definition, such symbols would be 'gestural'; for them to change from the visual to vocal mode requires another step. As a simultaneous change to the vocal mode would be unmotivated, I hold that *initial syntactic lexical items were atomic gestures,* not words. This fits well with

---

14 Determiners furnish more examples for F' in (24). Primates probably lack 'Definite', i.e. a category signalling previous reference, Also, Hauser (note 13) claims animals compute accurately only through 3. Yet Slavic grammar has special syntax for 1-2-3-4, counting recursively only above 4. Humans may thus have raised 'low counting' to 4.

primates' most advanced sense being vision – as shown for example by their rich mix of discrete and non-discrete visual categories.

M. Corballis (see note 3) provides other arguments that vision is the 'more highly developed' primate sense, reviewing interesting correlations between cerebral hemispheric dominance for language and for gesturing. Though the models he cites explaining the correlations are not straightforward, the gestural origins they suggest for language have not been successfully explained away.

Are there actually reasons to doubt that linguistic mutation first led to gestural syntax, prior to development of Duality? Corballis's summary finds two objections to his (and my) position:

> Language evolution emerged not from vocalisation, but from manual gestures, and switched to a vocal mode relatively recently in hominid evolution, perhaps with the emergence of H. sapiens. This idea was suggested by … Étienne Condillac and revived in the 1970s by the American anthropologist Gordon W. Hewes. It has not found general favor among linguists or anthropologists, perhaps because *it lacks parsimony* and because *there is no direct evidence* that any of our hominid ancestors gestured rather than spoke.[15]

But parsimony cannot be invoked against gesture-based theory. The slightest awareness of language systems shows that (i) human speech uses complex combinatorial systems of (meaningless) phonemes, and (ii) humans have a second potential for recursive syntax. Moreover, sign language studies show that the two potentials are independent. It cannot 'lack parsimony' to state the obvious, that human language involves two fully independent non-primate properties. Nor can it then offend parsimony to postulate their emergence in one or the other order. And as for his second point, there will never be 'direct evidence' either way: neither that our extinct hominid ancestors gestured rather than spoke, nor the opposite.

In this view, *human language first consisted of a gestural Syntacticon*, whose members combined using recursive syntax and whose inventory of features was nearly the same as in language today. These few hundred items were based on discrete primate concepts, necessarily visual because we are primates. This early language had no purely semantic features nor open classes of lexical items. Nonetheless, its 'syntactic' gestures may well have been supplemented by non-syntactic gestures like those still used for non-discrete categories such as hunger (rubbing the stomach), beckoning, anger (shaking a fist), cold (clasping one's elbows), reprimanding (shaking a finger), bad smells, rejection, etc. Then as now, these non-syntactic gestures

---

15 See Corballis, 'Gestural Origins', p.140 (my italics).

could not in themselves escape the perceived 'here and now'; only an accompanying use of syntax could accomplish the Displacement function.

Let's now address the emergence of Large Lexicons. As the store of syntactic and non-syntactic atomic gestures approached its carrying capacity, c. 400-500 (today an approximate Syntacticon size), pressure grew to enlarge lexicons with some more flexible system of smallest elements. The new *'building blocks,' which had to be independent of meaning, were found in the (until then meaningless) vocal mode.* These 'sounded' items were phonemes (or perhaps syllables); since phoneme or syllable inventories in the world's languages number under 100, the order of magnitude of the smallest systemic elements, 10^2, remained as in the lexicon of atomic gestures.

I feel this required some independent mental mutation (25) but perhaps not; cf. Lieberman's extensive discussion.[16] This separate innovation, whatever its nature, furnished human language with Dual Patterning and paved the way for open lexical classes, Large Lexicons and fully developed Displacement.

(25) Phonology Mutation. Gestural lexical items were replaced by permanently stored sequences of inherently meaningless phonemes (or syllables).

This scenario controversially orders the Syntax Mutation (21) before the Phonology Mutation. Yet there are three independent linguistic indications that Computational Syntax precedes (25), assuming that only later evolved mechanisms can use aspects of earlier developments.

(i) As many linguists have argued, syntactic rules make essentially no use of phonological categories or constructs. This inaccessibility of phonology to syntax is rather natural if phonology is a later evolutionary development. This same ordering also explains why, in contrast, phonology is so often dependent on syntactic categories and configurations.[17]

(ii) Phonology is not the only property of Large Lexicons untouched by syntax. I referred earlier to works showing that Lexicons contain two subcomponents, a small Syntacticon of a few hundred 'closed class items' that constitute a syntactic system and a massive Dictionary of some 20,000 'open class items' whose lexical features f play no role in syntax. As discussed in section III, these features f, the mainstay of Large Lexicons, are largely products of human culture and so necessarily *postdate* the emergence

---

16 Philip Lieberman, Eve Spoke: *Human Language and Human Evolution* (New York: Norton, 1998).

17 Both these positions are extensively supported in Noam Chomsky and Morris Halle, *The Sound Pattern of English* (New York: Harper and Row, 1968).

of syntax; it would be bizarre if such Lexicons, packed with semantic features for culturally based concepts, came into mental existence earlier than the human syntax that expresses them. Rather, because the Large Lexicons created by the Phonology Mutation (25) come later, human syntax to this day does not integrate Dictionary features f into its combinatorial system.

(iii) The relation of the syntactic features F of the syntactic component to the Dictionary is quite different. The morphemes of the Syntacticon are used for all kinds of 'idiomatic usages' connected to open class cultural constructs: *come* 'have an orgasm,' *come to* 'recover consciousness,' *do gooder, do's and don'ts, go for it, have nots, make it, put off, put out, a wanna be,* etc. That is, the Dictionary uses items stored in the Syntaction for creating ever more 'open class items,' showing again that the Syntacticon develops earlier than the Dictionary, i.e. the Syntax Mutation (21) precedes the Phonology Mutation (25).

VI. Concluding Questions for Further Thought

I have proposed a model for linguistic evolution based principally on a natural language property Dual Patterning. Within it, I have distinguished an initial Syntax Mutation (21) from a later change that fully establishes Duality, a Phonology Mutation (25). This sequencing presupposes a gesture-based period of early language. Perhaps my most original claim is that *natural language's syntactic categories derive from the discrete aspects of primate vision*, which I claim are the only truly discrete categories in primates' sensory apparatus.

Properties of discrete vision and grammatical features should thus generally correspond. But one can formulate problems. When we find syntactic categories not in vision, we can (i) rethink our understanding of vision, (ii) question the analysis of syntax or (iii) attribute the categories to development since the Syntax Mutation (21).

An instance of (i) is the grammatical category Tense, not commonly understood as visual, but perhaps better understood in terms of visual memory.[18] The converse problem related to (ii) is posed by cross-linguistic differences in grammatical but not visual categories. (Generative grammar holds these differences are actually small.) For example, grammatical items for 'numeral classifiers' in many languages (Japanese, Mayan, etc.), while

---

[18] This issue is discussed in the longer version of this essay referenced in note 1.

indeed related to vision, are not used at all in many languages. In this area, grammatical analyses remain far from satisfactory.

Finally, if grammar doesn't use some category of discrete vision, *this essay's framework interestingly predicts its absence in primate vision*, as in (iii). For instance, discrete human vision clearly encompasses simple fractions (e.g. 1/3, 1/4) – it is hard to imagine they are learned through any other sense, and fractions have no special grammatical status. Then as predicted, exact fractions are doubtless beyond primates' grasp, suggesting that human mastery of them postdates the Syntax Mutation.

Charles Darwin's Centenary and the Politics and Poetics of Parenting: Inheritance, Variation, and the Aesthetic Legacy of Samuel Butler[1]

David Amigoni

Abstract: Variation, inheritance, and parenting were complex aspects of Darwin's theory in that they entangled the biological and cultural-intellectual domains. The celebration of Darwin's centenary in 1909 was actually an occasion that questioned Darwin's theories of inheritance and variation. The Mendelian William Bateson was one of those who questioned, and in doing this he styled Darwin as a 'poet,' and cited, for support, Samuel Butler, an earlier 'maverick' critic of Darwin. The essay explores the aesthetic legacy of Butler, and offers a reading from his novel, *The Way of All Flesh*, that both supports Bateson's critique of Darwinian inheritance while moving beyond it.

> It is easy to understand the difficulty felt by the fathers of evolution when we remember how much had to be seen before the facts could be well before them. (Samuel Butler, *Evolution, Old and New*, p.37)[2]
>
> *All flesh is not the same flesh: but there is one kind of flesh of men, another flesh of beasts, another of fishes, and another of birds.* [Corinthians 15.39] (William Bateson, epigraph, *Materials for the Study of* Variation, p.1)[3]

Introduction

Samuel Butler's controversial *Evolution, Old and New* (1879) critically assessed the relationship between Charles Darwin's 'new' theory of evolution by natural selection, and the 'old' theories (of Buffon, Lamarck and Erasmus Darwin) from which, Butler claimed, Charles Darwin's theory had descended. In the quotation above, Butler implicitly makes a point about the genealogy of intellectual ideas: theories are parented into existence, and subsequent growth. The substantive point of Butler's observation is that the 'fathers' of evolutionary theory had major challenges of 'seeing' with which

---

1 I am grateful to Nick Saul and Simon J. James for inviting me to speak at their conference; I am grateful to Cannon Schmitt and Marga Vicenda for providing me with the opportunity to present a version of this paper to the Institute for the History and Philosophy of Science and Technology at the University of Toronto, and for the valuable feedback from the audience.

2 Samuel Butler, *Evolution, Old and New: Or, the Theories of Buffon, Dr Erasmus Darwin, and Lamarck, as Compared with that of Charles Darwin, 1879, 3rd edn* (London: Fifield, 1911).

3 William Bateson, *Materials for the Study of Variation Treated with Especial Regard to Discontinuity in the Origin of Species* (London: Macmillan, 1894).

to contend: they had to observe a great deal of life and its history before the 'facts' of the theory of life's common descent could come dimly into view. To this end, evolutionary theories play variously, and with great complexity, on the opposition between likeness and difference among living and dead organisms.

Thus, when William Bateson, an English pioneer of Mendelian genetics, wrote his study of variation (1894) that sought to move beyond Darwin's theory of gradually accumulated variation, he selected an epigraph, drawn from Corinthians, that played on both likeness and difference: flesh was common to all human and animal life, but the flesh of humans, beasts, fishes and birds assuredly differed one from another between these genera? In the Preface to his study, Bateson drew attention to his epigraph, and in particular commented on its linguistic solemnity, and the division of intellectual labour in which his study of variation was to intervene: 'on the first page I have set in all reverence the most solemn enunciation of that problem that our language knows. The priest and the poet have tried to solve it, each in his turn, and have failed.'[4] The genetic scientist would set right the 'failures' of the priest and the poet, perhaps by seeking the building blocks of 'life' that were common, in the event, to all flesh. Bateson would go on to temper his sense of the poet's 'failure,' while being simultaneously aware of Butler's work with its explorations of relations between 'all flesh,' living and dead.

In assessing Darwin's legacy at the point of his bicentenary this chapter will focus on the conceptual play in certain key debates that emerged from the centenary discussion of his legacy: firstly, parenting both as an object of biological knowledge, and an intellectual force in the shaping of theory; secondly, the question of the representation of empirical evidence for variation, inheritance and common descent at a moment when Darwinian theories of natural selection were under pressure from a variety of sources including the new Mendelian genetics. Samuel Butler's 'playful' work on evolution complicates further the aesthetic dimensions of my argument about Charles Darwin's legacy: cited by Bateson as a prescient critic of Charles Darwin, I suggest ways in which Butler's fiction about inheritance between generations, *The Way of All Flesh* (1903), expanded evolutionary ideas 'poetically' in ways that were both seen, and yet not fully grasped, by the scientists in the midst of an intense debate.

---

4 Bateson, *Materials for the Study of Variation*, p.xii.

I

Darwinian theories of inheritance consisted of multiple conceptual strands; the strands crossed into one another, and became entangled. Firstly, inheritance concerned characteristics passed from parents to children, and increasing nineteenth-century awareness of the biological drives behind the way in which parents 'invest' protectively in their offspring. Darwin's theory of inherited characteristics was premised on the slow accumulation and passing on of infinitesimally small variations: variation set against the backdrop of, more or less, stability, in which 'nature' itself invests by selecting and preserving the good. Parents could contribute to this process by 'investing' time and care in their young: Darwin's selection theory made it crucial that they should do so. As August Weismann, the germ plasm theorist, noted in an essay exploring Darwin's theory of selection, published in the celebratory centenary work of 1909 – *Darwin and Modern Science*, a crucial source text for assessing Darwin's legacy at his centenary – selection was a harsh task master: a pair of elephants, notoriously slow breeding animals, are likely only successfully to protect two out of the average six young that they will produce in their life span into adulthood; the other four are most likely to perish without breeding.[5] Weissman's example draws implicit attention, perhaps, to the evolutionary fortunes of another slow-breeding animal, namely homo sapiens. Did human flesh fare better than elephant flesh in the struggle to invest in its transmission onwards as breeding material?

Thus, inheritance, variation, and selection, became intensely debated concepts of intellectual and theoretical discourse in the closing decades of the nineteenth century. Consequently, there is a need to think about the intellectual inheritance of biological paradigms, in which different generations of scholars and scientists played a role. This was often a disputatious process in which the evidence for selection and variation, and the language in which it was represented, became an arena of contestation. At times it concerned the way in which children of paradigm transformers such as Darwin sought to resist theoretical challenges in order to protect the inheritance of which they had become guardians. Darwin's biological theory of inheritance needed variation, of course: natural selection had to have variations from which to select. What I highlight is the way in which the new generation of scientists who came to do homage to Darwin in 1909 argued for a stronger theory of variation allied to a theory of particulate inheritance that challenged Darwin's theory of heredity. This generation of scientists was exemplified by William Bateson. In

[5] August Weismann, 'The Selection Theory', *Darwin and Modern Science: Essays in Commemoration of the Centenary of the Birth of Charles Darwin and of the Fiftieth Anniversary of the Publication of 'The Origin of Species'*, ed. by A. C. Seward (Cambridge: Cambridge University Press, 1909), p.20.

challenging the Darwinian paradigm, Bateson chose an intriguing path: he invoked both the maverick figure of Samuel Butler, and the office of the poet. For William Bateson, Darwin's value was analogous to that of a 'poet': according to Bateson, Darwin's legacy – what future generations could inherit from the great thinker – was an expansive, poetical view of life's workings. According to this view, Darwin's work matched the thought-provoking expansiveness generated by fictions and other aesthetic discourses.

Scientific exposition could and did resonate aesthetically. Anniversary and birthday celebrations relating to Darwin's theory of evolution were good opportunities for reflecting, poetically, on 'life's working,' and the variety of mechanisms and support systems through which it could be sustained: in his lecture on 'the coming of age of the *Origin*' – twenty-one years since its birth, in April 1880 at the Royal Institution – T.H. Huxley recalled his own role as public controversialist in 1859-60 seeking to keep Darwin's text alive as a 'vital' contribution to science. Huxley claimed that he 'acted for some time in the capacity of a sort of under-nurse, and thus came in for my fair share of the storms which threatened the very life of the young creature'.[6] It is a striking image of surrogate parenthood, and in Huxley's mind at least he was not 'Darwin's Bulldog' – rather, he was Darwin's under-nurse. 'Parental investment' in fragile infant forms needing protection and nurture – be they forms of organic life, or artefacts of culture, in this instance a text elaborating a system of evolutionary biology – can be seen in terms of the sheer variety of support systems that have evolved to ensure that life continues and flourishes in a harsh and unforgiving world in which natural selection operates. Huxley's image of surrogate parenthood participates in the aesthetic dimension of discourse in a fairly limited way. By contrast, Samuel Butler's famous posthumous novel about inter-generational inheritance and the relations between parents and children, *The Way of All Flesh*, radically aestheticised and re-drew the place of human-animal kinship in schemes of parenting and regeneration; this can further complicate our sense of the biological inheritance that emerged from Darwin's increasingly contested legacy.

II

In June 1909, some four hundred scientists gathered in Cambridge to celebrate the centenary of the birth of Charles Darwin, and the fiftieth anniversary of the publication of the *Origin of Species*. There were a number of events: formal speeches delivered in the Senate House, speeches delivered at a celebratory

---

6 T. H. Huxley, 'The Coming of Age of *The Origin of Species*', *Science and Culture and Other Essays* (London: Macmillan, 1881), pp.310-24.

banquet, later edited and published by Darwin's most eminent sons, Sir George and Francis Darwin.[7] Darwin's children played a big part in the celebrations. Inheritance was, of course, a theme in the presentations at the banquet; the Swedish physicist and chemist Svante Arrhenius said of Sir George Darwin, the mathematician and geophysicist, that he was 'a brilliant example of the heredity of intellectual greatness'.[8] This raised the important issue of inheritance between members of notable intellectual families, the very line of research initiated by Darwin's eugenicist cousin, Francis Galton. Latterly, the cause of eugenics was championed by his disciple the evolutionist-statistician Karl Pearson, who pushed it in a slightly different direction. Galton's theory of heredity had itself posed questions about the authority of Darwin's model of inheritance, given that Galton's view of 'hereditary genius' challenged Darwin's incremental, cumulative model through its account of extraordinary 'sports'. Pearson's commitment to biometrical statistics and correlation coefficients did, however, reinstate Darwin's assumed view of natural selection. Bateson responded forcefully against Pearson and his group. [9] The central point here is that Darwin's legacy – both familial and scientific-intellectual – was vigorously discussed during the centenary.

This puts into context a related question that was raised about Darwin – again, by one of his children, W(illiam). E. Darwin: what had his father, intellectually, passed on to the world? Was it a rich and varied legacy? In this respect, W. E. Darwin sought to refute the image of aesthetic aridity that Darwin's autobiography spread: the man who had become a machine for grinding out general laws, who had 'lost' a taste, an instinct, for poetry. Darwin's son pleaded a different case: referring to the 'grandeur' passage in the *Origin*, he argued that his and future generations inherited from Darwin a 'deep

---

7 *Darwin Celebration, Cambridge, June, 1909. Speeches Delivered at the Banquet Held on June 23rd*, ed. by George Darwin and Francis Darwin. Cambridge: *Cambridge Daily News*, 1909. See: The Complete Works of Charles Darwin Online, <http://darwin-online.org.uk/content/frameset?itemID=A279&viewtype=text&pageseq=1> [accessed 29 June 2009].

8 Speech by Dr Svante Arrhenius, *Darwin Celebration, Cambridge, June, 1909*, 8; <darwinonline.org.uk/content/frameset?itemID=A279&viewtype=text&pageseq=1 > [accessed 29 June 2009].

9 See Francis Galton, 'Hereditary Talent and Character', *Macmillan's Magazine*, 12 (1865), 157-66, 318-27; for Bateson's objections to Pearson, Weldon and the journal *Biometrika*, see his Preface to *Mendel's Principles of Heredity: A Defence, with a Translation of Mendel's Original Papers on Hybridisation* (Cambridge: Cambridge University Press, 1902). For contexts here I am indebted to the work of Daniel J. Kevles, *In the Name of Eugenics: Genetics and the Uses of Human Heredity* (1995), and James Moore, *Good Breeding: Science and Society in a Darwinian Age* (Milton Keynes: Open University, 2001), *Study Guide*, 50-54.

sense of the beauty and the poetry of the world, and of life'.[10] The inheritance of 'poetry in life' is something that I want to explore through the way in which Darwin's legacy was contested in 1909, and through Bateson's invocation of the legacy of one of his 'foes' from the late 1870s: Samuel Butler.

Samuel Butler spoiled the *Origin*'s twenty-first birthday party in 1879-80 when he published *Life and Habit* and *Evolution, Old and New*; he had a very different sense from Huxley of what had been inherited from Charles Darwin, and the role of previous generations in what Darwin had himself given birth to. Butler famously challenged Darwin's relative 'silence' on the intellectual inheritance from his grandfather Erasmus Darwin – and Buffon and Lamarck. Butler also became embroiled in a controversy about the authority of natural selection as the engine of evolutionary adaptation, a question that would be reinvoked when the mechanisms governing biological inheritance and variability were under discussion during the centenary celebration of Darwin's birth in 1909 among germ plasm and Mendelian theorists. Butler's early critique was distinctive because of the way in which he drew attention to Darwin's language, and to its ambiguity. Thus, Butler begins chapter 21 of *Evolution, Old and New* by citing Darwin's refutation of the 'imagined' conclusion drawn by some readers that '"natural selection induces variation"'. Butler contends that the 'imagined conclusion' about the inducement of variability by natural selection was indeed plausible rather than erroneous because 'his language concerning it is so confusing that the reader is not helped to see that it really comes to nothing but a cloak of difference from his predecessors'.[11] Thus, while Darwin struggled to write of the complex agencies at work in nature between variation and selection, Butler read Darwin's inscription of these relations as an effect of ambiguous language which concealed a silent conceptual struggle with predecessors.

In 1909 the presumption was that Darwin's writing was a model of clarity and lucidity. Joseph Dalton Hooker, in the introductory letter to the commemorative volume remarked that Darwin's writings were 'lucid and so little burthened with technical terms that they must be regarded as models for those who were asked to address themselves to the educated reader rather than the expert'.[12] This was indeed the brief of the centenary volume, as its editor A. C. Seward recorded. Despite Hooker's claim, Butler's

---

10 Speech by Mr W. E. Darwin, *Darwin Celebration, Cambridge, June, 1909*, 13-14; <http://darwin-online.org.uk/content/frameset?itemID=A279&viewtype=text&pageseq=1> [accessed 29 June 2009].

11 Butler, *Evolution, Old and New*, p.362.

12 *Darwin and Modern Science*, p.I.

allegations haunted the proceedings, as Darwin's language continued in practice to be an object of contention in the published birthday celebrations. A. C. Seward's Preface to the volume made the following point about the essay by Hugo de Vries: 'in regard to the interpretation of a passage in *The Origin of Species* quoted on page 71, it seemed advisable to add an editorial footnote; but, with this exception, I have not felt it necessary to record any opinion on views stated in the essays.'[13] In his essay on 'Variation,' de Vries, quoting from *The Origin*, remarked that:

> Darwin stated his view on this point very clearly: – "The doctrine of natural selection or the survival of the fittest, which implies that when variations or individual differences of a beneficial nature happen to arise, these will be preserved." In this sentence the words "*happen to arise*" appear to me of prominent significance [....]
> A distinction is indicated between ordinary fluctuations which are always present, and such variations as "happen to arise" from time to time. The latter afford the material for natural selection to act upon the broad lines of organic development, but the first do not. Fortuitous variations are the species-producing kind, which the theory requires; continuous fluctuations constitute, in this respect, a useless type.

Seward added this corrective footnote:

> [I think it right to point out that this passage from the *Origin* by Professor de Vries is not accepted as correct either by Mr Francis Darwin or by myself. We do not believe that Darwin intended to draw any distinction between *two types* of variation].[14]

De Vries had disturbed a powerful combination of defensive interests; Seward effectively created the field of Darwinian palaeobotany, and would go on to hold eminent offices in Cambridge and elsewhere (Master of Downing, Vice-Chancellor, President of the Geological Society, President of the British Association).[15] In addition, Francis Darwin, Charles Darwin's eldest son, had played an important role in defending the family integrity when both natural selection and Darwin's probity were deemed to be under attack by Samuel Butler. If Huxley imagined himself to be in the position of under-nurse to Darwin's new-born theory in the 1860s, then Francis Darwin was the defender of a theory that had rapidly aged into a venerated ancestor. De Vries was given a right of reply in a letter which was published as a

13 *Darwin and Modern Science*, p.v.

14 Ibid., p.71.

15 Seward, Sir Albert Charles, *Oxford Dictionary of National Biography* (2004), <http://www.oxforddnb.com/view/article/36026> [accessed 29 June 2009].

footnote: he conceded the impossibility of seeing the clarity of Darwin's presumed 'distinction' between different kinds of variability, while insisting that 'happen to arise' pointed sharply to a direction of travel that was confirmed by his writing on pangenesis in *Variation of Plants and Animals Under Domestication.*[16] There is a subtle conflict here over the interpretation of language, and the authority of models of inheritance were at stake: de Vries, along with Weismann, was seeking to eradicate the latent Lamarckism in Darwin's theory, appealing to his theory of heredity, pangenesis, as the opening that would cast terminal doubt on the theory of acquired characteristics. [17]

Butler's distinctive approach to the evolution question had been deeply out of favour in the scientific community during the 1880s and 90s. The dismissive tone was still echoing in 1909. J. Arthur Thomson, the theorist (with Patrick Geddes) of the evolution of sex, stated in his chapter on pre-Darwinian evolutionists – a sensitive topic, given Butler's charge – that 'To write, as Samuel Butler did, "Buffon planted, Erasmus Darwin and Lamarck watered, but it was Mr Darwin who said 'That fruit is ripe,' and shook it into his lap"… seems to us a quite misleading version of the facts of the case'.[18] It makes it even more astonishing then that Butler should have been rehabilitated in *Darwin and Modern Science* by William Bateson, author of the all important chapter on 'Heredity and Variation in Modern Lights'. In that essay, Bateson strikingly refers to the out-of-favour, indeed marginal Butler as 'the most brilliant, and by far the most interesting of Darwin's opponents'.[19] Bateson also quotes Butler's argument from *Life and Habit* that the origin of species is really the origin of variation.[20] Bateson thus records a moment when Butler found himself in agreement with Darwin, but which Bateson clearly chose to push in a particular direction, by reading for the origin of species in sports and mutations rather than slow, insensible accumulated variation.

Why was Butler important to Bateson's complex separation from Darwin's legacy? In part because there were historic links between the Butler and Bateson families based on schooling (Shrewsbury) and Cambridge University, links which overlapped with the Darwins.[21] Given his

---

16 *Darwin and Modern Science*, p.71.

17 Darwin would not make such a task easy: the pangenesis hypothesis set out in the *Variation of Plants and Animals Under Domestication* was actually preceded by a chapter outlining inheritance based on use and disuse.

18 *Darwin and Modern Science*, p.9.

19 Ibid., p.8.

20 Ibid., p.9.

21 See for the family genealogy and careers of father and son see Bateson, William Henry (1812-81), *Oxford Dictionary of National Biography*,

increasing antagonism towards Darwin's legacy, within Cambridge William Bateson became, for a time, something of an outcast figure: Butler's marginal and indeed maverick status was available for Bateson to identify with.

That said, their critiques of the Darwinian tradition were simultaneously shared and different. Butler was attached to a kind of universal Lamarckism: organisms and indeed inorganic matter were imbued with purposiveness and response to the environment. Bateson argued that discontinuous variation owed nothing to the environment. Yet, this was also the source of some odd affinity with Butler. For Butler, the interiority of the organism, its 'will' or 'cunning,' enabled it to develop, to evolve. To be sure, Bateson's rejection of environmental influence was founded upon experimentation, observation and statistical precision. At the same time it was also apparent to Bateson that there was a genetic alchemy in the interior of the organism that promised disclosure of the archaic secrets of heredity. Bateson was attracted to what he saw as Butler's 'natural penetration' into the identity of each living thing; quoting from Butler, this identity comprises '"*actually* the primordial cell which never died nor dies, but has differentiated itself into the life of the world, all living beings whatever, being one with it and members one of another"'.[22] In the preface to his book on Mendel and heredity (1902), Bateson argued that experimentation needed to be supplemented by an 'instinct for truth' of the kind that Butler's 'natural penetration' demonstrated.[23] This was a very different approach to Pearson's Darwinian biometrics with its population coefficients, and a 'grammar' of scientific method that refused on principle to look inwardly at the object, and only at what the data of inherited or non-inherited characteristics calculated.[24]

Above all – and Bateson's attraction to Butler's lyricism in *Life and Habit* seems to suggest this – the differences between Bateson and Butler, and indeed Bateson, Butler and Darwin might be smoothed out by the phenomenon of poetical language, which in turn could work in support of this 'instinct'. Samuel Butler, in an essay intervention on thought and language from the mid-1890s (a major topic in evolutionary speculation)

<www.oxforddnb.com/view/article/1685?docPos=2>; for the son, William (1861-1926) see <www.oxforddnb.com/view/article/30641/30641?back=,1685> [both accessed 29 June 2009].

22 Bateson included this quotation from Butler's *Life and Habit* (1878 edn, p.86) in a footnote to a reflection on the germ plasm thesis; *Darwin and Modern Science*, p.91.

23 Bateson, *Mendel's Principles of Heredity*, p.xii.

24 See Karl Pearson, *The Grammar of Science* (London: Scott, 1892), chs. 2 and 9.

contended, arrestingly, that 'the written symbol extends infinitely'.[25] There is no evidence that Bateson read this essay, but in his contribution to *Darwin and Modern Science* he also celebrated 'the creative power by which [Darwin] inaugurated a line of discovery endless in variety and extension'. For Bateson, in the same way that Shelley was 'a poet for poets: so Darwin was a naturalist for naturalists'.[26] Poetry gives thinkers the license to extend an argument, push it in new directions, and to exercise 'an instinct for truth'. For Bateson as he pondered heredity and variation as keys to 'the way of all flesh' – remember that his epigraph from the book of Corinthians to his great study of variation posited both the commonality and variability of flesh – Butler's novel of this title might have offered a wonderful opportunity for poetical extension, and instinctively driven quests for truth. James Moore reminds us that Bateson, by adopting Mendel's findings about the importance of grandparental characteristics in patterns of heredity, was able to argue against biometric models of inheritance precisely because of the reappearance of grandparental characteristics in offspring whose biological parents were bereft of the trait.[27] Bateson and the Mendelians might have been inspired by the instinctual poetical truth of a novel in which traits possessed by the benign yet highly skilled Old Mr Pontifex, born in 1727, reappear in his great-grand child Ernest Pontifex having bypassed the patriarchal tyrants, George and Theobald Pontifex.

Conclusion

Even so, Butler's narrative of inheritance, variation and likeness, and parental investments (both nurturing and damaging) takes us beyond a confirmatory story for Mendelians as they sought to push their own theory to supersede the Darwinian theory of heredity. Butler's poetry, and his 'instinct for truth' at once simplified and complicated all received stories about inheritance, descent, and parenting – the building blocks of evolution. This in turn further expanded the sense of what could be inherited from Darwin, intellectually: and how that expanding inheritance might feed later generations of thinkers and writers.

Everybody who has read Butler's novel recalls what happens to Ernest Pontifex's children in chapter 79 after Ernest's marriage to the lower class Ellen ends disastrously in her increasingly visible alcoholism.[28] After their

---

25 Samuel Butler, 'Thought and Language', *Essays on Life, Art, and Science* (London: Fifield, 1908), p.198.

26 *Darwin and Modern Science*, p.85.

27 James Moore, *Good Breeding*, p.53.

28 Samuel Butler, *The Way of All Flesh*, ed. by Michael Mason (Oxford: Oxford University Press, 1993); page and chapter references in the text will be to this edition.

separation, Ellen meets Bill, and Ernest effectively condones their bigamous marriage as they emigrate together to the United States. By this point, it has been revealed to Ernest that he is in fact bigamously married, for Ellen was already secretly married to his father's dismissed coachman, John. 'Of the children she has taken no notice,' Overton remarks, and the union with Bill remains childless; Ellen thus seems to exempt herself from the trade-offs around parental investment on two counts (p.360). So it is down to Ernest to settle the children's future. Readers recall, often with a degree of moral shock, that Ernest chooses to send his children away. He supports them financially, to be sure; in fact, chapter 78 is exclusively concerned with the success of Ernest's various financial speculations, which suggests that parental investment was soundly informed by financial investment. Thus, Ernest removes his children from their immediate biological inheritance and sends them to a respectable working class family in Gravesend, who present 'as fair a chance of coming to a good development as in those of any whom he knew'; as many readers have noted, there is a kind of engineered social and cultural solution to the danger of hereditary taint: the author, through Ernest, removes them from the hereditary cycle of oppression and abuse that characterised Ernest's own upbringing, and the line of terrible father-son relationships that the novel traces.

So far so good – but how did Butler utilise and complicate discourses of inheritance, selection, variation and parental investment in doing this? In the evolutionary discourse elaborated by Darwin, parental investment was no longer evidence of human ethical transcendence. Instead, Darwin was keen to point out the similarities between patterns of parental investments in the higher and lower animals. This is evident in Darwin's great treatise on human sexual selection, *The Descent of Man*:

> Monkeys are born in as almost helpless a condition as our own infants; and in certain genera the young differ almost as fully as much in appearance from the adults, as do our children from their full grown parents. It has been urged by some writers as an important distinction, that with man the young arrive at maturity at a much later age than with any other animal: but if we look to the races of mankind which inhabit tropical countries the difference is not great, for the orang is not believed to be adult till the age from ten to fifteen years.[29]

This reaffirms Darwin's imaginative strategy of dismantling the cultural barrier between humans and simians. Viewed from *The Way of All Flesh*, the passage enables us to read for Butler's implicit endorsement of, yet poetic expansion upon, Darwin's thought in the chapter in which Ernest divests

29 Charles Darwin, *The Descent of Man, and Selection in Relation to Sex*, 2 vols (London: Murray, 1871), p.13.

himself of his children to another carer. If readers remember the giving away of the children, then they are inclined to forget the depression that Ernest falls into after the break-up of his marriage and family life. The means by which he is coaxed out of it is fascinating. He is taken by Overton to a doctor who mistakenly assumes the rather underdressed people before him to be poor. He recommends a cheap remedy which he calls 'crossing'. 'Crossing' involves 'crossing' the depression out of oneself by visiting Regents Park Zoological Gardens. Of course, in one sense 'crossing' means erasure: but the word also points in the Mendelian direction of cross-breeding, hybridization. The doctor recommends the selection of 'a course of the larger mammals' (hippos, rhinos, elephants). He adds, 'the monkeys are not wide enough to cross, they do not stimulate sufficiently' (p.369). The narrative and identification strategies on which Darwin came to depend are thus expanded: and yet what Ernest realises is a curious and hybridised form of aesthetic and affective investment in the sight of other animals that seems, paradoxically, to suspend narratives of animals as restorative 'otherness,' enabling him instead to tap into an 'instinctual' grasp of common descent: 'As for the elephants, especially the baby elephant, he seemed to be drinking in large draughts of their lives to the re-creation and regeneration of his own' (p.370). We should recall Weismann's point in his essay on selection in *Darwin and Modern Science* that, for elephants, parental investment was a brutal challenge with limited rewards. Yet, in a curious kind of extension and reversal, the baby creature regenerates, indeed 'parents' Ernest. There is a sense in which elephant flesh and the evolutionary order to which it belongs has always already 'parented' Ernest. Of course, it takes a poetical sleight of hand to realise this. The image of the baby elephant regenerating humans is a powerful image of both investment and inheritance – the consequence of a common inheritance derived from 'the tree of life,' and perhaps an enactment of the communicative 'information' passed between humans and animals. Butler's theory of language, as I've argued elsewhere, did not prioritise human language over the embodied communicative action through which other organisms 'articulated' themselves.[30] In many ways, Butler's evolutionary reverie builds on Darwin's poetry, and before that the poetry of Corinthians, gesturing forwards, expanding infinitely. We should recall that, though temporarily on the margins of Cambridge science in 1894, William Bateson was a major scientific force by 1909 and investing in his own intellectual dynasty:

---

[30] David Amigoni, '"The Written Symbol Expands Infinitely". Samuel Butler and the Writing of Evolutionary Theory', *Samuel Butler: Victorian Against the Grain*, ed. by James G. Paradis (Toronto: Toronto University Press, 2007), pp.91-112.

Gregory Bateson (b.1904), the anthropologist and theorist of cybernetics and the 'ecology of mind,' was his son. He was also a follower of Samuel Butler.[31]

31 Gregory Bateson, *Steps to an Ecology of Mind*, 1972 (Chicago: Chicago University Press, 2000); in a striking echo of his father's words, Bateson celebrates Butler as 'the ablest contemporary critic of Darwinian evolution', p.xxi. The other 'personal inspirations' to whom Bateson appeals are Lamarck and the poet William Blake, who 'saw through his eyes, not with them'.

By An Evolutionist: Poetic Language in Chambers and Tennyson

Anna Barton

Abstract: Tennyson's poetic response to the evolutionary science of his day in *In Memoriam* is a familiar topic, but discussion has rarely focussed on the implications of evolution for Tennyson's poetic voice. Using Kristevan theories of gendered language, this essay offers the unborn and newly born children of poetic and scientific texts by Tennyson and Chambers as sites that both motivate and disrupt a paternal desire for meaning that will silence poetry, and argues that the way this desire is articulated, achieved or avoided might usefully contribute to a discussion of literature as constitutive of its own evolutionary progress.

As numerous readers of the poem have established, the Epilogue to *In Memoriam* incorporates Robert Chambers's theory of evolution whereby phylogeny is directly linked to ontogeny into a declaration of faith in the spiritual progress of the species towards apocalyptic reunion with the creator God.[1] Tennyson's Coleridgean blessing to his as yet un-conceived nephew or niece, son or daughter of his recently married sister and Edmund Lushington, is an imaginative leap of faith that restores Christian belief to a doubtful mourner, but it is also a leap of literary faith that restores textual authority to a doubting poetic.[2] Whereas, in Part V, Tennyson's mourner complains that 'words, like Nature, half reveal | And half conceal the Soul within' (V. 3-4), the Epilogue looks forward in anticipation of the 'crowning race' in whose hand 'Is Nature like an open book;' (132).[3] Between the

---

1 See, for example, W. R. Rutland, 'Tennyson and the Theory of Evolution', *Essays and Studies* 26 (1940), 22-23. The title of this essay also acknowledges an early essay by Walker Gibson, which explores the difference between the description of evolutionary theories in scientific texts and Tennyson's poetic descriptions of science: 'Behind the Veil: A Distinction Between Poetic and Scientific Language in Tennyson, Lyell and Darwin', *Victorian Studies*, 2 (1958), 60-68.

2 Tennyson's use of language in *In Memoriam* has generated a good deal of critical discussion. Debate has tended to focus on the relationship between *In Memoriam* and Victorian philology and the relationship between speech and text. See, for example, Elizabeth A. Hirsh, '"No Record of Reply": *In Memoriam* and Victorian Language Theory', *English Literary History*, 55 (1988), 233-57, William A. Wilson, 'Victorian Philology and the Anxiety of Language in Tennyson's *In Memoriam*', *Texas Studies in Literature and Language*, 30 (1988), 28-48 and P. G. Scott, 'Flowering in a Lonely Word: Tennyson and the Victorian Study of Language', *Victorian Poetry* 16 (1978), 371-81. Elizabeth Hirsh's essay also offers a persuasive application of Chambers's remarks about the evolution of language (p.237) to *In Memoriam's* focus on the acquisition of language.

3 All quotations from *In Memoriam* will be taken from *The Poems of Tennyson*, 3 vols, ed. by Christopher Ricks, 2nd edn (London: Longman, 1987).

beginning and the end of the poem, simile and object change places so that the words which were like the half-hidden truth of nature become, as James Secord puts it, 'the model for pure understanding'.[4] The poem advances the progress of mourner and reader beyond that of scientific knowledge, claiming text as a means to perfect comprehension by holding a like comprehension out as a promise to readers of nature: 'mastery over creation will be transcendent literacy. When nature can be read like a book, it will need no interpretation.'[5] For Secord, the end of *In Memoriam* describes a kind of textual vanishing point that scientific writing seeks and always already fails to achieve.[6]

Secord's book, which has done important work reintroducing *Vestiges of the Natural History of Creation* into the canon of evolutionary science, urges us to think of that canon as literary history and he concludes by considering the implications of Tennyson's poetic faith for himself as a reader of scientific text: 'The texts of science have no meaning apart from what readers make out of them, yet – ironically – they aspire to be a transcript of the truth of nature, needing no interpretation.'[7] Secord seeks to disrupt that aspiration by recovering part of what has been lost to a scientific progress which forgets, incorporates or discards its past. What, then, of poetry? Towards what event does an evolutionary poetic move? Beginning with the end of *In Memoriam*, this essay considers such an event as it is imagined within Tennyson's poetry. Using Kristevan theories of gendered language, it offers the unborn and newly born children of poetic and scientific texts by Tennyson and Chambers as sites that both motivate and disrupt paternal desire for meaning, and argues that the way this desire is articulated, achieved or avoided might usefully contribute to a discussion of literature as constitutive of its own evolutionary progress.[8]

---

4 James A. Secord, *Victorian Sensation: The Extraordinary Publication, Reception, and Secret Authorship of* Vestiges of the Natural History of Creation (Chicago: Univesity of Chicago Press, 2000), p.532.

5 Secord, p.532.

6 Secord identifies a similar metaphor at work throughout *Vestiges*: 'the physical act of turning the pages becomes a metaphor for scientific exploration: the description of the strata series is presented as "the leaves of the *Stone Book*," and subdivisions in the work are the "record of this period," "great natural transactions," "a new chapter in this marvellous history," or the "next volume of the rock series."' (p.100).

7 Secord, p.532.

8 As such, this essay develops out of recent re-examinations of the relationship between nineteenth-century poetry and science that recognise that 'Victorian poetry and science frequently employed the same metaphors, themes, images and ideological orientations' and assert that 'science, no less than poetry is a mode of discourse that, rather than being objective and value-free, is always to some extent conditioned by the social, linguistic, and cultural contexts in which it is produced.'

In *Desire in Language* (1980), Julia Kristeva describes this process in terms of a paternal drive towards the symbolic. Kristeva's post-Lacanian, post-Saussurean theory defines the paternal as 'the thetic function of naming, establishing meaning and signification,' a function that is primarily concerned with the formation of objective and transcendental reality.[9] Preceding or interrupting the symbolic, the semiotic, 'a distinctive mark, trace, index, the premonitory sign, the proof, engraved mark, imprint – in short a *distinctiveness* admitting of an uncertain and indeterminate articulation' is associated with the body and therefore the mother: 'Before recognizing itself in a mirror and consequently as signifiying, [the] body is dependent vis-a-vis the mother. At the same time instinctual and maternal, semiotic processes prepare the future speaker for entrance into meaning and signification (the symbolic)' (pp. 133, 136). Kristeva describes a linguistic family romance whereby the speaking subject enters the symbolic order through a rejection of the bodily, maternal, semiotic origins of articulation:

> But the symbolic (i.e. language as nomination, sign and syntax) constitutes itself only by breaking with this anteriority, which is retrieved as 'signifier,' 'primary process,' displacement and condensation, metaphor and metonymy, rhetorical figures, but which always remains subordinate – subjacent to the principal function of name-predicating. Language as symbolic function constitutes itself at the cost of repressing instinctual drive and continuous relation to the mother. (p.136)

She offers this description of the coming into meaning of language as part of a discussion of the nature of poetic language, which, she argues, is semiotic, characterised by a heterogeneousness that she associates with 'the first echolalias of infants [...] rhythms and intonations anterior to the first phonemes, morphemes, lexemes and sentences' (p.133). Conversely, symbolic language is linked with science, which seeks to restore the objecthood of man as signified phenomenon.

Kristeva's theory, in its association of the poetic with the infantile, provides a framework for rereading the description of phylogenic evolution in both *In Memoriam* and in those sections from Chambers's *Vestiges of the Natural History of Creation* that may have provided Tennyson with his source.[10] For both poet and scientist the child is synechdocal: at once

---

(Gowan Dawson and Sally Shuttleworth, 'Introduction: Science and Victorian Poetry', *Victorian Poetry*, 41 (2003), 1-10 (p.2).

9 Julia Kristeva, *Desire in Language: A Semiotic Approach to Literature and Art* (Oxford: Blackwell, 1980), p.138.

10 Tennyson claimed that he composed the sections of *In Memoriam* that deal with evolution before reading *Vestiges* in 1844 (Hallam Tennyson, *Alfred Lord Tennyson: A*

representative and part, both of the evolutionary progress of the species and, I suggest, the text's achievement of the symbolic. Chambers's earliest scientific work exhibits a concern with the establishment of likenesses between different natural processes and phenomena. In an early essay, published in *Chambers' Journal*, Chambers makes a series of observations concerning 'Animal Humanity': 'lambs resemble children very much in their sports'; adult monkeys discipline their children using 'exactly that kind of chastisement which has so long been in vogue among human parents and human teachers'; there is 'genuine human-like merriment – in many of the lower animals'; and an injured bird is helped by 'her fellows, exactly as a human being might have been in similar circumstances'.[11] These resemblances are noted by Chambers in order to allow him to take up a broadly ecological position, urging respect for the animal kingdom through an acknowledgement of their 'intellectuality and sentiment' and their value as the work of 'a divine Author'.[12] Light-hearted articles of this kind paved the way for *Vestiges*, by suggesting in mild, uncontroversial terms, the argument that would be asserted more firmly in his sensational treatise.

However, as Secord establishes, *Vestiges* also employs simile and metaphor freely.[13] Its metaphors are often domestic, housing the changing universe within the walls of the middle-class home and the middle-class womb. Reporting the recent but familiar claim of physiologists, 'that each animal passes, in the course of its germinal history, through a series of changes resembling the permanent forms of various orders of animals inferior to it in that scale,' he builds on this evidence to formulate his theory that evolutionary leaps happen within the womb:

> Are we to presume that, in this case, the simple engender the more complicated? Undoubtedly, this would not be more wonderful a natural process than one which

---

Memoir By His Son, 2 vols (London: Macmillan, 1897), I, p.223), but Christopher Ricks argues in his notes to the poem that the Epilogue, which so closely echoes Chambers, is likely to have been composed around 1844-45.

11 Robert Chambers, *Essays Familiar and Humorous*, 2 vols (Edinburgh: Chambers), I, pp.156-69.

12 Chambers, I. p.169.

13 Secord, pp.100-04. The use of metaphor in nineteenth-century scientific writing is addressed in detail by Gillian Beer, who recognises a centrality of metaphor to Darwin's argument in the *Origin* that has much in common with my own reading of Chambers: 'Darwin's aim is to discover analogies which can move beyond the provisional and metaphorical and prove themselves to be "true affinities".' (Gillian Beer, *Darwin's Plots: Evolutionary Narrative in Darwin, George Eliot and Nineteenth-Century Fiction*, 2nd edn (Cambridge, Cambridge University Press, 2000), p.74. For a lucid introduction to the use of metaphor in science more generally, see Janet Martin Soskice, *Metaphor and Religious Language* (Oxford: Clarendon, 1985).

> we never think of wondering at, because familiar to us – namely, that in the gestation of the mammal, the animalcule-like ovum of a few days is the parent, in a sense, of the chick-like form of a few weeks, and that in the subsequent stages – fish, reptile, etc. – the one may, with scarcely a metaphor, be said to be the progenitor of the other. I suggest, then, […] that the first step was an advance under favour of particular conditions, from the simplest forms of being, to the next more complicated, and this through the medium of the ordinary process of generation.[14]

In this passage Chambers seeks to make his theory palatable to a Victorian readership by asserting its close relationship to the familiar process of human gestation and by arguing that the development of the species is a simple extension of – indeed takes place during – the development of the foetus within the womb. Each embryo retells the evolutionary history that it perpetuates: the fish is the parent of the reptile; the reptile is the parent of the bird, which is the parent of the mammal and so on. In its description of the coming into being of an entire species out of the heterogeneous and indeterminate body of the unborn child, the text of *Vestiges* effects transformation and is itself transformed, performing the erasure of its own semiotic origin. At the crux of the argument, the reader is presented with a sequence of textual feints: 'etc.,' 'scarcely a metaphor,' 'may […] be said'. These might be described as semiotic traces that both interrupt and generate the argument, drawing attention to the symbolic drive that the text embodies. By stating that one developmental stage may, 'with scarcely a metaphor, be said to be the progenitor of the other,' Chambers demonstrates that the object of his text is a literal representation that is almost but not quite achieved. Metaphor, a rhetorical figure that Kristeva identifies as a remnant of linguistic anteriority, is simultaneously rejected and confessed in a moment that reaches towards the objective signification that is also the most radical proposal of Chambers' theory:

> *that the simplest and most primitive type, under a law to which that of like-production is subordinate, gave birth to the type next above it, that this again produced the next higher, and so on to the very highest.* (p. 222)

In the paragraphs that follow this assertion, Chambers returns to the domestic, asking the reader to compare the initial reaction of a person learning about the 'circumstances attending to the production of an individual of our race' for the first time with their own likely reaction to the startling facts of the gestation of a new species, concluding that this less

---

14 Robert Chambers, *Vestiges of the Natural History of Creation* (Leicester: Leicester University Press, 1969), p.205.

familiar process is 'as simply natural, and attended as little by any circumstances of a wonderful or startling kind, as the silent advance of an ordinary mother from one week to another of her pregnancy' (p.223). Secord argues that this household vocabulary again works to address fears about transmutation: 'to avoid accusations of undermining faith, nature's story is presented as unthreatening, uncoercive, and implicitly narrated within a domestic scene'. As Secord points out, the similarity insisted upon here is a rhetorical flourish that aims to check any alarm that Chambers's readership might feel, but its rhetoric is Aristotelian in that it relies on an understanding of the universe that is fundamentally symbolic, as the passage goes on to make clear:

> Yet, be it remembered, the whole phenomena are, in another point of view, wonders of the highest kind, for in each of them we have to trace the effect of an Almighty Will which has arranged the whole in such harmony with external physical circumstances, that both were developed in parallel steps – and probably this development upon our planet is but a sample of what has taken place, through the same cause, in other countless theatres of being which are suspended in space. (p.223)

Under the auspices of the Almighty or paternal will, a universal harmony is established that closes the gap between metaphor and object, sign and signified.

This symbolic, paternal impulse is also Tennyson's. The only way that Tennyson is able to end his long elegy for Arthur Hallam is by imagining an end to all poetry. Affirmation is marked by the end of this lengthy exchange between poet and reader, but the implications of this end are more troubling for poet than for scientist. As the mourner marks out the cyclical progress of days, months, seasons and years, his grief is most often registered as despair at being unable to reach, touch or see his dead friend; and the poem describes a series of gulfs that separate them. These material chasms often suggest a linguistic gap that the poetry seeks to bridge. In Part II, the roots of the 'Old Yew, which graspest at the stones | That name the underlying dead' (II.1-2) reach down towards the bodies buried beneath the ground and, Tennyson imagines, wrap around their bones. In doing so they connect the world of the living with the world of the dead and at the same time provide a link between bodies and names. Tennyson's desire to 'grow incorporate' into the tree is the desire of mourner and poet to embrace and to name. Tennyson 'seem[s]' to fulfil this desire, but the seeming nature of this moment marks it as an illusion. In the stanzas that follow, the poet struggles to make sense of things that seem and, by implication, are not as they seem. In Part VIII, Tennyson describes his elegy as a flower planted on

Hallam's tomb, but suggests that the truth of this simile may be simply a trick of his grief: 'So seems it in my deep regret,' (VIII. 17). The wish for Hallam's body to be brought for burial in England is described as an 'idle dream' of the mourners, to whom it 'sweeter seems I To rest beneath the clover sod' (X.9-12). Likewise, wondering at the changeful nature of his sorrow, Tennyson expresses doubt in his ability to read himself: 'doth she only seem to take I The touch of change in calm and storm [...]?' (XVI. 5-6). His faltering sense of self-knowledge points towards the possibility of more universal delusions. In Parts LV and LVI, through a performance of corrected misreading, Tennyson describes how evolutionary theory strips away that which seems to be the case. Nature's protection of individual species – 'So careful of the type she seems, I So careless of the single life;' (LV.7-8) – is dismissed: '"So careful of the type?" but no' (LVI.1) and 'Man, her last work, who seemed so fair' (LVI.9) is revealed to be no more than 'A monster then, a dream, I A discord' (LVI.21-2). In each of these examples, the word 'seems' indicates a misinterpretation or misrecognition that is also a failure to name. Set against this, Hallam, a 'noble type,' the prophet of a future race, is also described in terms of seeming and being: 'Not being less but more than all I The gentleness he seemed to be, I Best seemed the thing he was,' (CXI.11-13). This might be as much as to say that what you saw was what you got with Hallam; but, for Tennyson, the elision of semblance and substance that Hallam achieved represents a state of grace.[15] Hallam's nobility recalls that of Christ who is described as embodied language: voice as breath, speech as deed: 'And so the Word had breath, and wrought I With human hands the creed of creeds' (XXXVI.9-10). However, in the stanza it is not Hallam that is compared with Christ, but the poet himself: 'In loveliness of perfect deeds, I More strong than all poetic thought' (XXXVI.11-12). This Romantic definition of poetry as replica of divine creativity, like the image of the old yew, invites us to understand *In Memoriam* as a poem that aims to heal the rift that separates word and world so that, as it reaches its conclusion, it can look forward to a time 'when all that seems shall suffer shock':

[15] The connection between poetic language and fallen language is recognised by William A. Wilson: 'Absent from Tennyson's poem is the relationship between words and things enjoyed by Adam [...] Unmediated language was lost with Paradise, and for Tennyson and others in the nineteenth century this was not a *felix culpa*' (p.31).

O living will that shall endure
  When all that seems shall suffer shock,
  Rise in the spiritual rock,
Flow through our deeds and make us pure. (CXXXI.1-4)

The 'will' that remains is the 'Almighty Will,' creator of the harmonious universe, to which Chambers cedes authority in *Vestiges*. It is imagined flowing through deeds, which are also the poet's words, and lending them the purity that Kristevan (or fallen) poetic language lacks.

The epilogue therefore is not so much an afterword as an after poem. Tennyson's celebration of his sister's wedding, a celebration that insists on the poet's renewed confidence in the vital performativity of language is patterned by voices that lapse into permanent and meaning silence.[16] The mourner's 'dying songs' are transformed into a statue-like memorial to his love that is 'solid,' 'calm' and implicitly silent. The 'most living words of life' that are 'breathed' between the couple, performing their union, are succeeded by their signatures: 'Mute symbols of a joyful morn'. This progression from the spoken to the written word is often read in terms of Tennyson's reconciliation to his professional identity as print poet, but it might also be read as a troubled rejection of the poetic via the performance of a symbolic.[17] In each case, silence is represented as the apotheosis of language and each silence foreshadows and forestalls the silence that will fall at the end of the poem. As evening falls, the mourner remains with a few guests, recalling the newly departed bride and groom and knitting them together with a community of absent friends through genial conversation:

Discussing how their courtship grew,
  And talk of others that are wed,
  And how she looked, and what he said,
And back we come at fall of dew.

Again the feast, the speech, the glee,
The shade of passing thought, the wealth
Of words and wit, the double health,
The crowning cup, the three-times-three, ('Epilogue' 96-104)

16 For a relevant discussion of faith and meaning, see Henry Kozicki, '"Meaning" in Tennyson's *In Memoriam*, *Studies in English Literature*, 17 (1977), 673-94.

17 Eric Griffiths offers a fine reading of the relationship between the written and spoken word in Tennyson's poetry in *The Printed Voice of Victorian Poetry* (Oxford: OUP, 1989).

But this too ends in silence and light, amidst which the poet takes his leave:

And last the dance; – till I retire:
  Dumb is that tower which spake so loud,
  And high in heaven the streaming cloud,
And on the downs a rising fire:

And rise, O moon, from yonder down,
  Till over down and over dale
  All night the shining vapour sail
And pass the silent-lighted town, ('Epilogue' 105-112)

This sequence of departures betrays a sense of something approaching reluctance. The mourner's retirement from the dance, which precedes his retirement from the poem by only nine stanzas, allows the poet to rehearse and postpone that last event via a kind of poetic dragging of his feet. The inevitable upward course of the moon snags on the stammering repetition of 'down,' putting off the progress that it performs by saying again in order to avoid saying nothing. Resistance is futile, however, and the final stanzas accelerate, imagining months, then years, then ages into the future as they describe the conception of the poet's descendant, its development in the womb and its birth and growth into an adulthood that will advance the race towards its apex, where knowledge shall be looked on 'eye to eye': understood, not interpreted, seen, not read.

By imagining the birth of this child into the symbolic, striking its inchoate being into the 'bounds' of flesh and text, Tennyson usurps his brother-in-law by performing the paternal function in his place. As he does so, he is himself usurped by the child whose imagined development stands for and exceeds his own. Up until this point, it is the poet who has featured as the child of his own poem: 'An infant crying in the night: | An infant crying for the light: | And with no language but a cry' (LXIV.17-20); and then later: 'as a child that cries | But crying, knows his father near' (CCXXIV.19-20). In these familiar lines, Tennyson identifies his inability to put the grief he feels into words with an infantile state that is explicitly pre-linguistic. He is the 'baby new to earth and sky' who, in Part XLV, he describes in terms of a connection to the mother and an incapacity to name: 'What time his tender palm is prest | Against the circle of the breast, | Has never thought that "this is I",' lines that Kristeva echoes in her description of the semiotic body, which 'before recognizing itself as identical in a mirror and, consequently as signifying [...] is dependent vis-à-vis the mother'.[18] His cry, which must be read as one of despair and failure is, therefore also the

---

[18] Kristeva, p.136.

cry that for Kristeva is fundamentally poetic. The last stanzas describe a moment, future to the race, but present to the poem, when that cry is silenced. The lines gain a kind of purchase on the generations that will follow, securing for the poet a ghostly presence amongst the 'crowning race' that will succeed even his unborn nephew or niece. However, in so doing they incur a loss: a break with the poetic despair that holds up and holds together the circular progress of *In Memoriam*.

It was not just Tennyson who saw the future of the species in the faces of the next generation.[19] Secord argues that '[m]ore important than technical treatments of embryology, however, was Chambers's familiarity with everyday processes of birth and development'.[20] Secord also identifies a piece of writing by Chambers about his first son, Robert: a poem entitled 'Lines To a Little Boy' that was published in *Chambers Edinburgh Magazine* in 1840.[21] Even in this uncomplicated expression of a father's love, a paternalistic urge to establish meaning is never far from the surface. The opening lines of 'Lines To a Little Boy' name and possess the baby, repeatedly asserting the link between parent and child: 'My winsome one, my handsome one, my darling little boy I The heart's pride of thy mother and thy father's chiefest joy' (1-2). Having established that the boy is his father's son, Chambers goes on to describe him in terms of the meaningless sounds that he emits. Brought firmly within the compass of the father's, rather than the mother's, body – sitting on his knee or riding on his shoulder – the child is invited to 'lisp' and 'prattle' nonsense, and Chambers even delights in hearing the happy hum of his son's heart. The following stanzas admit the possibility of what *Vestiges* will later state definitively: that the progress of the species is not certain: 'they say that later years I Will sometimes change a parent's hope for bitter grief and tears' (9-10). However, Chambers attempts to deny this uncertainty through an expression of fidelity to his son, which also insists on a visceral connection between them: 'No – no, that time can ne'er arrive, for, whatsoe'er befall I This heart shall still be wholly thine or shall be not at all' (25-26). Chambers signals his physical investment in his son by suggesting that it is his own heart that beats inside his son's body and in doing so he inscribes the happy hum that he hears emanating from it at the beginning of the poem with his

19 I argue elsewhere that the compulsion to write into meaning, which I identify here as paternal and symbolic, can be traced in Tennyson's other poems about parents and children: *The Princess*, 'De Profundis', 'Aylmer's Field' and 'The Grandmother'. Anna Barton, *Tennyson's Name: Identity and Responsibility in the Poetry of Alfred Lord Tennyson* (Aldershot: Ashgate, 2008), pp.31-53.

20 Secord, p.96.

21 See Appendix for the text of the poem.

name. Finally Chambers's poem, like Tennyson's, imagines its 'little boy' into adulthood, insisting on the father's continued presence: 'A home as happy shall be thine, and I too shall be there, | The blessings purchased by thy worth in peace and love to *share*,' (45-6). However, the linear order of succession that these lines put in place encounter a disruption at the end of the poem, which seeks a return to the beginning, by imagining Chambers's son, now a father, holding his son on his knee. Like the seasons that pattern *In Memoriam*, moving the poem forward but also round and back on itself, the cyclical sense of time that Chambers invokes at the end of 'Lines To a Little Boy' describes a future that recalls its past in formal echoes and reflected images.

In 1889, Tennyson penned the fragmentary epigram, 'By An Evolutionist'. The final lines of this brief poem run as follows:

> I have climb'd to the snows of Age, and I gaze at a field in the Past,
> Where I sank with the body at times in the sloughs of a low desire,
> But I hear no yelp of the beast, and the Man is quiet at last
> As he stands on the heights of his life with a glimpse of a height that is higher.

The poem is written from the perspective of an old man and, rather than the form of the infant, it is an elderly body, grown beyond the base materiality of the flesh and looking towards death that is understood in terms of an evolutionary progress that is, in this case, progress into the next life rather than progress into adulthood. Here the summit of human achievement is explicitly associated with silence, which comes as a relief from the animal yelp that reduces all articulation to the primitive and the corporeal. Although, by the age of eighty, Tennyson was reconciled to the theory of evolution, the title he gives the epigram isolates the voice of the speaker from his own, suggesting that he cannot quite bring himself to share the evolutionist's delight in the coming quiet and inviting the reader to listen with him to the yelps and cries that 'make[...] what is known as "literature" something other than knowledge: the very place where social code is destroyed and renewed'.[22]

---

[22] Kristeva, p.132.

Appendix

Robert Chambers, 'Lines To A Little Boy'[23]

My winsome one, my handsome one, my darling little boy,
The heart's pride of thy mother and thy father's chiefest joy,
Come ride upon my shoulder, come sit upon my knee,
And prattle all the nonsense that I love to hear from thee:
With thine eyes of merry lustre and thy pretty lisping tongue,
And thy heart that evermore lets out its humming, happy song;
With thy thousand tricks so gleesome, which I bear without annoy
Come to my arms, come to my soul, my darling little boy!

My winsome one, my fairest one, they say that later years
Will sometimes change a parent's hope for bitter grief and tears:
But *thou*, so innocent! canst thou be aught but what thou art,
And all this bloom of feeling with the bloom of face depart?
Canst thou this tabernacle fair, where God reigns bright within
Perform like Judah's children, with the pagan rights of sin?
No – no, so much I'll cherish thee, so clasped we'll be in one,
That bugbear guilt shall only get the father with the son;
And thou, perceiving that the grief must *me* at least destroy,
Wilt still be fair and innocent, my darling little boy!

My gentle one, by [sic] blessed one, can that time ever be,
When I to thee shall be severe, or thou unkind to me?
Can any change which time may bring, this glowing passion wreck,
Or clench with rage the little hand now fondling round my neck?
Can this community of sport, to which love brings me down,
Give way to anger's kindling glance, and Hate's malignant frown?
No – no, that time can ne'er arrive, for whatsoe'er befall
This heart shall still be wholly thine, or shall be not at all;
And to an offering like this thou canst not e'er be coy,
But still will be my faithful and my darling little boy!

My winsome one, my gallant one, so fair so happy now,
With thy bonnet set so proudly upon thy shining brow,
With thy fearless bounding motions, and thy laugh of thoughtless
glee
So circled by a father's love which wards each ill from thee!

---

23 *Chambers Edinburgh Journal*, 424 (1840), 64.

Can I suppose another time when this shall all be o'er,
And thy cheek shall wear the ruddy badge of happiness no more,
When all that now delight in thee far elsewhere shall have gone,
And thou shalt pilgrimage through life, unfriended and alone,
Without an aid to strengthen and console thy troubled mind,
Save the memory of the love of those that left thee thus behind.
Oh let me not awake the thought, but in the present blest,
Make thee a child of wisdom – and to heaven bequeath the rest.

Far rather let me image thee, in sunny future days,
Outdoing every deed of mine and wearing brighter bays;
With less to dull thy fervency of recollected paine,
And more, to animate thy course of glory and of gain;
A home as happy shall be thine, and I too shall be there,
The blessings purchased by thy worth in peace and love to *share*,
Shall see within thy beaming eye my early love repaid,
And every ill of failing life a bliss by kindness made –
Shall see thee pour upon thy son, then sitting on thy knee,
A father's gushing tenderness, such as I feel for thee;
And know, as I this moment do, no brighter, better joy,
Than thus to clasp unto thy soul thy darling little boy!

Victorian Evolutionary Criticism and the Pitfalls of Consilience

John Holmes

Abstract: The emergent school of 'literary Darwinism' or 'biopoetics' was anticipated in the late nineteenth century by Herbert Spencer, Hippolyte Taine, Grant Allen and Constance Naden, among others. A close examination of their attempts to derive a critical approach to literature and the arts from evolutionary theory exposes a number of errors still current within evolutionary literary criticism today. For all its limitations, the work of their contemporary John Addington Symonds remains suggestive of alternative and more promising ways of bringing about the 'consilience' between the arts and the sciences demanded by Edward O. Wilson.

In *Consilience* (1998), the biologist Edward O. Wilson expresses a hope that the humanities, including the 'interpretation of the arts,' 'will draw closer to the sciences and partly fuse with them'. Wilson's method is to interpret the arts in the light of evolutionary psychology, an approach he identifies with the work of Joseph Carroll and Frederick Turner, among others.[1] Carroll and Turner, together with Wilson himself, are the leaders of an emergent school of criticism, self-styled 'literary Darwinism' or 'biopoetics'. Since *Consilience* was published, this school has been growing in numbers and self-confidence. Alongside monographs by individual critics, it has produced a number of substantial collections of essays, and has been taken up by leading public intellectuals including Steven Pinker and Ian McEwan.[2] It has even begun to split into factions, with Carroll in particular keen to distance his own adaptationist literary criticism from the 'cosmic evolutionism' of Turner and others.[3]

As yet, however, biopoetic critics have paid little attention to the history of their own discipline. Carroll discusses the nineteenth-century French critic Hippolyte Taine's model of evolutionary criticism, but he has very little to say about any other forebears.[4] Similarly, the foundational collection *Biopoetics* includes only a very brief descriptive summary of evolutionary aesthetics before the twentieth century, without passing much comment on it apart from to say that, as science then was not as well grounded as it is

---

1 Edward O. Wilson, *Consilience: The Unity of Knowledge* (London: Abacus, 1999), pp.10, 240-41.

2 See Steven Pinker, 'The Biology of Fiction', and Ian McEwan, 'Literature, Science and Human Nature', in *Human Nature: Fact and Fiction*, ed. by Robin Headlam Wells and Johnjoe McFadden (London: Continuum, 2006), pp.27-39, 40-60.

3 Joseph Carroll, *Literary Darwinism: Evolution, Human Nature, and Literature* (London: Routledge, 2004), pp.xii-xiii.

4 Joseph Carroll, *Evolution and Literary Theory* (Columbia: University of Missouri Press, 1995), pp.16-24, 32-40.

now, writers were 'forced to speculate' and their speculations seem 'quaint' today.[5] Yet a close examination of the theoretical assumptions and critical practice of Victorian evolutionary criticism (as opposed to its scientific foundations) exposes errors which remain uncorrected within current biopoetics.

Victorian evolutionary criticism begins with Herbert Spencer's tentative speculations in various essays published in the late 1850s. Its first systematic theorisation comes in his *First Principles*, published in instalments from 1860 to 1862. While Spencer was expounding his theory in England, Taine was working out an alternative model grounded in his reading of Darwin in his *History of English Literature*, published in France in 1863-64. The lead set by Spencer and Taine was followed by a number of critics and evolutionary theorists over the next thirty years, including H. M. Possnett and Edmund Gosse, as well as the critics discussed in this chapter.[6]

There are four strands in Victorian evolutionary criticism which are especially revealing for their parallels within evolutionary criticism today. The first is Spencer's project of interpreting the evolution of the arts in the light of comprehensive evolutionary principles. The second is the theory of physiological aesthetics, built up by Grant Allen and Constance Naden on foundations laid again by Spencer. The third is Taine's model of the literary critic as a natural historian. All three of these approaches, as articulated by the Victorian theorists themselves, involve false steps which have been repeated apparently unknowingly by modern biopoetic critics. Finally, while they do not always avoid these errors, John Addington Symonds's experiments in scientific criticism, gathered together in *Essays Speculative and Suggestive* in 1890, and the theoretical expositions that underlie them, give promising glimpses of fresh approaches to evolutionary criticism.

The Arts and the 'Law of Evolution': Herbert Spencer

In *First Principles*, Herbert Spencer defines the 'Law of Evolution' as 'a change from an indefinite, incoherent homogeneity, to a definite coherent heterogeneity'. Drawing on his essay 'Progress: its Law and Cause,' published in the *Westminster Review* in 1857, he takes his evidence for this law from the history of the arts as well as from elsewhere. The shift from incoherence to coherence is exemplified in the transformation of 'the tales

---

5 Nancy E. Aiken, 'Literature of Early "Scientific" and "Evolution" Aesthetics', in *Biopoetics: Evolutionary Explorations in the Arts*, ed. by Brett Cooke and Frederick Turner (Lexington: ICUS, 1999), pp.417-31 (p.419).

6 See Hutcheson Macaulay Posnett, *Comparative Literature* (London: Kegan Paul, 1886); Edmund Gosse, *A Short History of Modern English Literature* (London: Heinemann, 1897), pp.386-92.

of primitive times, like those with which the story-tellers of the East still daily amuse their listeners,' in which events are often both impossible and unconnected, into complex modern novels, in which they are 'the proper products' of 'characters working under given conditions' and 'held together by complex moral relations'. The complementary move from homogeneity to heterogeneity can be seen in 'the co-ordinate origin and gradual differentiation of Poetry, Music, and Dancing,' and in the development of literature itself, from texts such as the Bible or the Iliad which combine many different genres and functions, to 'its present heterogeneous development, in which its divisions and subdivisions are so numerous and varied as to defy complete classification'.[7]

Spencer's dabbling in the history of the arts is empirically inadequate. As A. T. W. Borsdorf pointed out in *Science of Literature*, an early critique of Spencer and Taine, both realistic and fantastical modes of literature have co-existed throughout literary history.[8] It is not clear that the current popularity of fantasy is an evolutionary throwback, nor for that matter that opera is. But while such historical criticisms are valid, to dismiss Spencer on these grounds alone, without examining his methodology more closely, is to miss his more fundamental mistakes.

Spencer's reasoning stumbles into three pitfalls in particular. The first of these is to derive a cosmological principle – the 'Law of Evolution' – from a biological process. For Spencer himself, this was not unwarranted. Teleological explanations of biological evolution were widely accepted in the late nineteenth century, while current theories in other sciences, such as the nebula hypothesis in astronomy, seemed to justify wide-ranging evolutionary parallels. But the cosmic evolutionism of certain modern biopoetic critics is less justified. Alexander Argyros's claim that 'evolution moves in the direction of temporal complexity' and Turner's account of 'the deepest tendency or theme of the universe,' beginning with 'unity in multiplicity' and 'complexity within simplicity' and ending with 'hierarchical organization' and 'self-similarity,' both bear out Carroll's charge that these critics take their lead illegitimately from Spencer, rather than from Darwin.[9]

---

7 Herbert Spencer, *First Principles* (London: Williams and Norgate, 1863), pp.169, 173-74, 210-11, 216. See also Spencer, 'Progress: its Law and Cause', *Westminster Review*, n.s. 11 (1857), 445-85 (pp.459-65).

8 A. T. W. Borsdorf, *Science of Literature: On the Literary Theories of Taine and Herbert Spencer* (London: Nutt, 1903), p.47.

9 Alexander J. Argyros, *A Blessed Rage for Order: Deconstruction, Evolution, and Chaos* (Ann Arbor: University of Michigan Press, 1991), p.148; Frederick Turner, 'An Ecopoetics of Beauty and Meaning', in Cooke and Turner, eds, pp.119-37 (pp.126-27); Carroll, *Literary Darwinism*, pp.xii-xiii

The second mistake in Spencer's reasoning is to take accounts of the origins of the arts based on very slight evidence, and then extrapolate from them as though they were gospel. Wilson emphasises continuity rather than evolutionary change, but he repeats the same fundamental error. Wilson seeks to ground a comprehensive model of how the arts work today in the premise that they were born of stone age humanity's 'need to impose order on the confusion caused by intelligence'. He admits that 'the biological origin of the arts is a working hypothesis,' not a fact, but as Steven Jay Gould observes, he nonetheless treats an 'unproven theory for the origin of aesthetic preferences' as if it were already established.[10] Furthermore, it is not at all clear, even if we were able to prove that the arts had a biological function thirty thousand years ago, and to identify precisely what that function was, that that knowledge would or should have much bearing on how we interpret individual artworks today. Modern biopoetics has produced some stimulating speculative work on the origins of art, Wilson's included. Yet these accounts remain ultimately myths, both inadequately empirical and too remote to give solid grounds to readings of historical rather than prehistoric texts.

Spencer's third mistake is to assume that there is a hierarchy of aesthetic forms that can be described in scientific terms. In Spencer's case, this entails identifying his own aesthetic preferences with the cosmic drive of evolution. Where Spencer's evolutionism is dynamic and frankly progressive, Wilson's evolutionary psychology is static, as the human nature which evolved slowly during the Pleistocene has not kept pace with more recent cultural change. Nevertheless, Wilson too appeals to evolution to underpin his own aesthetic value-judgements. The conservative structure of Wilson's evolutionary theory is matched by his reification of 'the masters of the Western canon, and those of other high cultures' into the spokesmen of human nature.[11]

There are two problems with both Spencer's and Wilson's arguments on literary value. The first, noted by Carroll, is that 'taking evolution as the basis for normative value judgments' can 'justify virtually anything, even values radically opposed to one another'.[12] Spencer and Wilson may agree about George Eliot, but their estimations of Homer are widely divergent. The second, equally fundamental, problem is that their arguments involve what Gould calls 'the logically fallacious transition from a factual claim of

10 Wilson, pp.250, 254; Steven Jay Gould, *The Hedgehog, the Fox, and the Magister's Pox: Mending and Minding the Misconceived Gap between Science and the Humanities* (London: Vintage, 2004), p.238.

11 Wilson, p.236.

12 Carroll, *Literary Darwinism*, p.xiv.

science to an intrinsically nonfactual issue that must be adjudicated'.[13] Whether challenging the canon or affirming it, they imply that there is an empirical basis in evolutionary theory for judging the aesthetic merit of a work of art or literature when, as Carroll points out, it is rather aesthetic responses themselves that are the fundamental empirical data for generalising aesthetic judgements.[14]

Physiological Aesthetics: Grant Allen and Constance Naden

In *First Principles*, Spencer simply assumes that his aesthetic judgements follow from his place in the evolutionary hierarchy. Elsewhere, he argues more directly that aesthetic judgement itself is a physiological process. In one of a series of essays on education, 'What Knowledge is of most Worth?,' published in the *Westminster Review* in 1859, Spencer argues that 'as all mental natures have certain general principles in common, there must result certain corresponding general principles on which alone art-products can be successfully framed'.[15] This argument seems to contradict Spencer's own thinking on the evolution of art in 'Progress: Its Law and Cause' and *First Principles*. If human nature – our common mental characteristics – determines how we respond to art, then how can aesthetic taste differ so much between modern Englishmen who applaud realist novels and Orientals still delighted by 'the story-tellers of the East'?

Spencer's disciple Grant Allen set out to solve this problem in *Physiological Aesthetics* (1877). Allen's argument is that different aesthetic standards are equally natural, but that they reflect different stages of evolution. According to Allen, good and bad taste are not 'mere matters of convention'. Rather, bad taste is the result of 'a coarse and indiscriminative nervous organisation, an untrained attention, a low emotional nature, and an imperfect intelligence'. Good taste, on the other hand, 'is the progressive product of progressing fineness and discrimination in the nerves, educated attention, high and noble emotional constitution, and increasing intellectual faculties'. It would therefore be unreasonable to 'expect a child or a savage to admire the poetry of Wordsworth, the landscapes of Turner, the sonatas of Beethoven, the duomo of Milan'. By Allen's logic, the different standards of beauty held by different races map onto different developmental stages in evolution, with 'the judgment of the finest-nurtured and most discriminative, the purest and most cultivated of our contemporaries'

13 Gould, p.238.

14 Carroll, *Literary Darwinism*, p.165.

15 Herbert Spencer, 'What Knowledge is of Most Worth?', *Westminster Review*, n.s. 16 (1859), 1-41 (p.32).

approximating most closely to 'an absolute aesthetic standard'.[16] Where Spencer's evolutionary aesthetics culminate in his own taste, Allen's lead more conventionally up to that of a Matthew Arnold or perhaps a Walter Pater.

Racist pseudo-science aside, the fundamental weakness in Allen's argument is that his logic is circular. Allen argues that we should accept as our aesthetic standard the good taste of those whose physiology is most refined. At the same time, he argues that we can see who is blest with this refined physiology by seeing who has the best taste. The same circularity lies hidden at the heart of Wilson's assertion that the canonical artworks of 'high cultures' are the truest and best expression of human nature. According to Wilson, as art expresses human nature, 'Creativity is therefore humanistic in the fullest sense. Works of enduring value,' he declares, 'are those truest to these origins'.[17] But if all art is the product of human beings, how can some artworks be more 'human' than others? And even if this were a meaningful distinction to make, to equate this most truly human art straightforwardly with the canon is to imply with Grant that some people's aesthetic preferences are more fully human than others' – those of savages, children and others excluded from the process of canon-formation.

Like Allen, the poet, evolutionary philosopher and feminist Constance Naden sought to develop Spencer's suggestion that there was a biological basis to aesthetics in an essay on 'The Evolution of the Sense of Beauty,' published in *Knowledge* in 1885. The contrast between Naden's argument and Allen's is again instructive for modern biopoetic critics. The racist evolutionary trajectories implicit in Allen's argument are affirmed openly and unashamedly by Naden. In constructing an evolutionary teleology of art culminating in classical sculpture, she identifies 'the primitive anthropoid ape' as its starting point. The evolution of aesthetics takes place as 'the projecting jaw, the receding forehead, the high cheek-bones, and the flat nose' of 'the lower races' are increasingly rejected as it is recognised 'that they are generally associated with deficient intelligence'.[18]

Naden's grotesque science discredits her argument. In other regards, however, she is more persuasive than Allen, because she is both less ambitious and less doctrinaire. Allen aims 'to prove that our existing likes and dislikes in aesthetic matters are the necessary result of natural selection'.[19] Like Wilson, he is a card-carrying adaptationist. Naden restricts

---

16 Grant Allen, *Physiological Aesthetics* (London: Henry S. King, 1877), pp.47-48.

17 Wilson, p.237.

18 Constance Naden, *Further Reliques of Constance Naden: Being Essays and Tracts for our Times*, ed. by George M. McCrie (London: Bickers, 1891), p.97.

19 Allen, p.viii.

herself largely to visual aesthetics and is content with the more modest claim that, while 'the love for light and colour is not a product of natural selection' itself, 'without natural selection it could not have been perpetuated.'[20] In the language of current evolutionary theory, aesthetics for Naden are a by-product of natural selection for acute perceptions and good health, not the direct product of natural selection in themselves. Nevertheless, whether it is Spencer attributing the modern preference for novels over fairy tales to the Law of Evolution, or Allen explaining the poetic appeal of the word 'buoyant' through its associations with good health, or Naden identifying the Greek ideal in sculpture as the product of the progressive weeding out of ape-like features from art, the Victorian physiological critics all move directly from a supposed artistic phenomenon to its ultimate cause, leapfrogging any influences in between.[21] Similarly, when Brett Cooke characterises Pushkin's short story 'The Snowstorm' in terms of 'the male constraint of female sexuality, which appears to be a crosscultural universal,' rather than in terms of the form that supposed 'universal' took in early nineteenth-century Russia, he too effaces and so loses the subtlety of the rich, thick-descriptive readings of texts and contexts yielded by historicist criticism.[22]

The Natural History of Literature: Hippolyte Taine

Troubled by this approach, Carroll compensates for it within his own critical model by admitting the impact of historical contexts alongside biological imperatives. In this, he is following Hippolyte Taine, the French critic who developed his own model of evolutionary criticism at the same time as Spencer and his followers were devising their physiological aesthetics in England. Criticism for Taine is an exercise in historical psychology. Where a psychoanalytic critic might seek to understand an author's psychology by looking in detail at his or her family and upbringing, what matters for Taine is not these specific experiences so much as the wider circumstances into which a given author happens to be born.

There are, Taine suggests, three main forces which shape an author's work. The first of these is race (*la race*). Like Wilson's human nature but more local, Taine's concept of race is the accumulated characteristics of a given people shaped by their environment, in practice an author's biologically predetermined tendency towards a particular temperament or character. The second shaping force is the author's own surroundings or

---

20 Naden, pp.83-84.

21 Allen, p.253.

22 Brett Cooke, 'Sexual Property in Pushkin's "The Snowstorm": A Darwinist Perspective', in Cooke and Turner, eds, pp.175-204 (p.191).

environment (*le milieu*), both natural – Taine is particularly fixated on climate – and social or political. The third is the epoch (*le moment*). This Taine defines as the 'acquired momentum' of the historical moment. Although the distinction between the social *milieu* and the *moment* is not very exact, the latter implies a diachronic perspective on the former. For Taine, these three categories – race, environment and epoch – not only provide the critic with the tools to understand the psychology of authors and their texts, but also explain the success or failure of a given artist or movement. If the racial temperament, the natural, social and political environment, and the epoch are in accord with one another, a great literary movement will flourish. If they run counter to one another, then new literary experiments will come to nothing.[23]

There are a number of obvious problems with Taine's model for historicist evolutionary criticism. As early critics such as Borsdorf and Leslie Stephen observed, we do not have the comprehensive scientific understanding we would need to determine how climate, for example, impacts on an author's psychology, nor how that translates into the form or content of a literary text. Race too is a problematic category, and not only politically. As Borsdorf noted, 'individual characters are not necessarily hereditary,' while, as both he and Stephen pointed out, all nations and races are mixed. Anticipating both the New Criticism and post-structuralism, Borsdorf objected too that the very project of looking for an author behind a text was misconceived, leading only to an imaginary past and a false portrait of the artist made up of those fragments of themselves which they had chosen to leave behind.[24]

These objections are real, and they leave Taine's project badly damaged, for all that it remains an early and suggestive model for a number of different current critical schools including historicism and ecocriticism. Furthermore, for all that Taine's criticism itself is often astute, it is not clear that his practice is grounded in his supposed method, nor that that method generates specific insights into individual authors or texts. This is equally true of many current biopoetic critical models. Wilson's reduction of 'myth and fiction' to some 'two dozen' archetypal structures, Robert Storey's and Peter Swirski's models of narrative respectively as an adaptive device for developing self-awareness and social cohesion and as a mechanism for thought experiments along the lines of game theory – none of these models appears able to generate anything other than predetermined insights into a

---

23 H. A. Taine, *History of English Literature*, trans. by H. van Laun (Edinburgh: Edmonston and Douglas, 1871), 2 vols, I, pp.10-14.

24 Borsdorf, pp.18, 24, 29, 32-35; Leslie Stephen, 'Taine's History of English Literature', *Fortnightly Review*, 20 (1873), 693-714 (pp 693-97).

narrow range of pre-selected texts, ironing out their distinctive literary qualities in the process.[25] As Harold Fromm observes in an otherwise sympathetic account of biopoetic criticism, 'There is always the danger of forcing a variety of artefacts through a critical grinder that makes them all come out looking like the same dust'.[26] Even Carroll's more nuanced model of biopoetic criticism is clearly tailored to a particular genre and set of preoccupations. Carroll's test case is *Pride and Prejudice*, a realist novel with a romance plot.[27] It is not clear that either his method or any other biopoetic approach yet devised would have much to say about lyric poetry, for example, yet the lyric is presumably as much a product of human nature as narrative.

For Carroll, Taine's concept of race remains valuable, providing it is 'transposed to the level of the whole species'.[28] But in attempting to align Taine with modern biopoetics, he exposes discontinuities as much as parallels between their two projects. Stephen criticises Taine for failing to take into account what he ironically calls the 'minutiae' of the differences within populations.[29] Yet Taine himself critiques his eighteenth-century predecessors, arguing that 'They thought men of every race and century were all but identical; the Greek, the barbarian, the Hindoo, the man of the Restoration, and the man of the eighteenth century, as if they had been turned out of a common mould; and all in conformity to a certain abstract conception, which served for the whole human race.' Like Stephen's rebuke to Taine himself, Taine's indictment of his predecessors who 'knew man, but not men' is an apt criticism of evolutionary psychology. Taine's point is precisely that generalisations about human nature – whether decided a priori as in the Enlightenment or derived from statistical norms as in evolutionary psychology – are of little use when it comes to interpreting the works of particular individuals chosen from among 'the infinite diversity and marvellous complexity of souls'.[30] Taine goes on to make generalisations of his own under the rubric of race, but this does not undermine the force of his criticism as, for all that it is a blunt and patently inadequate instrument, the function of race within Taine's critical model is to distinguish between

25 Wilson, pp.248-49; Robert Storey, *Mimesis and the Human Animal: On the Biogenetic Foundations of Literary Representation* (Evanston: Northwestern University Press, 1996); Peter Swirski, *Of Literature and Knowledge: Explorations in Narrative Thought Experiments, Evolution, and Game Theory* (London: Routledge, 2007).

26 Harold Fromm, 'The New Darwinism in the Humanities', *Hudson Review*, 56 (2003), pp.89-99, 315-27 (p.321).

27 See Carroll, *Literary Darwinism*, pp.187-216.

28 Carroll, *Evolution and Literary Theory*, p.17.

29 Stephen, p.695.

30 Taine, I, p.5.

groups of authors who might otherwise be lumped indistinguishably together as members of the 'human race'. To suggest that Taine's concept of race is equivalent to Wilson's 'human nature' is to ignore the reason why Taine invokes race in the first place, and to miss the pertinence of his critique of enlightenment universalism to biopoetics itself.

Alternative Avenues in Evolutionary Criticism: John Addington Symonds

The study of Victorian evolutionary criticism throws up a number of unresolved problems within modern biopoetics. Spencer's *First Principles* exposes the temptations of unwarranted cosmic evolutionism and of myths of the origins of the arts. The physiological aesthetics of Spencer, Allen and Naden show up the difficulties involved in extrapolating aesthetic principles from evolutionary science. Taine's *History of English Literature* bears witness to the challenge of turning a model of literary criticism as natural history into a method capable of generating detailed critical analysis, while his powerful critique of enlightenment universalism suggests that the very project of grounding a critical approach in the generalisations of a statistical science such as evolutionary psychology may be misguided.

If Wilson's ambitious project of consilience is to bear fruit, biopoetic criticism needs to address these criticisms and to avoid these failings. But Wilson's is not the only model of consilience, nor are these the only Victorian evolutionary critics. In his last book, *The Hedgehog, the Fox, and the Magister's Pox*, Gould counters Wilson by proposing 'a consilience of equal regard' that 'respects the inherent differences' between science and the humanities while seeking 'to emphasize and nurture the numerous regions of actual overlap and common concern'.[31] This less ambitious, more catholic, model of consilience opens the door to a range of alternative literary critical approaches informed by science. Among the Victorians, John Addington Symonds stands out as a precursor to this alternative approach to scientific criticism.

Symonds's most concerted piece of evolutionary criticism, 'On the Application of Evolutionary Principles to Art and Literature,' is both less and more scientifically rigorous than those of his contemporaries.[32] On the minus side, it makes questionable use of analogies, as Symonds invokes organic concepts of growth and decay, evolution and extinction, to account for the history of artistic movements such as Elizabethan and Greek drama, Gothic architecture, and Italian Renaissance painting. Carroll derides evolutionary models of cultural change – specifically Richard Dawkins's

---

[31] Gould, p.259.

[32] John Addington Symonds, *Essays Speculative and Suggestive*, 3rd edn (London: Smith, Elder, 1907), pp.27-52.

concept of the meme – complaining that 'the analogical theorist takes it for granted that the causal processes in one field will provide a neat and reliable pattern for processes in other fields.'[33] Yet unlike Spencer and arguably Dawkins, Symonds is acutely aware that he is using evolution as a metaphor, a critical tool that comes to hand, rather than the key to all mythologies. The use he makes of it is to generate a hypothesis which he can subsequently test to see how far it fits with literary and artistic history as a whole. In this he anticipates, and would presumably win the approval of, modern biopoetic critics such as Jonathan Gottschall and Carroll himself, for all that – again on the minus side – his pet theory may be weak and his own tests of it not convincingly impartial.[34]

Symonds's significance as an evolutionary critic lies less in his practice than in the theoretical premises that that practice imperfectly embodies. There are three suggestions within his work for the role of science in criticism which remain pertinent today. In his essay 'On Some Principles of Criticism,' Symonds follows Taine in arguing that the critic should 'become the natural historian of art and literature'. At the same time, however, he insists that that same critic 'cannot abnegate the right to judge' nor 'divest himself of subjective tastes which colour his judgment'. Of the three principles of criticism that he outlines, Symonds is clear that scientific criticism should supplement classical (or evaluative) and romantic (or subjective) criticism, not supplant them.[35] He thus opens the door to evolutionary criticism, but he does not hand over the keys. Biopoetics has pretensions to be a new 'paradigm' that will require the re-evaluation of 'all of our prior perspectives on the arts'.[36] In Jonathan Gottschall's words, 'the primary theoretical, methodological, and attitudinal struts that support the field [of literary studies] are suffering pervasive rot' and 'this scientific turn represents the only responsible and attractive correction of course.'[37] By contrast, Symonds advances a pluralist model of criticism that has space for such scientific approaches within it but does not preclude other lines of enquiry.

In the same essay, Symonds defines 'Scientific criticism' as proceeding by 'induction, historical investigation, morphological analysis, misdoubting the certainty of aesthetic principles, regarding the instincts and sensibilities of

33 Carroll, *Literary Darwinism*, p.xiii.

34 See Jonathan Gottschall, *Literature, Science, and a New Humanities* (New York: Palgrave Macmillan, 2008).

35 Symonds, pp.60, 62.

36 Brett Cooke, 'Biopoetics: The New Synthesis', in Cooke and Turner, eds, pp.3-25 (p.23).

37 Gottschall, p.3.

the individual with distrust' and 'seeking materials for basing the canons of perfection upon some positive foundation'.[38] Like Wilson, Symonds is proposing a consilience of science and literary criticism. But where Wilson aspires to a unity of knowledge, Symonds like Gould calls instead for a shared commitment to empiricism. Described in these terms, 'scientific criticism' looks more like a combination of historicism and New Criticism than like biopoetics. Indeed, the biopoetic critics' largely unquestioning dependence on evolutionary psychology for their hypotheses about literature looks unscientific by comparison.

Symonds's third hint at how the study of literature might proceed in the light of science again tallies with Gould's version of consilience. For Spencer, the modern taste in realist fiction is a product of the evolutionary process itself. For Symonds, it is rather a product of the place and role of science within our culture. In 'Realism and Idealism', Symonds observes that 'Science has rendered our sense of veracity acute'.[39] Science may not explain art, but it undoubtedly impacts on how we experience it. Readers taught to understand the world through modern science will read given texts differently from those brought up believing in the literal truth of a particular revelation. Different readers will experience different disjunctions between what they are reading and their own worldviews, reflecting whether they are able and how far they are willing to suspend their disbelief in the alternative perceptions or imaginations they encounter. These disparities between reading experiences might lead us to ask, for example, what role texts that do not tally with science play in a culture that accepts the validity of scientific knowledge, for readers brought up in that culture, or for those on its margins. Because of the philosophical and religious questions it raises, evolutionary theory would be central both to the context in which we ask questions of this kind and to any answers we might give. Taking a lead not from Wilson or Spencer, but rather from Gould and Symonds, the most productive course for modern scientific criticism may be to ally itself not to evolutionary psychology at all, but to reader-response theory – not to pursue a Darwinian critical method, but to aim for a greater awareness of ourselves as readers of literature in a Darwinian age.

---

38 Symonds, p.61.
39 Ibid., p. 127.

H. G. Wells's *The Time Machine* and the End of Literature

Simon J. James

Abstract: As a writer of fiction, polemic and popular science a generation after Darwin, H. G. Wells sought to place the aesthetic at the service of evolutionary imperatives. Like many of his Victorian contemporaries in the decades following the publication of the *Origin of Species* and the 1870-71 Education Acts, Wells worried over the place of imaginative literature in mankind's evolutionary future. He feared in particular that the pursuit of aesthetic perfection for its own sake would risk not only literature's extinction, but also humanity's. *The Time Machine* dramatises the biological and cultural consequences of mankind's neglect of its own future survival, and the consequent extinction of artistic culture in this imagined degeneration.

Victorian writing, especially scientific discourse, was fond of Galileo's image of Nature as itself a book. Wells himself uses the image in his very first book, 1893's *Text Book of Biology*, suggesting that 'Zoology is, indeed a philosophy and a literature to those who can read its symbols.'[1] The popularity of the trope in the nineteenth century is perhaps due to its recurrence in Thomas Carlyle's enormously influential *Sartor Resartus* (1833-34):

> We speak of the Volume of Nature: and truly a Volume it is, – whose Author and Writer is God. To read it! Dost thou, does man, so much as well know the Alphabet thereof? With its Words, Sentences, and grand descriptive Pages, poetical and philosophical, spread out through Solar Systems, and Thousands of Years, we shall not try thee. It is a Volume written in celestial hieroglyphs, in the true Sacred-writing; of which even Prophets are happy that they can read here a line and there a line. As for your Institutes, and Academies of Science, they strive bravely; and, from amid the thick-crowded, inextricably intertwisted hieroglyphic writing, pick out, by dextrous combination, some Letters in the vulgar Character, and therefrom put together this and the other economic Recipe, of high avail in Practice. That Nature is more than some boundless Volume of such Recipes, or huge, well-nigh inexhaustible Domestic-Cookery Book, of which the whole secret will in this manner one day evolve itself, the fewest dream.[2]

---

1 Quoted in John Hammond, '*The Island of Doctor Moreau*: A Swiftian Parallel', in *The Wellsian: Selected Essays on H. G. Wells*, ed. by John S. Partington (Haren: Equilibris, 2003), pp.45-67 (p.46).

2 Thomas Carlyle, *Sartor Resartus: The Life and Opinions of Herr Teufelsdröckh in Three Books*, ed. by Roger L. Tarr and Mark Engel (Berkeley: University of California Press, 2000), pp.189-90.

The same metaphor is also employed both by Charles Lyell in *Principles of Geology* (1830-33), whose reimagining of time was of course crucial to Darwin's formulation of evolution, and subsequently by Darwin himself in the *Origin*:

> For my part, following out Lyell's metaphor, I look at the geological record, as a history of the world imperfectly kept, and written in a changing dialect; of this history we possess the last volume alone, relating only to two or three countries. Of this volume, only here and there a short chapter has been preserved, and of each page, only here and there a few lines. Each word of the slowly-changing language, being more or less different in the interrupted succession of chapters, may represent the apparently abruptly changed forms of life.[3]

For Carlyle, the book of nature cannot be fully read because mankind has not yet developed the esoteric knowledge required to do so. In H. G. Wells's *The Time Machine* (1895), the Time Traveller is such a Carlylean academician, a palaeontologist who travels into the year 802701, seeking evidence of his own present in the records held there. In Darwin's development of the metaphor, however, the fragmentary nature of the reading is the fault of the text, not of the reader: the whole story may not have been fully written down and time has in any case erased some of the writing forever. The Time Traveller may be an intellectual in his own time, but not only has the environment of the future made his literacy archaic, but he does not even have a grammatically coherent text to read:

> Suppose you found an inscription, with sentences here and there in excellent plain English, and interpolated therewith, others made up of words, of letters even, absolutely unknown to you? Well, on the third day of my visit, that was how the world of Eight Hundred and Two Thousand Seven Hundred and One presented itself to me![4]

In a writing career that would last for fifty years after *The Time Machine*, Wells feared that real literature would itself have to take part in the human species' struggle for existence, or would cease to be of any importance in the order of nature. Writing, and the novel in particular, must lead the imagining and building of a utopian future for mankind. Art produced purely for art's sake, on the other hand, might risk becoming obsolescent and even contributing to humanity's evolutionary degeneration. Max Nordau's *Degeneration*, first published in English also in 1895, imagines a utopian

---

3 Charles Darwin, *The Origin of Species*, ed. by Gillian Beer (Oxford: Oxford University Press, 1996), p.251.

4 *The Time Machine: An Invention*, ed. by Patrick Parrinder (London: Penguin, 2005), pp.43-44.

future in which art and literature no longer have any place. As dancing was once 'a State function of the highest order' and is now 'no more than a fleeting pastime for women and youths,' so too might literature pass from being a part of culture into becoming only its relic.[5] For writing to survive, it too must be adapted, and Wells sought to compel literature to do just that, namely, to fulfil a role as a 'powerful instrument of moral suggestion,' and not just a beautiful irrelevance.[6]

In order to achieve this, literature must leave the study and enter the laboratory. Wells was arguably both the first writer in the canon of English Literature to have been trained formally as a scientist (a word not in use until the second half of the nineteenth century), and at the same time, according to Roslynn D. Haynes, 'the last great literary writer to have been so strongly influenced by science'.[7] Science for Wells is not an essentially disinterested practice, but in a more Huxleyan fashion, should be fundamentally instrumental, a form of thinking that is a spur to doing. So too Wells's aesthetic ultimately places cultural production at the service of a larger *telos*.

Writing in 1892, J. A. Symonds had argued that the study of culture was not an end in itself but an Arnoldian preparation for life, a 'training [...] consciously carried on by the individual'.[8] Symonds, however, is unusual in including science within his definition of 'culture'. He argues for artistic culture's superiority, as being more genuinely disinterested, while science is limited by being more narrowly functional – but also that science should nonetheless be liberally tolerated and not despised by its counterpart. Presciently, Symonds foresees a division between these two cultures of the arts and the sciences: 'for the future it is probable that there will always be two differently constituted orders of minds, the one inclining to the purely humanistic, and the other to the purely scientific side of culture' (p.109). In 1887, John Morley also predicted that 'after the severity with which science was for so many ages treated by literature, we cannot wonder that science now retaliates, now mightily exalts herself, and thrusts literature down into

---

5 Max Nordau, *Degeneration*, ed. by George L. Mosse (Lincoln: University of Nebraska Press, 1993), p.543.

6 'The Contemporary Novel', in *An Englishman Looks at the World* (London: Cassell, 1914), pp.148-69 (p.159).

7 Roslynn D. Haynes, *H. G. Wells: Discoverer of the Future: The Influence of Science on his Thought* (New York: New York University Press, 1980), p.ix, p.1.

8 J. A. Symonds, 'Culture: Its Meaning and Uses', *New Review*, 7 (1892), 105-15 (p.106). See 'Literature and Science', in Matthew Arnold, *The Complete Prose Works of Matthew Arnold*, 10 vols, ed. by R. H. Super (Ann Arbor: University of Michigan Press, 1974), X, pp.53-73; C. P. Snow, *The Two Cultures and The Two Cultures: A Second Look*, ed. by Stefan Collini (Cambridge: Cambridge University Press, 1993).

the lower place.'[9] Wells claims the authority bestowed by his scientific training for his anti-aestheticism, and by 1931's *The Science of Life*, he was making his retaliation against literature explicit:

> The self-love of the ignorant has demanded that the man of science, in play and story, should be caricatured, ridiculed, and misrepresented. From Laputa to the Pickwick Club, British Literature, for example, spits and jeers at its greater sister, and to devote a life to science and the service of truth is still to renounce most of the common glories and satisfactions of life for a hard and exalted mistress.[10]

In his utopian text *Mankind in the Making* (1908), Wells recommended instead a synthesis of the artistic and literary imaginations:

> For the lack of a sufficient literature we specialize into co-ordinated classes. A number of new social types are developing, ignorant of each other, ignorant almost of themselves, full of mutual suspicions and mutual misunderstandings, narrow, limited, and dangerously incapable of intelligent collective action in the face of crises. The medical man sees nothing beyond his profession; he misunderstands the artist, the divine, and the engineer. The engineer hates and despises the politician, the lawyer misses the aims of the medical man, the artist lives angrily in a stuffy little corner of pure technique; none of them reads any general literature at all except perhaps a newspaper. Each thinks parochially in his own limits, and, except for his specialty, is an illiterate man. It is absolutely necessary to the progress of our civilization that these isolations should be overcome, that the community should become aware of itself collectively and should think as a whole. And the only thing that can overcome these isolations and put the mass of intelligent men upon a common basis of understanding, is an abundant and almost universally influential contemporary literature.[11]

The Time Machine attempts just such a synthesis of the literary and scientific, here evolving the new genre of the 'scientific romance' from biology's redrawing of the limits of human physiology, and geology's expansion of the Victorian perception of time.[12] In narrative theory from Aristotle onwards, discrete 'events' are made into a 'plot' by connections of both time and causality.[13] Time travel radically destabilises both of these elements, as the

---

9 John Morley, 'On the Study of Literature' (1887), in *Studies in Literature* (London: Macmillan, 1891), pp.199-200.

10 *The Science of Life*, 2 vols (New York: Doubleday, 1931), II, p.19.

11 *Mankind in the Making* (London: Chapman and Hall, 1908), pp.359-60.

12 Gillian Beer, '"The Death of the Sun": Victorian Solar Physics and Solar Myth', in *The Sun is God: Painting, Literature and Mythology* (Oxford: Clarendon, 1989), ed. by J. B. Bullen, pp.159-80.

13 Aristotle, *Poetics*, Section 3, B, iv, in *Classical Literary Criticism*, ed. by D. A. Russell and Michael Winterbottom, trans. by M. E. Hubbard (Oxford: Oxford University Press,

enormous scale of geological time makes local causal connections appear tiny and inconsequential. Lyell's own version of the 'book of nature' speculates towards a new fantastic genre of narrative reshaped by geology's new understanding of temporality:

> How fatal every errour as to the quantity of time must prove to the introduction of rational views concerning the state of things in former ages, may be conceived by supposing that the annals of the civil and military transactions of a great nation were perused under the impression that they occurred in a period of one hundred instead of two thousand years. Such a portion of history would immediately assume the air of a romance; the events would seem devoid of credibility, and inconsistent with the present course of human affairs. A crowd of incidents would follow each other in thick succession. Armies and fleets would appear to be assembled only to be destroyed, and cities built merely to fall in ruins.
>
> In the vast interval of time which may really have elapsed [...] one or more races of organic beings may have passed away, and yet have left behind [...] no trace of their existence. The imagination is as much perplexed by such errors as to time, as it would be if we could annihilate space, and by some power, such as we read of in tales of enchantment, could transfer a person who had laid himself down to sleep.[14]

When the Time Traveller awakes in the future, he makes a series of misreadings of the book of nature before eventually realising that mankind has evolved into two separate smaller species, the delicate Eloi and the simian Morlocks. In 1874, the future poet laureate Alfred Austin had wondered whether imaginative literature produced in its readers anything higher than mere pleasure, fearing that too many books published in the wake of the 1870-71 Education Acts possessed too little value:

> We entertain the profoundest veneration for works of the imagination, and we hope we should be the last to under-estimate their value. But we venerate and value them on one condition: that they raise man not only from the slough of despond, but from the mire of selfish aims, of ignoble desires, cynical beliefs and purely material views of existence.[15]

Austin proposes semi-seriously that unless literature does seek to improve humanity's future in this way, printing should be suspended for ten years in order to improve man's future evolutionary prospects:

1989), pp.51-90 (pp.62-63); E. M. Forster, *Aspects of the Novel* (Harmondsworth: Penguin, 1962), pp.93-94.

14 Charles Lyell, *The Principles of Geology*, ed. by Martin Rudwick, 3 vols (Chicago: University of Chicago Press, 1990), I, pp.78-79, pp.159-60. See Stephen Jay Gould, *Time's Arrow, Time's Cycle: Myth and Metaphor in the Discovery of Geological Time* (Cambridge, MA: Harvard University Press, 1987), pp.85-86.

15 Alfred Austin, 'The Vice of Reading', *Temple Bar*, 42 (1874), 251-57, (p.252).

> Let reading continue to be a part of his life, but a subsidiary part to thinking, seeing, observing, and energising. [...] That such reading as at present prevails has, by reason both of its quality and quantity, led to a deterioration of the human species, physically, mentally, and morally, we entertain no doubt; nor do we see how, unless the vicious habit be somehow corrected, the race can escape from being ultimately divided into two sections, the members of one of which will be little removed from invalids, and the members of the other scarcely distinguishable from *cretins.* (p.257)

It is hardly likely that the seven-year old H. G. Wells had read it, but Austin's prophecy presciently connects anxieties over mass literacy and the correct uses of literature with fears of the kind of cultural and evolutionary apocalypse dramatised in *The Time Machine.*

The Time Traveller 'had always anticipated that the people of the year Eight Hundred and Two Thousand odd would be incredibly in front of us in knowledge, art, everything. Then one of them suddenly asked me a question that showed him to be on the intellectual level of one of our five-year-old children' (p.25). The explorations of Victorian philologists into the science of language revealed language also to have a history, thus revealing its nature to be contingent, rather than permanent.[16] If the Eloi no longer possess any abstract nouns, the Time Traveller worries whether the abstractions represented by those nouns, such as 'art,' can still be said to exist. His 'confident anticipations of a profoundly grave and intellectual posterity' (p.26) are displaced by the realisation that natural selection has eliminated parts of speech, books, art, anything that is no longer of any advantage. The Time Traveller is troubled by how little affected the Eloi are the day after the Morlock attack: they have no memory; with no memory they have no capacity for art, and the cultural residuum has become fossilised into the 'ancient monument of an intellectual age' (p.65) that is the Palace of Green Porcelain.[17] The Eloi and the Morlocks show not only mankind's physical shape degenerating, but also humanity's mental life and civilisation as a whole; and natural selection has shown as little respect to the cultural corpus inherited from previous generations as it has to the shape of the human form.

---

16 See, for instance, Max Müller, *Lectures on the Science of Language: Delivered at the Royal Institution of Great Britain on April, May and June, 1861*, 2nd rev edn, (London: Longman, 1862), p.386.

17 In the *National Observer* text of *The Time Machine*, the Time Traveller despairingly adds, '"What need for education when there is no struggle for life? What need of books, or what need of stimulus to creative effort?"', *H. G. Wells: Early Writings in Science and Science Fiction*, ed. by Robert M. Philmus and David Y. Hughes (Berkeley: University of California Press, 1975), p.77.

Darwin's notion of beauty in *The Descent of Man* had placed the aesthetic as a by-product of the reproductive imperative: if art is a kind of display, beauty is thus ultimately a matter of sexual selection:

> When, however, it is said that the lower animals have a sense of beauty, it must not be supposed that such sense is comparable with that of a cultivated man, with his multiform and complex associated ideas. A more just comparison would be between the taste for the beautiful in animals, and that in the lowest savages, who admire and deck themselves with any brilliant, glittering, or curious object.[18]

For the Eloi, however, the taste for the beautiful no longer plays a part in reproduction. 'In all the differences of texture and bearing that now mark off the sexes from each other, these people of the future were alike. And the children seemed to my eyes to be but the miniatures of their parents' (p.29).[19] Since the Eloi have substituted the competitive testing of sexual selection for constantly 'making love in a half-playful fashion' (p.41), there is no longer any need for such selective adaptations as aesthetic beauty. If Darwin's surmises about the evolutionary purpose of language and of artistic culture are correct, once a species reaches the fatal equilibrium that precedes degeneration, there is little purpose for either.[20] As mankind has degenerated, its grip on language has been surrendered.[21] In 1994, Steven Pinker hypothesised:

> I would guess that most other human 'cultural' practices (competitive sports, narrative literature, landscape design, ballet), no matter how much they seem like arbitrary outcomes of a Borgesian lottery, are clever technologies we have

---

18 Charles Darwin, *The Descent of Man, and Selection in Relation to Sex*, ed. by James Moore and Adrian Desmond (London: Penguin, 2004), p.246. See also p.208 and Darwin's thoughts on different varieties of aesthetic pleasure in the 'M Notebook', in *Metaphysics, Materialism and the Evolution of Mind: Early Writings of Charles Darwin*, ed. by Paul H. Barrett and Howard E. Gruber (Chicago: Chicago University Press, 1980), pp.6-45 (pp.11-13). Cf. *The Science of Life*, 2 vols (London: Cassell: 1931), II, pp.1390, 1416-32.

19 'Angry Old Buffer' lamented of New Women and Aesthetes in 'Sexomania':
A New Fear My Bosom Vexes,
Tomorrow there may be no sexes!
Unless, to save all Pother,
Each One Becomes the Other.
Punch, 27 April, 1895, p.203. See Linda Dowling, 'The Decadent and the New Woman in the 1890s', *Nineteenth-Century Fiction*, 33 (1978-79), 434-53.

20 See E. Ray Lankester, *Degeneration: A Chapter in Darwinism* (London: Macmillan, 1880), pp.28-33. <books.google.com> [accessed 31 August 2009].

21 See 'The Beginnings of Mind and Language', *Mankind in the Making*, pp.113-56.

> invented to exercise and stimulate mental modules that were originally designed for specific adaptive functions.[22]

Such functions originally possessed by *homo sapiens* are clearly no longer of any use to the Eloi. The Time Traveller, attempting 'a Carlyle-like scorn of this wretched aristocracy in decay' (p.62) classifies the Eloi as the descendants of aristocrats or effete bourgeois on whom the proletarian-descended and technically-minded Morlocks take grisly class revenge by eating them.[23]

The Eloi are described as possessing features of 'Dresden-china' (p.24) not only for their actual pallor and fragility, but also for the association with class privilege and conspicuous consumption.[24] The class of the Eloi also associates them with high culture: descended from the 'Haves, pursuing pleasure and comfort and beauty' (p.48), hyper-evolved Wildean aesthetes or inhabitants of Morris's Nowhere without the need to work, listless consumers not energised producers: 'The little people displayed no vestige of a creative tendency. There were no shops, no workshops, no sign of importations among them' (p.41).[25] (As Linda Dowling notes, Morris's Nowhere produces no literature, and seems distrustful of it, thus perhaps

---

22 Steven Pinker, *The Language Instinct: The New Science of Language and Mind* (London: Penguin, 2000), p.473. For Pinker's assault on the high cultural canons of his own time, see *The Blank Slate: The Modern Denial of Human Nature* (London: Lane, 2002), pp.401-07; and Jon Adams's essay in this volume.

23 In the *National Observer* text, he speculates: 'since this time-journey of mine, I have fancied that there is such a split going on even now in English society, a split that began some two hundred and fifty years ago or more. I do not mean any split between working people and rich [...] but between the sombre, mechanically industrious, arithmetical, inartistic type, the type of the Puritan and the American millionaire and the pleasure-loving, witty, and graceful type that gives us our clever artists, our actors and writers, some of our gentry, and many an elegant rogue. Conceive such types drifting away from one another each into its own direction.' (*Early Writings*, p.86).

24 John Glendening's thought-provoking reading of *The Time Machine* makes a connection with Wilde's famous fondness for his china: 'The Track of the Sphinx: H. G. Wells, The Modern Universe, and the Decay of Aestheticism', *Victorians Institute Journal*, 32 (2004), 129-66, (p.140). In a version of the predecessor of *The Time Machine* 'The Chronic Argonauts' that no longer survives, 'art and literature are cultivated in a very dilettante manner' by the Eloi's ancestors (Bernard Bergonzi, *The Early H. G. Wells: A Study of the Scientific Romances* (Manchester: Manchester University Press, 1961), p.38.

25 See also Steven Mclean's note in the Penguin edition of *The Time Machine*, pp.100-01. Stephen Derry, 'The Time Traveller's Utopian Books and his Reading of the Future', *Foundation*, 65 (1998), 16-24; Gowan Dawson, *Darwin, Literature and Victorian Responsibility* (Cambridge: Cambridge University Press, 2007), p.216.

implying literature's eventual extinction.)[26] The Time Traveller thus speculates, correctly at last, that as mankind has become eventually over-civilised and comfortable, its need for the evolutionary adaptation of art has atrophied alongside its need for physical strength:

> In a state of physical balance and security, power, intellectual as well as physical, would be out of place. For countless years I judged there had been no danger of war or solitary violence, no danger from wild beasts, no wasting disease to require strength of constitution, no need of toil. For such a life, what we should call the weak are as well equipped as the strong, are indeed no longer weak. Better equipped indeed they are, for the strong would be fretted by an energy for which there was no outlet. No doubt the exquisite beauty of the buildings I saw was the outcome of the last surgings of the now purposeless energy of mankind before it settled down into perfect harmony with the conditions under which it lived – the flourish of that triumph which began the last great peace. This has ever been the fate of energy in security; it takes to art and to eroticism, and then come languor and decay.
>
> Even this artistic impetus would at last die away – had almost died in the Time I saw. To adorn themselves with flowers, to dance, to sing in the sunlight: so much was left of the artistic spirit, and no more. Even that would fade in the end into a contented inactivity. We are kept keen on the grindstone of pain and necessity, and, it seemed to me, that here was that hateful grindstone broken at last! (*Time Machine*, pp.32-33)

The 'mere beautiful futility' (p.58) celebrated by Wilde and aestheticism eventually proves an evolutionary liability. The pursuit of pleasure without social responsibility leads to 'intellectual degradation' (p.62). Herbert Spencer had warned in 1843 that under conditions of too much leisure:

> Every powerful spring of action is destroyed – acuteness of intellect is not wanted – force of moral feeling is never called for – the higher powers of his mind are deprived of their natural exercise, and a gradual deterioration of character must ensue. Take away the demand for exertion, and you will ensure inactivity. Induce inactivity, and you will soon have degradation.[27]

Pinker adds that:

> Although language is an instinct, written language is not. [...] Most societies have lacked written language, and those that have it inherited it or borrowed it from

---

26 Linda C. Dowling, *The Vulgarization of Art: The Victorians and Aesthetic Democracy* (Charlottesville: University of Virginia Press, 1996), p.70.

27 Herbert Spencer, Letter IX, 'The Proper Sphere of Government', in *Political Writings*, ed. by John Offer (Cambridge: Cambridge University Press, 1994), pp.1-57 (p.49).

> one of the inventors. Children must be taught to read or write in laborious lessons [...] and people do not uniformly succeed.[28]

The childlike Eloi have no-one to teach them to read, and time has obliterated writing both from the metaphorical book of nature, and literally from the world itself. Wells's time traveller had hoped to find in 802,701 a future library that holds the world's accumulated knowledge. However, the only books that still exist are found in the (to use Fredric Jameson's term) 'prophetic archaeology' of an abandoned museum containing cultural relics and the remains of other species who have, like *homo sapiens*, lost the struggle for existence:[29]

> The brown and charred rags that hung from the sides of it, I presently recognized as the decaying vestiges of books. They had long since dropped to pieces, and every semblance of print had left them. But here and there were warped boards and cracked metallic clasps that told the tale well enough. Had I been a literary man I might, perhaps, have moralized upon the futility of all ambition. But as it was, the thing that struck me with keenest force was the enormous waste of labour to which this sombre wilderness of rotting paper testified. At the time I will confess that I thought chiefly of the *Philosophical Transactions* and my own seventeen papers upon physical optics. (pp.67-68)

This is an unsettlingly self-conscious moment to find within the pages of a book, a technology proclaiming its own obsolescence. The late-Victorian cultural elite had worried that the proliferation of cheap books would make literature valueless; here, no longer possessing any evolutionary value, literature and other monuments of culture have indeed been discarded. Art, then, that does not serve a purpose any greater than being merely beautiful for its own sake proves an evolutionary disadvantage.[30]

Matthew Arnold, in 'Dover Beach' (1867) had listened anxiously to the withdrawing roar of the 'sea of Faith'.[31] The Time Traveller too, voyaging to the time of the end of life on earth thousands of years after Arnold's vaunted panacea of culture has followed religion into extinction, listens to the sea:

---

28 Pinker, *Language Instinct*, p.199. See also Walter J. Ong, *Orality and Literacy* (London: Routledge, 1988), pp.28-29, 171-72.

29 Fredric Jameson, *Archaeologies of the Future; The Desire Called Utopia and Other Science Fictions* (London: Verso, 2005), p.99.

30 The history of applying a Darwinian understanding to literary criticism is summarized in Joseph Carroll, *Literary Darwinism: Evolution, Human Nature, and Literature* (New York: Routledge, 2004), vii-xxii. See also John Holmes's essay in the current volume.

31 Matthew Arnold, 'Dover Beach', *The Poems of Matthew Arnold*, ed. by Kenneth Allott (London: Longman, 1965), pp.240-42.

> The machine was standing on a sloping beach. The sea stretched away to the south-west, to rise into a sharp bright horizon against the wan sky. There were no breakers and no waves, for not a breath of wind was stirring. Only a slight oily swell rose and fell like a gentle breathing, and showed that the eternal sea was still moving and living. (p.82)

'Human history becomes more and more a race between education and catastrophe,' Wells famously wrote in 1919's *The Outline of History*.[32] His scientific romances evaluate culture and morals as a function of biology, only as enduring as biological reality will allow them to be. Biology, the parent, must always dominate over and even ironise culture. Without a literature consciously adapted to the environment of the future, mankind has become extinct: it is Wells's own work that he saw as remedying the failings of Victorian art and preventing such an imagined future from ever occurring.

---

[32] *The Outline of History*, 2 vols (London: Newnes, 1919), II, p.758.

Mind in Modern Fiction: Literary and Philosophical Perspectives after Darwin

Patricia Waugh

Abstract: This essay argues that the engagement with Darwin in contemporary literary studies has been skewed by an unhelpful critical stand-off between so-called Literary Darwinism and Postmodernism or contemporary literary theory. The argument here is that the relevance of Darwin's ideas to the representation of mind in modern fiction has therefore been under-explored. Darwin's emergent model of mind fundamentally challenged the Cartesian tradition in philosophy of mind, influencing later philosophical movements such as phenomenology, pragmatism, and the development of modern psychology and neuroscience from William James to Antonio Damasio and the work of the most recent generation of cognitive neuroscientists. British modernism, evolving in the shadow of, and often in contradistinction to, the logical positivist hegemony, constructs representations of mind indebted to Darwin's biological model (one that is explicitly repudiated in logical empiricism and logical positivism). This essay explores Darwinian 'integrative' modernist models of mind that anticipate the enactive mind of current cognitive science, and also engages their antitheses: the 'hyper-reflexive' consciousnesses which represent an intentional *reductio ad absurdum* of Cartesian and dualist models of mind and body.

Diversification and development of the biological sciences has brought with it a new oracular expansionism in scientific claims. In the last thirty years, supported by a steady flow of popularising books mixing respectable science with scientistic speculation, Darwinism has extended its reach into politics, ethics, epistemology, anthropology and even the law. But, if philosophers and social theorists have now established a canon of work engaging with the ethical and epistemological challenges of this naturalist turn, literary cultures – and literary criticism especially – have been slow to respond. The literary critical debate has been skewed by an unhelpful stand-off largely involving postmodernists and converts to a broadly ultra-Darwinist stance (mostly led by the American literary critic and 'theory'-baiter, Joseph Carroll). From the perspective of literary studies, some cautious patrolling of scientistic excursions into literary territory might be expected, but what is surprising is the extent of the general *indifference* of the majority of literary critics and theorists to the implications of this latest paradigm shift in scientific knowledge.

Along with this indifference runs a general literary unawareness of ways in which, far from threatening humanistic values, the 'naturalist' turn may be seen as supportive of, and providing complementary evidence for, the kinds of aesthetic experience advocated in the manifestos of many of the major modern creative artists and movements. Modernist aesthetics are often explicitly articulated as a resistance to scientific and philosophical positivism, with its neglect of the senses, affect and emotion, and its subsequent fragmentation and instrumentalisation of knowledge and human experience.[1] The turn from positivism to a more expansive naturalism opens up an orientation towards the biological body which acknowledges a more central role for the sensory and the affective organs in the fundamental processes of rational cognition. Literature might therefore be viewed as an extension of, and providing a unique perspective on, the symbolic and material processes involved in ordinary thinking.

Indeed, recent work in cognitive neuroscience has proposed a view of mind and consciousness remarkably close to that of the modernist poetics of Eliot, Woolf, Lawrence, Joyce and others. Throughout his critical writing from 1917 to 1961, for example, Eliot, the major architect of modernism in Britain and America, repeatedly drew attention to the significance of emotion and the senses in thinking, and to the role of poetry in demonstrating their centrality to human reason, judgement and knowledge. His essays on Dante, the Metaphysical poets, Hamlet, Massinger, Milton, and Marvell contributed to the construction of one of the sustaining myths of modernism (and of twentieth-century literary criticism until the early 1960s): that, in the seventeenth century, and particularly after the poetry of Donne, a pre-modern era came to an end which had evidenced a 'very high development of the senses,' and when thought was an experience that modified sensibility. Before this moment of 'dissociation of sensibility,' the human 'intellect was immediately at the tips of the senses'; the poet could 'feel his thought as immediately as the odour of a rose' and in this era, 'every precise emotion tends towards intellectual formulation.'[2] Accordingly, the Eliot-Pound account of modernism justified itself as the programme for a new and revolutionary poetics which might retrieve a mode of *embodied thought* and, in its New Critical versions, even go some way towards redeeming the original Fall into modernity.

Given that one of the most significant legacies of Darwinism has been the challenge to mind-body dualisms and the stimulation of the current

---

1 Many examples might be offered here, but amongst the most distinguished and influential are: T. S. Eliot, 'The Metaphysical Poets' (1921), *T. S. Eliot: Selected Essays*, 3rd edn (London: Faber and Faber, 1951), pp.281-92; Virginia Woolf, *Moments of Being*, ed. by Jeanne Shulkind (Sussex: University Press, 1976); Roger Fry, 'An Essay in Aesthetics', *Vision and Design* (1920), ed. by J. B. Bullen (London: Oxford University Press, 1981), pp.12-28; 'The Five Senses' of Lawrence's *Fantasia of the Unconscious* (1922), in *Psychoanalysis and the Unconscious and Fantasia of the Unconscious: Cambridge Edition of the Works of D. H. Lawrence*, ed. by Bruce Steele (Cambridge: Cambridge University Press, 2004), pp.xlii-xlvii.

2 T. S. Eliot, *Selected Essays* , 3rd edn (London: Faber, 1951), pp.209-10, 287, 135

understanding of mind as an emergent property of the body, it is worth reflecting on some of the ways in which Darwinism has impinged upon the construction and representation of mind in modernism. Much has been written on the kinds of experimental techniques, developed by Marcel Proust, James Joyce and Virginia Woolf, for representing consciousness, but reflections on the intellectual sources for their models of mind focus largely on Sigmund Freud, with some attention to figures such as Henri Bergson and Bertrand Russell, and occasionally William James. Yet the organicism of modernist poetics is grounded, like the Darwinian view of mind, in a thoroughgoing rejection of dualism and a demonisation of the Cartesian as the source of that dissociation of mind and body from which, uniquely (in Richard's famous formulation in the *Principles of Literary Criticism* (1924)), literary studies might save us. An evolutionary science of mind, such as the 'emergence' thesis at the core of contemporary neurobiology and neuroscience, a thesis that offers so fundamental a challenge to mind-body dualisms, should surely have something important to offer to literary studies.

Mind cannot be a *tabula rasa* in biological terms; at the very least, theories have been inscribed into the human genetic makeup through the trials and errors of our ancestors across space and time. Nor is mind, on this model, a proposition to a world completely 'other' than it: presumably we survive because there is a fit between our theories of the world and what the world is. Knowledge is both tacit and explicit, theory and practice; and, if we explore the world through our capacity for symbolic representation, then cultural practices such as literature *and* the natural sciences are both ways of negotiating knowledge of the world. Biology is not simply genetics, although the current obsession with the latter might lead one to forget this. Darwin did not lay down the terms of his theory of evolution by natural selection in a laboratory, but in his struggle to understand living organisms as they negotiate the environments which might sustain them. The descriptive language of the developmental biologist, for example, must be responsive, like the poet's and novelist's, to different possibilities of 'voice ,' to the perspective of an earthworm or bat as well as a human creature. Darwinism suggests that we humans are creative agents who learn through embodied practices and know and negotiate the world through making models which allow us to imagine and construct alternative worlds as tools for testing our beliefs in the real one. In this sense, its philosophical temper is pragmatist rather than positivist.

At the end of the *Origin of Species* (1859), Darwin envisions the likely demise of dualism: 'in the distant future I see open fields for more important researches. Psychology will be based on a new foundation, that of the necessary acquirement of each mental power and capacity by gradation.

Light will be thrown on the origin of man and his history.'[3] In his later and equally seminal work, *The Expression of the Emotions in Humans and Animals* (1872), he first explored the idea that the emotions are fundamentally adaptive and basically evaluative in orienting the organism to its environment, suggesting that emotions are primary evaluations of the world which orient our subsequent knowledge of and investment in it. Contemporary philosophers such as Bernard Williams have called on such evolutionary accounts of mind to voice reservations concerning the impossibly stern call of Kantian ethics: for the categorical imperative requires us to ignore our projects, prejudices and interests in a way which, for Williams, would seriously jeopardise our own flourishing and survival as human animals.[4] To know the world is always to value and evaluate it: knowledge and value, like reason and feeling, and mind and body, are fundamentally bound up with each other. Darwinism has offered a reorientation towards and acceptance of the biological body which has also suggested possible ways of modifying and redressing the exclusive and austere rationalism of mainstream traditions of ethics and philosophy of mind from Plato through Stoicism, Descartes, Kant and the neo-Kantians, and twentieth-century behaviourism. For each of these major traditions assumes a model of rationality and of ethical thinking where emotion, understood as raw bodily experience, or the 'feeling of a feeling,' is regarded as the major source of human error, deception, unreliability and faulty judgement.

Darwinism and Philosophical Models of Mind

In 1641, Descartes published his *Meditations on First Philosophy*: in writing it he famously set off on a sceptical journey that not only questioned the reality of all external and material things, including the earth, the air, the sky, but also entertained the possibility that his own body, 'hands, eyes, flesh and blood might similarly be the figment of a deluded imagination.[5] Earlier, in the *Discourse on Method* (1637), he had affirmed that he was 'a substance whose whole essence or nature is to think and whose being requires no place and

---

3 Charles Darwin, *The Origin of Species*, ed. by Gillian Beer (Oxford: Oxford University Press, 1996), p.394.

4 Bernard Williams, *Ethics and the Limits of Philosophy* (London: Routledge, 2006).

5 See especially Meditation VI, 'Concerning the Existence of Material Objects and the Real Distinction between the Mind and the Body', in René Descartes, *Meditations on First Philosophy*, ed. by John Cottingham (Cambridge: Cambridge University Press, 1996).

depends on no material thing'.[6] But in the very act of doubting, Descartes was able to affirm the reality of at least one entity: his existence as a thinking – and imagining – thing and, from this Archimedean point, to re-establish the existence of God and rebuild that material world whose reality had been brought into doubt. Cartesian assumptions persist into the twentieth century. Gilbert Ryle's *The Concept of Mind* (1939) insisted that a fully developed third-person, scientifically behaviourist account of human actions would eliminate any need to posit a Cartesian *cogito* with its 'ghost in the machine' theory of mind. But his philosophical behaviourism simply extended a tradition seeking to ground its model of mind in a rational account that eliminates the sensory and feeling body, or in a scientific narrative that evacuates consciousness as a space of interiority involving thoughts, feelings, sensations and impressions grounded in the activities of the body.

So, if Darwinism offers an account of human consciousness more congenial and sympathetic to the concerns of the literary humanities than the mainstream analytical philosophies of mind, are there good reasons for literary critics to be suspicious of evolutionary accounts of consciousness? George Levine has recently argued that we need Darwin, whose thinking is organicist, Romantic, and deeply expressive of what might be thought of as a 'feeling for the organism,' as an antidote to the current *neo*-Darwinist reductionism which threatens the integrity and foundational values of the literary humanities: those of creativity, autonomy and personal ethical independence.[7] Daniel Dennett's book, *Consciousness Explained* (1991) presented a scientific account of mind that claimed neither to suffer from the limited externalism of the behaviourist account nor the uncertain *aporias* of non-scientific forms of internalism, such as subjective introspection, where the 'I' is both object and subject of its own activities. Consciousness must be reconceived as a 'virtual machine,' a sort of evolved software or computer programme that runs off the hardware of the brain. There is, Dennett tells us, no 'Cartesian theatre': feelings or sensations or 'qualia' are simply complex dispositions of the brain, and the 'self (otherwise known as the Audience in the Cartesian theatre, the Central Meaner, or the Witness) turns out to be a valuable abstraction, a theorists' fiction rather then an internal observer or boss'. In principle, a suitably 'programmed' robot, with a silicon-based computer brain, 'would be conscious, would have a self'. By

6 René Descartes, *Philosophical Writings*, trans. and ed. by E. Anscombe and P. T. Green (Indianapolis: Bobbs-Merrill, 1971), p.32.

7 George Levine, *Darwin Loves You: Natural Selection and the Re-enchantment of the World* (Princeton: Princeton University Press, 2008).

thinking of our brains as information-processing systems, we 'can dispel the fog' and close the gap between consciousness and brain, mind and body.[8]

The model of mind engaged in Dennett's account is a synthesis built from two main intellectual sources. The first is Norbert Wiener's conception of the cyborg, where human and machine are equivalent components, feedback systems which might be engineered to produce maximally efficient transmissions of information. The second is the theory of genetic transmission: explaining how genetic information contained in DNA governs the making of proteins in living organisms. Just as Wiener's cybernetics made it possible to conceive of the mind as an information-processing machine, and the machine as a kind of mind, so the idea of DNA as a code, and heredity as mode of information-transmission, offered a way of understanding natural selection as the gradual transfer of information through a *mindless* and algorithmic process of replication which eventually builds minds. Daniel Dennett's evolutionary account of mind as an emergent property of the blind causality of natural selection pictures the mind as a parallel-processing machine, an updated version of Swift's famous writing machine in the Academy of Lagado, but now a multitrack processor where 'at any point in time there are multiple drafts of narrative fragments at various stages of editing in various places in the brain.'[9] If Darwin's theory of natural selection had made the notion of a creator or designer redundant, here too the evolutionary account of mind has no need of intention or what Dennett calls 'skyhooks': science will displace the final remnants of a Cartesian tradition in which mind is still conceived within the terms of an ultimately theological framework. Information, mind, machine and organism start to seem interchangeable.

A closer look at these revolutionary moments in the philosophy of mind, from Cartesian dualism, to Ryle and behaviourism, and finally to Dennett and neo-Darwinism, reveals how much they share: the extirpation of the visceral body as integral to rationality; the disconnection not only from the sensorimotor activity of the body, but also from the negotiation through purposive bodily and mental behaviour with an environment that is vital to sustain life. Cognition is conceived as the representation through internal symbols by a pre-wired brain running according to established procedures of an external world that is also pre-given and inertly available for internal manipulation. In all of them, the body is a machine driven by natural necessity. In neo-Darwinist versions, the primary drivers are the genes, regulating behaviour and deceiving their 'vehicles' into believing in the

---

8 Daniel C. Dennett, *Consciousness Explained* (Harmondsworth: Penguin, 1993), p.431.

9 Dennett, p.135.

delusion of their own rational control, while the real engines of the machine, the 'replicators,' are hidden from view and left in peace to get on with their mindless task of intelligent replication. Either way, mind is reduced to a pre-written narrative script or exists only as a transcendental apperception, a disembodied ghost in a machine.

The Integrative Mind in Modern Fiction: Darwinism and Embodied Consciousness

In the remainder of this essay, I will suggest that the anti-Cartesianism of modernist fiction, like modernist poetics, is enabled and supported by the Darwinian model of an emergent human consciousness, but in one or other of two modes. The first, which I am calling the 'integrative,' represents consciousness as embodied, enactive and distributed through engagement with a world which it constantly monitors, responds to, evaluates, negotiates and helps to constitute; the second, which I am calling the 'hyper-reflexive,' operates as a kind of thought-experiment that allows the reader to enter a consciousness conceived in entirely Cartesian terms, an alienated, disturbed and purely introspective consciousness, existing in a dead, hypertrophied or glassy world which provides neither depth nor any grounded sense of the real.

Traditionally, the conceptual periodisation of modernist fiction has identified its technical modes and philosophical outlook with some variety of self-reflexivity: the idea that whatever else, and compared to realism, all modernist fiction is introspective or self-begetting or self-conscious. Its inwardness is defended as a necessary strategy in a world where scientistic and behaviourist models of mind prevail and where literature is therefore the only arena where interiority might be preserved; where the most private self might find expression and access 'the deeper unnamed feelings which form the substratum of our being'.[10] Accordingly, the novel, in particular, allows us to experience *what it feels like* to be inside another mind looking out at and accessing and constituting a world through sensations, impressions, feelings, thoughts, and half-formed ideas, words, images or 'mentalese'. This is why, Eliot, for example, insisted that the knowledge of the world derived from literature is always indirect, but is nevertheless a kind of knowing.[11] From Henry James's idea of figural narration as mediated through a reflector or watcher at one of the many windows in the house of fiction, each seeing a slightly different world, to Lukàcs's negative view of modernism as solipsistic and withdrawn, the established view of modernist

---

10 T. S. Eliot, *The Use of Poetry and the Use of Criticism* (London: Faber and Faber, 1933), p.155.

11 T. S. Eliot, 'Dante', *Selected Essays*, pp.237-281.

fiction is that consciousness is represented as introspective, through techniques which are themselves self-reflexively foregrounded as a formal antidote to any reductively heresiacal tendencies towards paraphrase.[12] Virginia Woolf's 1919 essay 'Modern Fiction,' for example (written shortly before her first full experiment in fictional consciousness, the novel *Jacob's Room*), announces that 'for the moderns' as opposed to the materialists, Bennett, Galsworthy and Wells, 'the point of interest' in the future of fiction 'lies very likely in the dark places of psychology'.[13]

'Dark places' seems far removed from Darwin's 'open fields,' but I want to suggest that 'open fields' conjures an alternative image of mind, more outward-focussed than the inwardness of 'dark places,' and much more appropriate as a way of describing the embodied, enactive and highly responsive fictional consciousnesses that negotiate environments, landscapes, other minds, streets, meeting places, bars, hotels, and convivial gatherings in modernist novels such as *Ulysses* (1922) or *Women in Love* (1920) or *Mrs Dalloway* (1925). 'Dark places' resonates with a Freudian sense of the unknown evolutionary depths and origins of mind, and literary criticism has thoroughly explored this sense of modernist consciousness as introspectively plumbing the unconscious depths, probing self-deceptions, desires, obscure drives and self-dramatisations that hide characters from themselves and subvert their best laid plans. But what has been rather more neglected is the idea of consciousness conveyed in Darwin's image of 'open fields': mind as open to and adapted for the negotiation of social transactions and survival in a changing world; mind as embodied, purposive as well as unconsciously driven; orienting itself to its environments through affect and feeling, sensory response and reflection and in constant dialogue with its environment, including other minds, as well as itself.

Open the first page of *Mrs Dalloway* and the mind that is encountered is hardly solipsistic or withdrawn. In the very first sentence, Clarissa thrusts *open* her windows and steps out onto the London street, immediately negotiates traffic and shops and passers-by, registering the sounds of engines, the cries in the street, the smells of flowers; at one moment, she is transported back, through memory and imagination and the warm feel of sunlight, to the visceral sensations of adolescent passion; at the next, she is suddenly knocked back into the present as the boom and leaden circles, the

---

12 Henry James, 'Preface to *The Portrait of a Lady*', in *The Art of the Novel*, ed. by R. P. Blackmur (New York: Scribner, 1948), p.45; Georg Lukács, *The Meaning of Contemporary Realism*, trans. by John and Necke Mander (London: Merlin, 1963).

13 Virginia Woolf, 'Modern Fiction', *The Common Reader: First Series* (London: Hogarth Press, 1925), in *Virginia Woolf: Selected Essays*, ed. by David Bradshaw (Oxford: Oxford University Press, 2008), pp.6-13, 11.

sound waves of Big Ben, spin out into the air and assault the ears; immediately, as they fade, she is propelled towards the future, as the interim lull is registered viscerally as a sense of foreboding, of some terror lurking, waiting, biding its time before it bursts upon the scene. But this is not simply the presentation of an introspective consciousness, it is a view through a mind thoroughly immersed in, and engaged with, and negotiating the world through a process of constant and imperceptible adjustment, extension, and distribution of itself; a mind splitting itself off spatially, at one moment, so as to externalise itself empathetically and take up residence in the perspective of an onlooker in the present (Scrope Purvis), and, in the next, transporting itself temporarily so that the past is brought vividly if transiently into the present (Clarissa, Peter and Sally at Bourton, in their youth). This more embodied and enactive understanding of mind is the model now being developed in the work of the most recent contemporary cognitive neuroscientists such as Damasio, Edelman, Varela, Maturana and Andy Clark.[14]

Elsewhere Woolf narrates the experience of consciousness actually emerging – as in *The Waves* – in an almost textbook illustration of Damasio's narrative of the precarious and fragile autobiographical self gradually attaching itself to a more substantial core consciousness that emerges as the continuous adjustment and shifting of 'feeling of a feeling'. Damasio insists that the affective responses of the body are 'integral to the processes of reasoning and decision-making'; feeling prompts continuous bodily readjustment to the world, protecting the organism's equilibrium through periods of change, and providing for consciousness an ever-changing appraisal and map of the organism's relation to its world.[15] Running through the high modernist fictions of writers such as Woolf and Joyce, as well as the aesthetics of Eliot and Richards and Fry, is a preoccupation with reincorporating the body into language and thought as part of a process of *affective cognition,* of resisting the Cartesian pressure, as the individual self registers and responds to changing environments and other minds. As William James first recognised, in his *The Principles of Psychology* of 1890, it is emotion, the continuous flow of visceral, sensorimotor and

---

14 Andy Clark, *Supersizing the Mind: Embodiment, Action and Cognitive Extension* (Oxford: Oxford University Press, 2008); Antonio Damasio, *The Feeling of What Happens: Body, Emotion and the Making of Consciousness* (London: Vintage, 2000); Gerald Edelman, *Wider than the Sky: the Phenomenal Gift of Consciousness* (Harmondsworth: Penguin, 2004); H. R. Maturana and Francesco Varela, *The Tree of Knowledge* (Boston: Shambhala, 1987).

15 Antonia Damasio, *Descartes' Error: Emotion, Reason and the Human Brain* (New York, 2000), p.xiii.

affective response that anchors us in and endlessly and minutely adjusts and appraises our relation to the world, that makes us aware, more than anything, of the way in which mental life, and knowledge, and flourishing in the world are 'knit up with the body'.[16]

A masterly representation of consciousness negotiating its environment, constantly readjusting itself, engaging feeling and sensation, memory and desire, is Joyce's story 'The Dead' (1914). Its protagonist, Gabriel, is ironically presented as a character entirely unaware of the ways in which he shares the sentimentalism and narrow assumptions of the community he disdains. Separating himself from the crowd at the Misses Morkans' annual dance, he feels himself culturally superior, cosmopolitan, a subtle ironist in a roomful of earnest parochials. Repeatedly, Joyce represents him, through a technique of continuous kinaesthetic foregrounding, as physically positioned in space, constantly looking down or up or away from the gathering: he is positioned minutely in relation to the doorway, the window, the stairs and furniture, doorways, lights and shadows. Projecting a posture of cool indifference towards the moth-eaten brocade of custom, Gabriel is nevertheless shown burning with feeling as his true narcissistic investment in cutting the right kind of figure is gradually – and brilliantly – exposed. Things begin badly: he is immediately wrong-footed by Lily, whose rebarbative response to his intrusive questioning stings him. Blinded by a performative screen of patriarchal bluster, he is entirely unable to fathom the source of her displeasure with him, but it puts him out of sorts and he never regains equilibrium. As his turbulent and contradictory feelings mount, so the base grammar of his figural narration shifts increasingly into the subjunctive mood. He reaches out to the cool of the window as if to magically recover his own 'coolness' through its touch. Several times before the end of the story, he looks through its dark pane at the snowy spaces of elsewhere. This proleptic mode of consciousness now takes over, imagining and projecting alternative scenarios, a dramaturgy of his affective life raised to heroic proportions. He sees himself presiding magisterially and heroically at the funeral of his aunt, recovering his place at the centre of the narrative and the community, even though the story is of her death and burial. Even at the crucial moment when he looks up the darkened stairway and first (and literally) misrecognises his wife as a stranger, even here, her story is confiscated from her, and she is framed aesthetically in a hypothetical painting, 'Distant Music,' with Gabriel imagining himself as the artist. His only moment of ease and total attunement to the spatio-temporal co-

16 William James, *Principles of Psychology*, 2 vols (New York: Holt, 1950), II, p.452.

ordinates of the present occurs when he takes the patriarchal seat at the head of the table and takes up the knife ceremoniously to carve the ample meats and supervise their distribution. A story of haunting by the past, 'The Dead' is narrated through the proleptic impulse of Gabriel's consciousness as he negotiates a complex emotional and existential relationship with his environment. For this is how consciousness works: not through the internal manipulation of the symbolic representations of what is inertly outside of it, nor through the algorithmic hard-wiring, the 'consciousness explained' of Dennett's neo-Darwinism.

The Alienated Mind in Modern Fiction: Darwinism and Hyper-reflexive Consciousness

But what happens in fictional writing that intentionally runs with the idea of a Cartesian-style '*inward turn*'? Such writing produces a parodic or absurd effect that perpetuates the Cartesian picture in the mode of *reductio ad absurdum* to reveal its pathologisation, diminution and destruction of subjectivity as a feeling of being. This *hyper-reflexive* modernism might be seen to begin with the Dostoevsky of *Notes From Underground* (1864), though it emerges in the 1960s as the dominant of the 'post-modern' and appears in a number of contemporary guises, most emphatically perhaps in the 'blank' fictions of American writers such as Brett Easton Ellis. I've borrowed the term 'hyperreflexivity' from the clinical psychologist Louis Sass, whose work in phenomenological psychiatry develops the concept as the key to understanding schizophrenia as a consequence of overinvestment in the Cartesian model.[17] His work defies the mainstream psychiatric tradition, established with Bleuler's physicalist account of *dementia praecox* (1911) as a degenerative condition of the brain, or the Freudian account of psychosis as extreme regression. On the contrary, Sass reads schizophrenia in terms close to those of modernist 'dissociation of sensibility,' as a pathology of over-intellectualisation, a consequence of construing mind as a conceptual tool entirely unrelated to the body. In Sass's account, once the mode of the hyper-reflexive becomes 'operant,' what is ordinarily taken-for-granted becomes disrupted, hyperaware and self-conscious: the implicit or tacit sense of existential feelings of being, the solid sense of selfhood, of belonging to and being anchored in a world, comfortably inhabiting the body as the medium of relations with and perspectives on it. World and

---

17 Louis A. Sass, *Madness and Modernism: Insanity in the Light of Modern Art, Literature, and Thought* (Cambridge, MA: Harvard University Press, 1992), 'Affectivity in Schizophrenia: A Phenomenological View', *Hidden Resources: Classical Perspectives on Subjectivity*, ed. by Dan Zahavi (Exeter: Imprint Academic, 2004), pp.127-47.

mind begin to disintegrate. The body comes to be experienced as an alien thing, as carrion, as something dragged around, and the world too appears dead, flat, screened off, offering no affordances, no hierarchy of salience, no selective pressures or evaluations.

Sass quotes extensively from one of the earliest detailed accounts of schizophrenia, the *Autobiography of a Schizophrenic Girl*, first written in the 1950s by a recovered schizophrenic, Renee, and republished in 1968. Describing her first intimation of the coming illness, Renee notes the 'appearance of those elements which were always present in later sensations of unreality: illimitable vastness, brilliant light, and the gloss and smoothness of material things'. She begins to feel as though she has *lost touch* with the real. Subsequently, she describes the sensation as living in a condition of the sublime, but where her eventual disconnection becomes so intense that she experiences her body as dead and thought as coming from outside her own head, while each external object appears as if 'cut off by a knife, detached in its emptiness, in the boundlessness, spaced off from other things. Without any relationship with the environment, just by being itself, it began to come to life.'[18] In order to escape this world of stasis and emptiness, which she curiously refers to throughout as 'Enlightenment,' she begins to build an alternative world, an hallucinated space of escape from containment, an imaginary and autonomous and perfectly ordered aesthetic world. When the body effectively ceases to be, because it has ceased to be a feeling body, then Descartes's insistence that his self would still continue to be, because he would continue to reason, would seem to be far from the case.[19] Schizophrenia becomes a kind of bizarre and disturbing expose of the madness lurking in the Cartesian picture. As introspective withdrawal increases, a paralysing *akrasia* of the will sets in; the most trivial or routine action in the world outside requires a constant and explicitly formulated, enormous and energy-draining, effort. It is as if the actual world has to be rebuilt from scratch, like a child struggling to achieve its first movements or earliest perceptions, or a novelist making the first mark on the page to give birth to a world.

Fredric Jameson's seminal essay, 'Postmodernism, or the Cultural Logic of Late Capitalism' (1984), famously argued that the postmodern condition is characterised by a waning of affect, a disruption of temporality marked by the feeling of living in an illuminated and hypperreal and perpetual present,

---

18 Marguerite Sechehaye, *Autobiography of a Schizophrenic Girl*, trans. by Grace Rubin-Rabson (New York: Grune & Stratton, 1951), pp.19; 57.

19 René Descartes, *Meditations*, see especially Meditation VI.

and the sense of dwelling in an aestheticised or sublime space.[20] Jameson insists in the essay that he is not talking about *clinical* schizophrenia and is simply using the idea of the disease to encapsulate the mood of the postmodern. Yet he too quotes extensively, like Sass, from the *Autobiography of a Schizophrenic Girl.* For as in the experience of the schizophrenic, writers working in the mode of the hyper-reflexive, and postmodernists in particular, often remove from their fictional worlds those tacit and invisible threads that serve to bind selves to historical worlds and which weave the texture of the realist novel, or the integrative mode of modernism, with their sense of psychological depth, temporal and spatial anchorage, and close and multi-perspectival observations of behaviour. But although the hyper-reflexive emerges as the dominant of postmodernism, it was always one of the two major trajectories of modernist writing. Kafka's 'Metamorphosis' (1913) is a famous early example, with its representation of the strangely flattened consciousness of Gregor Samsa whose absence of mental perturbation as he surveys his bodily transformation into a giant beetle produces a kind of ghastly comic effect. For the world built through Gregor's consciousness, one where the appalling becomes the banal, and where bureaucratic rumination has already reduced Gregor's relation to his affective body as if to that of a form of alien or insect life, is a world uncannily familiar to the twentieth century urban dweller. Beckett would go on fully to exploit the comic potential of this mode of Cartesian splitting and, like Kafka, to flag up the disabling *akrasia* which follows such fundamental disconnection. Watt's tortured skewing of his body as he attempts to walk is a comic development of Gregor's plight: 'Watt's way of advancing due east, for example, was to turn his bust as far as possible towards the north and to fling out his right leg as far as possible towards the south' and so on ad nauseam, many times over.[21]

Some Paradoxes of the Fictional Representation of Mind

The stark distinction between an 'integrative' and a 'hyper-reflexive' model of mind in modern fiction is heuristic, of course. In practice, almost all modern novelists, including those oriented towards the exploration of an 'integrative consciousness,' are aware that any attempt to represent embodied thinking purely linguistically, even in aesthetic language, has to struggle against the ingrained tendency to think of thinking as purely internal mental representation or the manipulation of internal symbols which either stand in for an absent pre-given exteriority or 'real' or constitute interior

20 Fredric Jameson, 'Postmodernism, or the Cultural Logic of Late Capitalism', *New Left Review*, 146 (1984), 53-92.

21 Samuel Beckett, *Watt* (London: Calder, 1953), p.28.

worlds of their own. The very attempt to represent consciousness *in writing* threatens to subvert even the most anti-Cartesian commitment by reconstituting consciousness itself as an *explicit* effect of the materiality of the text. Discussing how the loss of a background or tacit sense of embeddedness produces a sense of the world as representation, Merleau-Ponty, in *The Phenomenology of Perception* (1945), offers the activity of writing itself as an example. So if we turn our focus too obsessively onto the activity itself, as the hand gripping the pen, or the finger touching the keys of the keyboard, or the nib scratching the surface of the paper, then the foregrounding of the medium and the act of representation begins to displace the flow of absorption in the world that is being built through the words. The activity, the writing, then becomes inhibited, the imaginary world vanishes into black marks on a white surface; awareness dawns that consciousness is never an entity locked away in a skull but always distributed and prosthetically extended. But the intrusion of that awareness of its extension also threatens to disrupt its flow and to produce a disturbing and alienating awareness of the material processes and conditions of its extension.[22]

Solipsistic dissolution is the spectre that, somewhere, haunts all modernist writing. The problem of how to represent an experiencing mind is the problem of preserving a sense of the tacit flow of feeling and consciousness that anchors the individual in an environment, but also where, in order to build such a picture, what is normally tacit must of necessity be explicitly constructed and selected and, therefore, carrying the potential to disturb the 'flow' by intruding the act of representation and the ontological awareness of the condition of fictionality, the status of the 'as if'. In Ian McEwan's *Saturday* (2005), having described what will become the axiomatic moment of the novel, as Perowne swings his car into a side street to avoid the Anti-Iraq War march, the narrator interjects: 'a second can be a long time in introspection'. He then describes the impossible challenge faced by any modern writer who seriously tries to represent as it is, in all its sheer *economy*, the workings of human consciousness:

> The assertions and the questions don't spell themselves out. He experiences them more as a mental shrug, followed by an interrogative pulse. This is the pre-verbal language that linguists call mentalese. Hardly a language, more a matrix of shifting patterns, consolidating and compressing meaning in fractions of a second, and blending it inseparably with its distinctive emotional hue, which itself is rather like a colour. A sickly yellow. Even with a poet's gift of compression, it could take hundreds of words and many minutes to describe. So that when a flash of red

22 Maurice Merleau-Ponty, *Phenomenology of Perception*, trans. by Colin Smith (London: Routledge, Kegan and Paul, 1962).

> streaks in across his left peripheral vision, like a shape on his retina in a bout of insomnia, it already has the quality of an idea, a new idea. Unexpected and dangerous, but entirely his.[23]

Saturday is a novel with complex intertextual relations with Woolf's *Mrs Dalloway*: both are one-day London novels that begin with their protagonists, Jamesian-style, gazing through windows; both foreground technologies and transport, cars and aeroplanes, set sanity and insanity side by side, investigate suicide and suicide-bombing, war and terror and the building of human defences against them. Each writes in the knowledge of Darwinian theories of mind, of William James and Bergson for Woolf, and of Damasio, Edelman and Varela for McEwan. Curiously too, *Saturday* ends with a direct echo of the last line of Joyce's 'The Dead ,' as Perowne, like Gabriel, looks fondly at his sleeping wife before returning to the window: 'There's always this, is one of his remaining thoughts. And then: there's only this. And at last, faintly falling: this day's over' (p.279). But *Saturday* does not close with a sense of potential annihilation that characterises the final moments of these earlier modernist fictions. McEwan appears to share his neurosurgeon-protagonist's materialist optimism, opening up the mind as Perowne does the brain, believing that science might, one day, explain not only its material wiring but also the hard-wiring for our normative behaviours, for what we ought to believe in, and what we ought to do. How might evolutionary science explain Perwone's seemingly perverse but apparently ethically autonomous altruism, his professional service in saving the man who has threatened the lives of his own kin (and therefore his own selfish genes)?

Is our desire to believe that science might one day offer more, by way of explanation for the human condition and the workings of the human mind, than in the crude accounts of evolutionary psychology, simply the contemporary expression of a perennial human coping strategy? A necessary defence against the fear of physical threat that we share with animals, and against the existential dread that human consciousness itself creates? This sense of existential dread running below the surface of engaged minds and lives is far more integral to the structure of feeling of Woolf's novel, even though McEwan's begins with the immediate terror (and Terror) that a suicide-bomber is hovering over London. Almost the last words of *Mrs Dalloway* are spoken by Lady Rosseter, who asks, '"what does the brain matter compared to the heart,"' followed by Peter Walsh thinking to himself, 'What is this terror? what is this ecstasy…What is it that fills me

---

23 Ian McEwan, *Saturday* (London: Cape, 2005), p.81.

with extraordinary excitement? It is Clarissa, he said. For there she was.'[24] In McEwan, the brain is rather more in evidence than the heart, opened up with the technical equipment of modern science and the skill of the neuroscientist. Terror and violence lurk, but are managed through the rationally applied skills and the cool compassion of the liberal professional. When it is opened up, the brain is revealed as a thing more beautiful than sublime. The watcher remains safely behind the window looking out or, equipped with the technical expertise of the surgeon, looking in. Terror and ecstasy give way to professional values and domestic harmony. In this novel in particular, McEwan presents consciousness by drawing explicitly on the latest theories of evolutionary psychology and cognitive neuroscience (which are acknowledged in the appendix to the hardback edition). But in his equally complex and self-conscious engagements with seminal modernist texts, he clearly signals his intention to do something new and closer to consciousness now, to write a fully materialist representation of the workings of the embodied and enactive mind, and to extirpate perhaps what he sees as the last traces of Cartesianism lurking in these earlier fictions. Yet his attempts to monitor and represent exhaustively this moment-to-moment functional rhythm of consciousness in its every visceral pulsation, impression, thought and imagining, produces a peculiar effect of *hyper-realism* which, paradoxically, in bringing into such explicit focus what is normally tacit and lived, produces something of the hyper-reflexive effects examined earlier. Ironic too, that the effort to be more realistic produces instead a feeling of the *hyper-real*, like those neo-realist paintings that look so much more real than photographs, so that we know that they can't be. Perhaps Woolf does in the end intentionally allow space for a lurking dualism, and even a notion of the soul: if only, perhaps, to guard against the unavoidable reductionism of a metaphysics of materialism that assumes, prematurely, that it has all the answers. And it is surely McEwan's novel, and not Woolf's, that comes closer to falling, whether knowingly or inadvertently, into the condition of the hyper-reflexive. How would we explain that in neo-Darwinist terms?

24 Virginia Woolf, *Mrs Dalloway*, ed. by Claire Tomalin (Oxford: Oxford University Press, 1992), p.255.

E-Volutionary Fictions: The Darwinian Algorithm in Literature and Computer Games

Alistair Brown

Abstract: Whilst Daniel Dennett argues that evolution offers a universal acid that underpins human culture, this essay contends that the algorithm offers a better explanandum that can be applied across disciplines. By comparing literature against two self-evidently algorithmic narratives, those of Richard Dawkins's 'The Blind Watchmaker' program and the computer game, *The Sims*, it concludes that literature and certain computer games cannot be said to operate through comparable algorithmic processes, even though both share narrative traits. This therefore undermines the assumption that evolutionary theory can be applied with universal efficacy to all products of human culture.

Evolutionary Orthodoxy

Though it is clear that evolution has not eliminated religious belief, it is also evident that the orthodoxy of religion inexorably shifted post-Darwin. Today, one's atheism or faith is significantly defined by one's perception of the validity of evolution. But has Darwinism changed the orthodoxy of literature to a similar degree? Though there is a clear Darwinian influence on the metaphors and narratives of fiction, is it possible to conceptualise literature and literary criticism *without* the evolutionary framework? For Darwinism to change the orthodoxy of literature requires not the superficial use of new images, but a fundamental adjustment in the structural frameworks in which literature (as religion) is both constructed and understood. This essay contests the idea that Darwinism changes the structural orthodoxy of literature through a double approach. Firstly, it questions whether it really is evolution in and of itself that offers the optimum empirical approach to literature, or whether evolution is the manifestation of a deeper, or more reduced, causal factor that might be better applied to cultural analysis: the algorithm. Secondly, it asserts a provocative comparison between literature and an alternative 'narrative' form, the computer game. In one strikingly realistic simulation of life – *The Sims* – the algorithm specifies its form at a very structural level, such that neither creation nor interpretation can occur without reference to an algorithmic framework. But although algorithms clearly underpin all digital games only some, like *The Sims*, can be said to use evolutionary algorithms. This problematises the idea that evolution generally offers a universal explanation for all cultural products, although the algorithm as an abstract concept might. Additionally, if computer games and literature are not comparable cultural artefacts, this complicates the notion that evolution can explain every aspect of human culture with equal efficacy, because an evolutionary algorithm explains the game of *The Sims* successfully, in a way it cannot do so well in relation to the novel.

Evolution or Algorithm?

Eager Darwinists such as Daniel Dennett have enlarged Darwinism as 'reductionism incarnate, promising to unite and explain just about everything in one magnificent vision,' thereby encompassing both the natural and the cultural worlds.[1] Dennett is disingenuous, though, in his claims for Darwin's dangerous idea. Though claiming evolution as a universal, interdisciplinary acid, Dennett spends the second chapter of his book explaining that evolution is itself reducible to a more powerful, fundamental acid, that of the algorithmic process:

> the algorithmic level *is* the level that best accounts for the speed of the antelope, the wing of the eagle, the shape of the orchid, the diversity of species, and all the other occasions for wonder in the world of nature. [...] No matter how impressive the products of an algorithm, the underlying process always consists of nothing but a set of individually mindless steps succeeding each other without the help of any intelligent supervision.[2]

Algorithms certainly seem better candidates for a universal acid at some level. Algorithms describe the interactions of the smallest molecular components that build up to the larger physical entities. Such entities include the neurons that constitute the brain, which must theoretically be describable in algorithmic terms (though Roger Penrose has argued that quantum mechanics mitigates against this).[3] Algorithms must ultimately regulate the most complex and abstract entities, such as mind. Mind's most potent productions, such as literary texts, must therefore be the products of algorithmic processes.

In principle, this chain of reasoning is valid. The problem lies in the size of the explanatory links. Even if algorithms can ultimately describe the neuronal (and embodied) cognition that constitutes consciousness, they may not offer the best mode for explaining the mind that they make manifest out of their reduced fundament. In such materialist theories of mind (and, by extension, of culture), the algorithmic acid floats at such a reduced level of complexity as to say little meaningful about the sensational experience of consciousness – that is, qualia – or its most advanced products such as art,

---

1 Daniel C. Dennett, *Darwin's Dangerous Idea: Evolution and the Meanings of Life* (London: Penguin, 1996), p.82.

2 Dennett, p.59.

3 Roger Penrose, *The Emperor's New Mind: Concerning Computers, Minds, and the Laws of Physics* (Oxford: Oxford University Press, 1999).

music or literature. This is the unresolved gap between the 'easy' and 'hard' problems of consciousness. The former is the monitoring of discrete electro-chemical states in the embodied brain, or the production of cybernetic models of mind; the latter is determining how these procedures in themselves simple and discrete can give rise to a manifest conscious and subjective experience – the experience of literature, for example. At issue is not the empirical validity of the scientific framework but the levels of translation between one domain (algorithms that govern neuronal firings or computer programs) and another (mind). There appears to be a threshold, such that any number and degree of objective measurements cannot tell us what it is like to be a bat, to use Thomas Nagel's famous example.[4] Alternatively, the chemistry of ink blots tells us how liquids coalesce on the page to form words according to a mindless molecular algorithm; but the same algorithmic formalism cannot explain the meaning of those words. Similarly, just as it can tell us about the adaptive function of the eagle's wing, evolutionary psychology can tell us much about the adaptive cognitive purpose served by narrative. The central dilemma is not whether evolution is right in principle as an explanation of literature, but whether it is best or most helpful in practice. As David Amigoni points out, it is unclear how arguments that literature adapts us for social life through its safe simulation of human motivations and behaviour constitute a radical shift in the orthodoxy of literary understanding.[5] Rather, a Darwinian terminology is invoked to legitimate the Aristotelian ideal that literature can help to make sense of the world by evoking sensations that make us want to avoid the fate of the characters within it. But if evolution cannot successfully explain literature at the scale of the large (from the top down), does the algorithm that at some unseen level underpins all of physics, evolution and mind included, do a better job of explaining it from the scale of the very small, building up from below? Or are both perspectives necessary but not in themselves sufficient (or sufficiently useful) explananda of literary objects, which thus preserves space for conventional literary critical analysis of textual characteristics?

For Darwinism to be fully successful as a universal acid that explains literature, it is surely insufficient for the process to be one way. Newton's second law symbolically models motion in the physical world and has predictive value, but reciprocally each spaceship that accelerates in the way Newton predicts accrues credibility to that original hypothesis. Similarly, Darwinism should not only catalyse new images and inspire new creative

---

4 Thomas Nagel, 'What is it Like to be a Bat?' *Philosophical Review*, 4 (1974), 435-50.

5 David Amigoni, 'A Consilient Canon?: Bridges to and from Evolutionary Literary Analysis,' *English Studies in Canada*, 32 (2006), 173-85.

narratives, but literature should correspondingly modify or validate our view of evolution. But it would surely be arrogant to argue that the *Island of Doctor Moreau* endorses evolutionary theory to the same degree as the Galapagos finches. The reason literature fails to transfer in this way is that evolution is not in itself substrate neutral; the domain to which it is being applied and from which it receives its data determines the quality of that science's explanation. The algorithm, however, is substrate neutral, since the test for an algorithm is whether it can run on any universal Turing-type calculating machine regardless of medium: the digital computer, graph paper, the natural world.[6]

Rather than the top-down perspective offered by evolution which is a legitimate but low quality explanation, a higher quality explanation that can translate to literature, and in which certain qualities of literature reciprocally enhance our understanding of that explanation, is the algorithm. This argument integrates two branches of recent literary epistemology: the first being neuroscientific or cognitive literary criticism (such as that exemplified in Mark Turner's *The Literary Mind*), and the other being the advent of automated story-telling programs designed to simulate the creative mind and in so doing reveal the structural, algorithmic principles of narratives (such as James Meehan's Tale-Spin, which generated Aesop-style fables).[7] However, without detailed reference to neuroscience or artificial intelligence, it is still possible to explore the algorithm as a literary explanation by contrasting conventional literature with another narrative form which has quite clearly incorporated algorithms to its foundations, in the structural modes of its modern writing. Evolutionary narratives, driven and regulated by evolutionary-type algorithmic principles, are to be found most strikingly not in conventional literature, but in the computer program.

E-Volutionary Narratives

Before entertaining the provocative claim that computer games constitute evolutionary 'literature,' consider one exchange between evolutionary theory and a computer program that transgresses less rigid generic and epistemological boundaries. In this example, the program is modelled upon

---

6 For example, John Conway's stunningly simple Malthusian algorithm for regulating population size, the aptly titled 'Game of Life,' works on computers or even a chess board, and appears to apply to natural population groups as well. The game was first described by Martin Gardner 'Mathematical Games: The Fantastic Combinations of John Conway's New Solitaire Game "Life",' *Scientific American*, 223 (1970), 120-23.

7 Mark Turner, *The Literary Mind: The Origins of Thought and Language* (Oxford: Oxford University Press, 1996). For the numerous examples of story generators, including discussion of Tale-Spin, see Margaret A. Boden, *The Creative Mind: Myths and Mechanisms* (London: Routledge, 2004).

evolutionary algorithms, but reciprocally verifies evolution in a mutual feedback. Richard Dawkins' 'Blind Watchmaker' software is a clever experiment designed to show the role of cumulative selection in the evolution of complex 'life' forms.[8] The program works through random mutation in nine variables analogous to genes. Starting with a basic tree structure, the variables govern features such as the length and number of branches, and the width of the tree. After each random mutation, the user non-randomly selects which tree to 'grow,' recursively carrying mutations through to subsequent levels of development. The selection pressure might be such that the user chooses whichever looks most like a butterfly, although with a more detailed program it could just as easily remove the controversial role of the human user (an Intelligent Designer?), so that the criteria are resource selection (choosing whichever form requires least processing power or memory), or genetic selection (choosing whichever tree most faithfully inherits variables from the previous generation). Because complex structures which strongly resemble real organic forms emerge from the simple underlying algorithms this ensures a reciprocal reinforcement between science and program, making the software offer data as well as a quaint analogy. Dawkins imagines displaying the generated 'insects' on a screen in his garden in summer, and using real-life bees as the natural selectors. The computer program is built on the neo-Darwinian algorithm of random mutations in the variables accumulating to organised, non-random forms, but the images that these principles manifest on screen validate evolutionary theory through a different order of symbolism: the resemblance of their shape to real trees and insects, such that the digital versions can be selected by their organic counterparts. The mutual exchange between the computer program and the summer garden is possible because of the substrate neutrality of the evolutionary algorithm: chance modifications in genetic chemistry or digital variables non-randomly accumulate to give rise to complex forms which just happen to be incarnated in pixels or cells. Since the algorithm is describable on any universal Turing machine, one could just as accurately – if not so efficiently – 'run' the algorithm using matchsticks or graph paper.

But whilst this is undoubtedly a simulation that reinforces the validity of the model on which it is predicated, the idea that it is a 'narrative' is a

---

8 Richard Dawkins, *The Blind Watchmaker: Why the Evidence of Evolution Reveals a Universe Without Design* (New York: Norton, 1987). Though the original program is now hard to obtain, there are numerous versions of the Blind Watchmaker freely available online, such as Mark Jones, 'Blind Watchmaker Applet,' *Syracuse University Department of Physics* (2001) <http://physics.syr.edu/courses/mirror/biomorph/> [accessed 5 June 2008].

teleological interpretation in which the human observer perceives a metaphorical connection between the pixellated shape and the biological spider. However, certain computer games inherit the same fundamental mechanics of evolution as 'The Blind Watchmaker,' with combinations of algorithms in themselves simple and mindless recursively producing emergent complexity – that is, evolutionary symptoms. As suggested above, the problem of algorithmic reductions, like the easy problem of consciousness, is that they describe the world at such a reduced level of complexity as to say almost nothing meaningful about the gestalt of sensational experience, the hard problems of qualia, such as literature. The critical difference with the computer game as opposed to literature is that meaning here is not contingent on a reader's personal disposition. The narrative arises in the first instance from the underlying programming alone, rather than any intervention by the encultured player. One could just as easily set a robot to play the game as use a human 'reader'. At the same time, and like a novel, it is designed by a mind; the processes that underlie it are coincidental tools deployed by an author, who does not think primarily of algorithmic rules, but imagines creating a virtual model in a complex way, and then considers which algorithms might best bring this vision about. Thus in the computer game we can separate more clearly the hard problems – how the mind is creative and why minds enjoy the computer game – from the easy ones – how that creativity is deployed in code, and how that code is processed or structured to generate specific outputs. Additionally, if computer games have a narrative aspect that makes them proximate to 'literature,' this implies that computer games are the true inheritors of evolutionism in a wholesale, orthodox sense, as opposed to the theoretically valid but practically meaningless sense that Darwinism, computer games, consciousness and literature are all underpinned by the reducing universal acid of the algorithm.

Literary Games

The Japanese roboticist Mashahiro Mori coined the term 'uncanny valley' to describe the counterintuitive effect that as a robot looks more human, our responses to it become increasingly sympathetic; however, at a point of very strong resemblance, we become repulsed by it.[9] Similarly, as graphics approach photo-realism so that we are less able to distinguish between the virtual and real world, game designers have realised that games require not

---

9 Mashahiro Mori, 'The Uncanny Valley,' trans. by Karl F. MacDorman and Takashi Minato, *Energy*, 7 (1970), 33-35 <http://www.androidscience.com/theuncannyvalley/proceedings2005/uncannyvalley.html> [accessed 2 June 2008].

just visual realism but a behavioural one if it is to be immersively enjoyed.[10] Oddly, the more real a game seems, the more uncanny it becomes when that environment does not behave in the way expected given its apparent visual fidelity. In an era of vast processing power, the skill of games design relies on knowing just which aspects of the world to model to preserve a holistic realism, whilst permitting the virtual environment to behave in a directed way such that it provides a different experience to reality and thus constitutes escapist entertainment. Recognising that this makes games designers not unlike novelists, Janet Murray suggests that, 'If there is to be a Charles Dickens or Charlotte Brontë of the digital medium, then Will Wright is surely one of his or her key antecedents.'[11] Wright's most influential creation, *Sim City*, models urban economics in a strikingly realistic way (reportedly it was used by professional council planners).[12] His latest project, *Spore*, is overtly Darwinian, taking the player from his or her development of cells, through multicellular organisms, tribes, civilisations and into space.[13] It is one of his earlier successes, *The Sims*, though, which best crosses Darwinian and literary paradigms.[14]

One of the most popular PC games to date, *The Sims* delivers virtual entertainment from the mundane activities of daily life: cooking, shopping, sleeping, exercising, working. The games player designs virtual houses, and controls avatars with various personalities to live in them. At the setup stage of the game, users create their characters through sliders which control a prototypical genetic makeup according to five parameters: Neat, Outgoing, Active, Playful, and Nice. Within the game proper, over the period of a day the player must carefully balance co-dependent properties such as tiredness, hygiene, hunger, sociability; the continually shifting states of these are

---

10 Darren Waters, 'What Exactly is a Next Generation Game?' *BBC News Online* <http://news.bbc.co.uk/1/hi/technology/6937058.stm> [accessed 2 June 2008].

11 Janet Murray, 'From Game-Story to Cyberdrama', *Electronic Book Review* (2005) <http://www.electronicbookreview.com/thread/firstperson/autodramatic> [accessed 3 June 2008].

12 Will Wright, *Guardian Unlimited* (2007) <http://www.guardian.co.uk/technology/2007/oct/26/willwright> [accessed 3 June 2008].

13 *Spore* (Guildford: Electronic Arts, 2008) [on DVD].

14 *The Sims Deluxe Edition* (Electronic Arts: Surrey, 2000) [on DVD]. I am using the first version of the game rather than its sequel, *The Sims 2*, for two reasons. Firstly, being less technically complex, the earlier program offers a more transparent window into its underlying, algorithmic processes. Secondly, since the sequel's release in 2004, there have been so many expansion packs and patches that it is difficult to consider a common and definitive version of the game (whilst there were add-ons produced for the first *Sims*, as I discuss later in this essay, they were not so wholesale revisionist as in the later version).

represented by bar graphs. After eight game hours at work, the virtual Sim will generally be tired, bored, and hungry, and the play revolves around prioritising the Sim's daily needs in accordance with their genetic predispositions. An extrovert Sim will want to socialise with other Sims, but this will make them more tired; getting some sleep will improve wakefulness but not alleviate hunger. The first key to successful gamesmanship is learning to satisfy these needs concurrently for the minimum expenditure of time. For example, a skilled player might organise a dinner party, thereby increasing fun and reducing hunger simultaneously. Over several days and weeks of simulated play, the player must decide when and how to compromise these immediate needs in order to make his Sims progress up the career ladder. By improving the character's personal skills through studying (if in a white-collar profession) or exercising (if in a manual job), each Sim will be left more hungry, tired and bored than before. However, if their intelligence or strength improves sufficiently they may get promoted, thus bringing in more money with which to purchase the devices of modern living – the widescreen television, the microwave, the dishwasher – which increase the speed with which the Sim's daily desires and chores can be fulfilled. The Dish Duster Deluxe dishwasher, for example, cuts the time spent on washing up dishes, thereby improving the hygiene attribute more efficiently.

The second art of the game is in ensuring the domestic environment matches the innate predispositions of each Sim: a Sim with a 'genetic' predisposition for Neatness will want better dishwashers; a playful Sim will want colourful furniture; an active Sim will require workout equipment. If done successfully, the cycle of success loops back to its beginning, as labour-saving devices and elaborate entertainment technologies increase the spare time available for practising the skills required at work, with the ultimate result being that the Sim can move even further up the career and financial ladder, thereby being able to purchase household appliances which alleviate their needs more rapidly.

Glossily presented, the colourful graphics and user interface screen (in both senses of the word) the fact that the game of *The Sims* is essentially reducible to a feedback loop: incremental shifts in the early conditions of the Sims' daily lives can reward the player by translating to large effects over the course of each recursion through the system. In this system construct, we see the same three structural affinities with evolution as present in the 'Blind Watchmaker'. Firstly, a 'genetic' foundation for a character which directs future development (a less physical Sim will be best adapted for an intellectual job, and will quickly complain and eventually 'die' if in the wrong sort of work). Secondly, environmental selection pressures play the key role in survival: changing their local environment to make it more comfortable facilitates the success of those Sims who are 'genetically' best adapted to that environment, and hampers those who are not. Finally, these two combine in a cumulative mechanism by which small and successfully adaptational changes at a previous stage translate recursively to dramatic changes later on, showing evidence of emergent complexity.

The Sims' expressions of these underlying, algorithmic processes draw attention to the deficits of evolutionary effects in the novel. Whilst the novel may be driven by the sense of an ending, the truly evolutionary narrative, however, should expand the timescale whilst doing away entirely with hints of teleology.[15] Evolution has no teleological intention, and comparably *The Sims* does not have any end, aim or goal as such; it is simply a perpetual *Bildungsroman* of domestic development and self-improvement. Additionally, in the supporting website, the game's objects are continually adapted – or 'modded' – by other players, so that each time the program is reopened, the potential design space of the game world has altered.[16] In its scale – which is not quite the same thing as size or length of time it takes to read or play, but is more like its Darwinian range of potential developments – *The Sims* is more evolutionary than a novel can be. Unlike the novel, the development of a character takes place not simply because of an a priori selection of events made by the author, but because of internal unpredictability within, and the changing context outside, the system. Because the game is constituted of variable components which are individually simple but combinable in infinite ways, with the components continually being modded by other players, each time the game is played the conclusion will be radically different – given a few early decisions and mods, Sims can emerge to be middle-class academics with antique furniture, or rock stars slobbing in a garage. This same alternative of ending given slightly different design decisions made along the way, is more in keeping with evolutionary selection than the novel's story can possibly be, in which a character dies or marries, and will do so again and again no matter how many times it is read.

Is *The Sims* Literature?

This contrast means that whilst *The Sims* is structurally evolutionary, and bears considerable psychological affinity to the *Bildungsroman*, it is arguably an over-extension of generic categories to propose the game as literature in any strong sense (an epistemological monotheism as problematic as the

15 Frank Kermode, *The Sense of an Ending: Studies in the Theory of Fiction* (New York: Oxford University Press, 1967).

16 The largest such expansion site is *The Sims Resource* <http://www.thesimsresource.com> [accessed 5 June 2008].

consilience of evolution and literary criticism). Certainly, computer games cover such a broad range that they cannot constitute literature according to any systematic ontology. Indeed, the great potential of the game as opposed to the book is that it is not tied to any formal demands of length or the left-to-right sequence of words across the page; rather the computer game can produce any structural relationship between its elements – graphics, sound, storyline, avatar – limited only by the machine's processing power. In the first-person shooter game, for example, story is often of only background relevance, with it mattering little whether the enemy are alien invaders or irradiated mutants. The narrative is the mere occasion for the game's internal structure, and as such the game is distinct from conventional literary narratives, in which story, form and structure are inseparable from each other. It is also fair to argue that whilst a set of evolutionary algorithms is common to 'The Blind Watchmaker' and *The Sims*, this does not mean that evolutionary algorithms systematically underlie programs or games, though clearly algorithms of some sort always do. In a game of paddle tennis, for example, the algorithm is about tangents and velocities, but has little to do with evolution.

Both challenges – that computer games are not systematically literary, and that they are not systematically structured on evolutionary algorithms – are legitimate. It is not possible to make the strong claim that computer games and literature are essentially commensurate, though there is certainly a level of parity not always acknowledged by academic literature departments. Neither is it possible to assert that games are always evolutionary. However, the aim of this essay is not to build up a poetics of computer games, but to challenge the view that evolution is a universal acid that axiomatically changes the orthodoxy in literature, just as it does other aspects of science and culture, such as religion. Both games and literature are underpinned by a common principle that is substrate neutral, the algorithmic processes that, in their different forms, underlie games and literature which emerge from a common structure, the brain. However, this suggests that evolution is itself also reducible to this algorithmic symbolism, and it is a misnomer to call evolution as a whole the universal acid, the algorithm being the better candidate. If it is legitimate to interpret *The Sims* as an evolutionary program, then it should be conceded that the contrast with the novel implies that evolution in and of itself as the bundling of certain algorithms is not a *universally* strong acid, because it changes the structural orthodoxies in some fields (such as particular computer games) more than in others (conventional literature, or other computer games). If we can talk about literary games, some can be described in terms of evolution at a deep structural level, with a degree of translation between expression and underlying process that we cannot automatically do with conventional literature. This is particularly true of *The Sims* as a model of evolutionary principles which, albeit in a less powerful way than the 'Blind Watchmaker,' reciprocates by providing validatory data for the science, in a way not true of the novel. In *The Sims* we see how the recursive pressures applied to simple variables can create a complex alternative environment that mimics real life, just as evolution predicts.

However, this building from simplicity to complexity explains why, although *The Sims* is a scientific model of evolutionary construction, it is also a literary-type model in its phenomenological effects on the individual player, if not in its specific modes of producing meaning. The structure of *The Sims* is fundamentally simple according to evolutionary algorithms, but the fact that it is enjoyable *despite* this simplicity is precisely what makes playing it a powerful philosophical experience. As Noah Wardrip Fruin has explained, though people begin playing with a preconceived idea of what constitutes a person, they quickly realise that a person in *The Sims*' virtualisation is simply comprised of eight continually-altering bar graphs. However, rather than being disappointed with this simplicity, by foregrounding the underlying model of personhood the game is highly immersive.[17] Given the importance of avoiding the 'uncanny valley,' the fact that *The Sims*' graphics are cartoonishly drawn is important in forcing us to encounter and interpret its realism at the behavioural rather than visual level. Rather than concealing the variables behind visually mimetic processes (such as detailed graphical representations of moods displayed on animated faces) the game instead presents the variables explicitly as bar graphs, drawing attention to itself at the meta-level. We start playing believing that *The Sims* must be imperfect models of ourselves, because they look quite functionally cartoonish. However, by overtly representing its internal processes rather than concealing these beneath a high definition graphical render, the game reveals that we are ourselves, beneath the skin, also essentially the product of particular simple mathematical or empirical factors which are common to both virtual and real life: time as a limited resource; the way additional capital allows one to save manual labour time; the way in which innate predispositions determine how we establish a balance between time and money that is best for our biological bodies. But rather than the shock of reductionism incarnate, this is a moment of positive revelation, because we realise that although in a theoretical sense the rules explain the game – which amounts to little more than stabilising ever-changing bar graphs – and

---

17 Noah Wardip-Fruin, 'Expressive Processing 8.3: *The Sims*,; *Grand Text Auto* <http://grandtextauto.org/2008/03/13/ep-83-the-sims/> [accessed 3 June 2008] (para. 5 of 19).

that algorithms expressed by these graphs must therefore also explain the human life on which the game is modelled, both the game as media or life as existence are so complex that the algorithms are not *on their own* the key to interpretation. They are necessary to the principle, but not sufficient to explain the practice, of enjoying the game of life.

Darwinism can certainly explain general patterns of behaviour in response to literature as a cultural artefact (why telling stories helps us survive as a social species), but cannot easily explain the detail of how novels make manifest worlds at the level of consciousness. Conversely, the algorithm can explain the detail of where literature arises from neurologically, or what computer game subroutine is taking place. But there remains a disconnect between the easy problem and the hard problem: between Darwinism's bird's eye view from above, and the algorithm's bottom up view, there is a conceptual gap. This middle perspective is taken by the human. Once we realise that life is governed by this underlying simplicity of algorithmic rules, we also recognise that they combine in unpredictable ways that provides for an immersive playing or reading experience; no matter how simple or algorithmically mindless the principles that regulate life are, experience appears manifest to our consciousness. Even though superficially computer games and the novel seem to operate quite differently, from a phenomenologically broad perspective a parallel emerges. Both make available to us a pattern or model of how human existence is constrained by particular rules of nature which are realised in human biology, behavioural prototypes, or characterisations. However, in their modelling of the world in a simplifying way both the novel and game allow access to higher complexity; they make the world at once more understandable, atomised into fewer components, whilst simultaneously preserving the multi-dimensional way in which we experience, live, read or play it. For a games designer like Will Wright, creativity lies in recognising the fewest possible algorithmic principles that can be programmed into the model, whilst allowing the model's complexity to emerge and hence behavioural realism to pertain: from so simple a beginning Sims most wonderful and wondrous can be evolved. Similarly, George Eliot recognises that the power of the realist novel lies not in a total representative correspondence between world and book but in cutting out the 'roar that lies on the other side of silence,' allowing what is of significance to be heard more clearly – as happens in her epic social simulation, *Middlemarch*.[18] The

[18] George Eliot, *Middlemarch*, ed. by David Carroll (Oxford: Oxford University Press, 1998), p.182.

ends of both successful games and novels, then, are the experience of subjective complexity emerging from empirical simplicity.

Value Judgements and Functional Roles: Carroll's Quarrel With Pinker

Jon Adams

Abstract: In recent years, a small band of literary scholars have sought to interpret the production and consumption of fiction as an evolved product of our 'adapted minds'. These so-called Darwinian Literary Critics seek to apply the insights of evolutionary psychology to the study of literary texts. The approach seems promising. However, a quarrel between these critics and the evolutionary psychologist Steven Pinker over reserving a space for 'high art' reveals a troubling divergence between their respective agendas, and an attempt by the Darwinian Literary Critics to employ biological norms to ratify evaluative claims.

Introduction

If literary study risks looking inferior in relation to other disciplines, it's at least in part because literary study has cleaved away from the sciences even as the sciences have become the dominant mode of intellectual enquiry. The ensuing rift, canonised as Snow's two cultures, has proved a fruitful source of anxiety for literary critics. On the one side, there's lots to say about how very different and irreconcilable the two cultures are; and on the other, there's lots to say about how that rift might be healed or doesn't really exist at all. For the positive thesis here – the possibilities of reconciling literary study with the sciences – the advantages are manifold: achieving methodological consistency would at once re-invigorate the study of literature and simultaneously set that study on an epistemically respectable footing.

And so goes the aim of so-called Darwinian Literary Criticism – an increasingly well-publicised if not yet influential branch of literary study which seeks to analyse literature through the lens of evolutionary psychology. The intentions of the Darwinian Literary Critics seem laudable indeed. Their agenda is optimistic, progressive, and intellectually ambitious; promising a future of meaningful interaction between scholars of literature and scientists at the cutting-edge of research into human cognition and evolutionary development. In some ways, theirs is a direct response to E. O. Wilson's calls for disciplinary unity as set out in his 1998 programme for 'consilience,' although the first and still the major text on the topic – Joseph Carroll's *Evolution and Literary Theory* – was published several years prior to Wilson's *Consilience*.[1] Carroll is very much the founding father of Darwinian

---

1 Joseph Carroll, *Evolution and Literary Theory* (Columbia: University of Missouri Press, 1995); E. O. Wilson, *Consilience: The Unity of Knowledge* (London: Little, Brown, 1998).

Literary Criticism, and his work sets the tone for those critics who have followed his lead – most notably Jonathan Gottschall and Brian Boyd.

In the years since Carroll's *Evolution and Literary Theory* was published, evolutionary psychology has received a good deal of attention in the press, largely as a result of popularisations by the likes of Matt Ridley, Daniel Dennett, Geoffrey Miller, and – perhaps most successfully – Steven Pinker.[2] In *How The Mind Works* and *The Blank Slate*, Pinker has sought to explain to a non-specialist audience how human cultural norms – including the production and consumption of art and literature – can be understood as products of evolved mechanisms conferring adaptive value. Taking narrative fiction as a human universal, present to some degree in every culture, Pinker's account seeks to explain storytelling in terms of its *functional* value as a means of 'rehearsing' for real life situations.

Pinker is one of the most successful popularisers of our time, and reaches an audience considerably larger than anything Carroll (as a literary critic writing for an academic audience) could hope for. Pinker's aim is to steer his readership around to the viewpoint of evolutionary psychology. (As a quantitative measure of that influence, Angela Cassidy has plotted mentions of 'evolutionary psychology' in the press against the publications of Pinker's books, and found a significant correlation.)[3] Thus, the more people read Pinker, the greater the chance that the perspective of evolutionary psychology (and with it the evolutionary view of literature) will be normalised, and the more likely that the evolutionary approach to literary criticism will be widely adopted.

So it perhaps comes as a surprise to find Carroll at odds with Pinker, complaining that Pinker's analysis is simplistic, and rejecting the account of literature he provides. Looking more closely at Carroll's objections to Pinker reveals evidence of a deeper (and potentially intractable) problem with

2 Matt Ridley, *The Red Queen: Sex and the Evolution of Human Nature* (London: Viking, 1993); Daniel Dennett, *Darwin's Dangerous Idea: Evolution and the Meanings of Life* (New York: Simon and Schuster, 1995); Geoffrey Miller, *The Mating Mind: How Sexual Choice Shaped the Evolution of Human Nature* (London: Heinemann, 2000); Steven Pinker, *How the Mind Works* (Harmondsworth: Penguin, 1997); *The Blank Slate: The Modern Denial of Human Nature* (New York: Penguin, 2002).

3 Cassidy plotted mentions of 'evolutionary psychology' in UK print media against the publication of Pinker's *How The Mind Works*. Graphs were created displaying the incidence of 'Pinker' and the incidence of 'evolutionary psychology'. Her results show an expected peak coincident with the publication of *How The Mind Works*, but even after reviews of the book have subsided, press interest in evolutionary psychology stands at a greater level than before. See Angela Cassidy, 'Popular Evolutionary Psychology in the UK. An Unusual case of Science in the Media?', *Public Understanding of Science* 14 (2005), 115-41; and 'Evolutionary Psychology as Public Science and Boundary Work', *Public Understanding of Science*, 15 (2006), 175-205.

reconciling evolutionary psychology's interest in fiction, and literary criticism's interest in evolutionary psychology. It reveals also that the attempt to naturalise the study of literature is – at least in its present form – bound up with some troubling views on the relation of literary value to moral character.

The Quarrel

The Darwinian Literary Critics use evolutionary psychology to do two things. The first is to explain the existence of literature, the second is to explain the content of literature – *why* we read and write, and *what* we read and write about. These aims are not unrelated, but the second is the more problematic of the two, for it seeks to treat fictional characters as organisms competing to survive and reproduce – exactly as in the evolutionary struggle in the world without. (Gottschall's recent work on Homer is exemplary of the method.)[4] Proponents of Darwinian Literary Criticism usually seek to characterise any resistance to their approach as symptomatic of a wider resistance to the theory of evolution.[5] As a rhetorical strategy, this allows them to play the role of victims to a dominant and entrenched elite of poststructuralist anti-realists, and has the advantage – especially in the US – of aligning them with the atheistic Darwinians against the Creationist religious right. But accepting the broader theory of evolution in no way entails accepting evolutionary psychology – notable dissenters have been Stephen Jay Gould and philosopher of biology John Dupré.[6] Indeed,

---

4 Gottschall understands literary study as a facet of 'the ancient project of seeking the nature of the human' (in *The Literary Animal: Evolution and the Nature of Narrative*, ed. Jonathan Gottschall and David Sloan Wilson (Evanston, IL: Northwestern University Press, 2005), p.xix). This is, of course, a perfectly noble reason to study literature. Not everyone will concur, but it is difficult to cite reasons for why the study of literature is valuable apt to persuade someone who thinks the project is a waste of time. Be that as it may, it's significant that Gottschall's primary aim in studying fiction is to 'explore the behavior and psychology of human beings' (in Gottschall & Wilson, p.xix): right from the start, the literary text is a conduit to other, larger questions.

5 Adopting an embattled position is something common to evolutionists of late, though the tactic is perhaps less sympathy-winning than its proponents hope. Reviewing *The Literary Animal*, Travis Landry takes issue with the victim stance: 'one repeatedly gets the sense that these naturalist critics would be better served without their 'us against the world' mentality. The empirical chest thumping and emblazoned rhetoric [...] may ultimately alienate the very literary critics these scholars hope to convert' (Landry, 'The Taming of *The Literary Animal*', review of *The Literary Animal*, ed. Gottschall and Wilson, in *Evolutionary Psychology* 4 (2006), 49-56, (p.56).

6 See, for example, Gould's essays 'Evolution: the Pleasures of Pluralism' and 'Darwinian Fundamentalism' in *The New York Review of Books*; and Dupré's *Human*

contrary to the impression of consensus created by the wealth of popular books on the topic, evolutionary psychology is a far from orthodox approach. (That being said, none of what follows stands or falls on the truth of evolutionary psychology's claims.) Unusually, it seems to collect the majority of its proponents from *outside* the disciplines in which it nominally inheres: neither biologists nor psychologists seem especially enamoured of the theory. Stranger still is that the Darwinian Literary Critics have been openly critical of the evolutionary psychologists on whose substantive work they rely. If they are as embattled as they make out, it surely makes sense not to bite one of the few hands that is willing to feed them. Pinker has been attacked more than most (though Geoffrey Miller's focus on sexual selection is also unpopular – criticised by Boyd and Carroll), and as might be expected, the criticisms centre on how Pinker deals with art – and specifically literature. [7]

Pinker's Account

Pinker works with an inclusive, quite loose definition of art. In *How The Mind Works*, the emphasis is on art as a 'pleasure technology'.[8] Pinker means to compare the enjoyment we receive from art to the enjoyment we receive from indulgences such as strawberry cheesecake – which deliver concentrated doses of fats and sugars that would have been scarce in the Pleistocene (our felt pleasure being the brain's evolved method of offering a reward for making the effort to source such ingredients). Art plays on similar effort-reward mechanisms, but unlike cheesecakes, doesn't actually deliver the 'calories'. By tricking the brain into thinking it has endured or undergone some experience it has not, art is a means of shorting the brain's neurochemical reward circuits 'without the inconvenience of wringing bona fide fitness increments from the harsh world'.[9] The stimulus is fake, but the response is real.

Nature and the Limits of Science, (Oxford: Oxford University Press, 2001), 'Against Maladaptationism: or What's Wrong with Evolutionary Psychology', in *Knowledge as Social Order: Rethinking the Sociology of BarryBarnes, ed. by Massimo Mazzotti (London:* Ashgate, 1998), pp 165-80, and his review of *How the Mind Works* in *Philosophy of Science* 66 (1999), 489-93.

7 Miller, pp.258-91; Boyd in Gottschall and Wilson, p.157; Carroll, *Literary Darwinism: Evolution, Human Nature, and Literature* (New York: Routledge, 2004), pp.xx-xxi.

8 *How the Mind Works*, p.525.

9 *How the Mind Works*, p.524.

An additional and more functional role for art is as an instructional device.[10] Tailored for narratives, this account suggests that the ability to engage with and create fiction offered our ancestors access to imaginative simulations in which they had the chance to mentally rehearse their responses to crises and threats, 'so they will be prepared if they ever find themselves in similar straits'.[11] 'Fictional narratives,' claims Pinker, 'supply us with a mental catalogue of the fatal conundrums we might face someday and the outcomes of strategies we could deploy in them.'[12]

Crucially, in neither function is any *qualitative* distinction made between art and mere entertainment: from the adaptive perspective, Shakespeare is on a level with pornography (both entertain, and in their own ways, instruct). Nor is Pinker concerned with developing a discriminating aesthetic theory.[13] There is no special place for high art here. Indeed, in his capacity as a psychologist, we might well expect Pinker to be *more* interested in popular fiction than specifically literary fiction. It's one of the few things modern psychology has in common with Freud and Jung: Freud had said he would be choosing for analysis 'not the writers most highly esteemed by the critics, but the less pretentious authors of novels, romances and short stories, who nevertheless have the widest and most eager circle of readers of both sexes'.[14] Similarly, Jung notes that 'Literary products of highly dubious merit are often of the greatest interest to the psychologist.'[15]

To some extent this is because the psychologist looks to fiction not for the story it tells the reader, but for the story it tells *about* the reader. What makes some literature worthy of further study also makes it less typical of fiction generally, and so less interesting as data for the psychologist looking to formulate general patterns. There's no reason to expect that the interest evolutionary psychology has in literature will line up with the interest that literary critics have in evolutionary psychology. Literature exists as a vast and

---

10 'Functional' intended in terms of adaptive utility – that is, doing actual useful work rather than exploiting a cognitive loophole. The idea is explored further in Pinker, *The Blank Slate*, pp.400-20.

11 *How the Mind Works*, p.542

12 *How the Mind Works*, p.543.

13 This is evident from the sources he uses to support his case in *How The Mind Works*: quotations from Woody Allen (p.542), National Lampoon's *Animal House* (p.548), Bob Dylan and Lou Reed (p.535), are given equal prominence as Shakespeare (p.528), Dryden (p.539), Joyce (p.541), and Kafka (p.541).

14 Sigmund Freud, 'Creative Writers and Day-Dreaming', *Art and Literature: Jensen's Gradiva, Leonardo da Vinci and Other Works*, trans. by James Strachey, ed. by Albert Dickson (Harmondsworth: Penguin, 1990 [1907]), pp.131-41 (p.137).

15 Carl Gustav Jung, *The Collected Works of C. G.* Jung, 18 vols, ed. by Herbert Readet al., trans. by R. F. C. Hull. 2nd edn, XV: *The Spirit in Man, Art, and Literature* in (Princeton, NJ: Princeton University Press, 1966), pp.87-88.

ready-made dataset, a durable record of human interests tapering back for several thousand years. The evolutionary psychologists hope that analysis of literature's content might provide evidence in support of their claims that human nature is always and everywhere the same. In respect of the aimed-for consilience, the gains for evolutionary psychology are obvious, but the reciprocal gains for literary criticism are not at all clear.

Carroll's Objection

It's perhaps out of a growing awareness of this troubling redundancy that Carroll's objection to Pinker first surfaces in a short essay from 1998 called 'Steven Pinker's Cheesecake for the Mind'. When the essay is republished in a 2004 collection of Carroll's work, it is less derisively retitled as 'Pinker, Dickens, and the Functions of Literature' (although the text remains the same).[16]

Carroll's chief complaint is that Pinker's account trivializes art and literature – by comparing art to cheesecake, Pinker seems to be denigrating it. It's a criticism echoed by Brian Boyd, who thinks the 'cheesecake' metaphor is intentionally provocative, and offers a limited account which explains the consumption of art but not its production. Boyd says: 'The limitation of Pinker's treatment of art lies [...] in an insouciance that results from his ignoring the link between artist and audience.'[17] Elsewhere, Peter Swirski, who is also seeking an evolutionary role for literature, calls Pinker's account 'so grossly distorted as to verge on nonsense'.[18]

What these critics don't like about Pinker's thesis is the epiphenomenal role it accords to art and literature.[19] Pinker is well aware that he has caused offence, and suspects this is because 'most people involved with the arts want them to be an adaptation because they feel it would somehow validate or ennoble the arts'. He thinks this is an error probably generated by the overlap between the (descriptive) biological and the (honorific) vernacular usage of 'adaptation,' and, pointing out that genocide may have adaptive value, notes that '[b]iological adaptations need not be praiseworthy by human standards. Quite the contrary'.[20] Rather than an adaptation and

---

16 Carroll, *Literary Darwinism,* pp.63-68.

17 Boyd, in Gottschall and Wilson, pp.154, 155.

18 Peter Swirski, *Of Literature and Knowledge: Explorations in Narrative Thought Experiments, Evolution, and Game Theory* (London: Routledge, 2007), p.75.

19 As Boyd complains: 'Art as cheesecake seems an indulgent extra. But if art were so superfluous, how could it not have been selected against?' (in Gottschall and Wilson, p.154).

20 Pinker, 'Toward a Consilient Study of Literature', review of *The Literary Animal,* ed. by Gottschall and Wilson. *Philosophy and Literature* 31 (2007), 162-78, (pp.169,170).

evolutionary goal, art may be simply a stimulus-simulator which works because it tricks the mind into releasing real rewards for fake efforts. Art need have no more *adaptive* function than masturbation. Such a situation is perhaps fine for cultural studies, but intolerable for literary study (especially a version of literary study organised around the evolutionary paradigm). And it is this indifference to literary merit that to Carroll seems such a failing in Pinker's account. Here's Carroll on *How The Mind Works*:

> To illustrate and decorate his text, [Pinker] has collected a substantial number of relevant quotations from literature, but there is no evidence that his familiarity with most of the works he quotes extends very far beyond those quotations. His literary taste and judgement seem those of undergraduate who is extraordinarily bright but who is much more sensitive to computers than to poems, plays, or novels.[21]

This 'insensitivity' to aesthetic value, which is of a part with what Boyd called Pinker's 'insouciance,' is important, for it is something essential to literary criticism, but entirely irrelevant to psychology. Indeed, what difference does it make to the veracity or explanatory power of Pinker's account if he has the literary tastes and judgement of an undergraduate? That the criticism seems to miss the mark so wide is telling. When we push the claim, we find in Carroll's quarrel with Pinker a vital asymmetry in their hopes for what consilience will achieve.

As a psychologist, Pinker wants to corroborate (that is, to lend inductive weight to) his claims for the universality of human nature. The Darwinian Literary Critics are happy to amass data in support of this goal, but apparently want to preserve within this a special place for literary as opposed to non-literary fictions, and they want evolutionary psychology to back them on this claim. Carroll wants consilience to preserve and support his value system. As it multiplies out, that minor difference in agendas has increasingly troublesome ramifications.

The Problem with Ringfencing High Art

In the autobiographical introduction to *The Literary Animal* ('Jon's Story'), Gottschall claims that his decision to view Homer through the lens of evolutionary psychology was regarded as a radical move. But in another respect, it is deeply conservative: few (if any) texts can rival the canonical centrality of the *Odyssey* and *Iliad.* There are plenty of lesser epics that would have done the same job. In fact, if human nature really is always and

---

21 Carroll, 'Steven Pinker's Cheesecake For The Mind', *Philosophy and Literature*, 22 (1998), 500-07 (p.501).

everywhere the same, pretty much anything written after Homer would have also done a fair job; for that matter, anything written *before* Homer as well. In light of the scientific posturing of the Darwinian Literary Critics, exclusive attention upon (and certainly the retention of any special place for) canonical literature (and high art generally) is in want of an explanation.

The question, then, is why use Homer to do what *Neighbours* could achieve? Why use Flaubert when Stephen King makes the same case? Why use Titian when Larry Flynt achieves the same goal (with considerably more success)? The answer, of course, is that the evolutionary psychologists themselves do not limit themselves to examples from high art (as was evident in Pinker's broad-brow cultural references). The decision to focus only on literary texts looks arbitrary in the face of evolutionary psychology's much wider claims for pan-human cultural unity. The point is that *any* sufficiently large selection of enduringly popular texts will achieve the same goal (and if your sample doesn't work, then it's a point against the claim for pan-human similarity).[22] Harold Bloom's Western Canon will do fine; but so will the top-grossing movies of the nineteen-eighties; so will any given bestsellers list from now back to when such lists began. We can find Carroll admitting just this:

> There is no work of literature written anywhere in the world, at any time, by any author, that is outside the scope of a Darwinian analysis. To be susceptible to a Darwinian analysis, an author does not have to be a Darwinian. [...] If Darwinism gives a true account of the human mind, and if the human mind produces all literary texts, all literary texts are susceptible to a Darwinian analysis. They are susceptible, that is, to an analysis of the constraining psychological structures that regulate the production of imaginative artifacts.[23]

How then to justify the focus on exclusively canonical texts? Carroll's comments here preface a close reading of Austen's *Pride and Prejudice*, a (canonical) choice he defends by analogy with the use of model organisms in

---

22 Sufficiently large to minimise the effect of outliers, and enduringly popular because such texts have appealed to the most individual humans for the longest period of time. It's easy to write a text that disagrees with the expectations of evolutionary psychology, but the evolutionary psychologist would not expect such a text to find a wide readership. As Denis Dutton points out, citing John Cage's musical silences and Duchamp's urinal: 'such modernist experiments [...] will never for long capture the attention of the vast public for art, whose abiding concerns involve the same themes they always have from the archaic Greeks to this afternoon's soap operas; love, death, adventure, and triumph over adversity': Denis Dutton, 'Sociobiology and Art', *Philosophy and Literature*, 23 (1999), 451-57 (p.452). Note that Dutton works with an *inclusive* definition of art.

23 Carroll, in Gottschall and Wilson, p.79.

biology.[24] In choosing *Pride and Prejudice* he is (in effect) making an arbitrary selection. *Any* book would have worked, because *every* book ever written is amenable to such analysis (just so long as every book was written by a human, and all the ones we currently have certainly were). Choosing this particular text is, Carroll claims, no more problematic than biologists choosing to work on a particular model organism. Some choose fruit-flies, others choose mice. He chooses *Pride and Prejudice.*

Yet that claim is undermined when we find Carroll reading books which *don't* agree so well with evolutionary psychology. Carroll doesn't want to make it explicit in his reading of Austen, but the implication is that the strong and detailed consonance with the expectations of evolutionary psychology constitute a point in favour of the novel. He does not say outright that what he is proposing is a system for demonstrable (objective) value judgements – which from I. A. Richards to Roman Jakobson had been the grail of 'scientific' approaches to literature, and something Northrop Frye called 'the donkey's carrot of literary criticism' – but the implication is clear. [25] More worryingly, the project seemingly embeds an attempt to employ evolutionary thinking to substantiate and corroborate value judgements according to their degree of agreement with biological norms.

Realism and Value Judgements

This comes to the surface in Carroll's comparative readings of three Palaeolithic fictions – Björn Kurtén's *Dance of the Tiger: A Novel of the Ice Age* (1980), Jean Auel's *Clan of the Cave Bear* (1980), and William Golding's *The Inheritors* (1955).[26] As comparisons, they must have been irresistible: the subject matter all three share – the conditions of life for Palaeolithic hominids – is also the very subject matter from which evolutionary psychology launches its claims to explain human ethology. Each novel offers

---

24 The analogy is telling. Model organism selection is by no means an unproblematic affair. Model organisms are certainly not chosen at random. They are, rather, chosen because they prominently display the types of features that the biologists are interested in or are better able to study (*Drosophila* and *Arabidopsis*, for example, both have unusually large chromosomes). The problem for the biologists working with such organisms is the problem of extension: how far can these findings located with the help of a particular (and particularly unusual) organism be extended to other species? After all, the choice of organism was influenced by the fact that some of its features were in the first place unusually prominent.

25 Northrop Frye, *The Anatomy of Criticism: Four Essays* (Harmondsworth: Penguin, 1990 [1957]), p.20.

26 Björn Kurtén, The *Dance of the Tiger: A Novel of the Ice Age* (Berkeley: University of California Press, 1995 [1978]); Jean Auel, *The Clan of the Cave Bear* (New York: Bantam Books, 1981); William Golding, *The Inheritors* (London: Faber, 1955).

a different account of life in the Pleistocene, not all of which are consistent with the account demanded for evolutionary psychology to hold true. Thus they give Carroll chance to provide a practical exemplar of Darwinian Literary Criticism, whilst simultaneously describing three competing theories of hominid evolution (only one of which he favours).[27] Yet for all the appealing economy of means, it's perhaps this overlap with the empirical grounds upon which Carroll's theoretical orientation rests that tampers with his literary critical reading – and ultimately leads him to expose a naturalistic fallacy at the heart of his critical approach.

Of the three, he likes only Golding. Golding's novel is art, and, moreover, 'one of the great fictional works of the twentieth century.'[28] The other two are not art; they are, in fact, 'bad art'. Tellingly, Carroll compares the two books he doesn't like to modern-day cultural productions with low artistic value: Kurtén's feel for dialogue 'would not be out of place in a tea party among the Teletubbies' (though of course, it *would*: any dialogue would be out of place in the *Teletubbies* as they don't speak natural language), and Auel's *Clan of the Cave Bear* is a 'soap opera' whose 'chief antecedent is [...] the 1960s cartoon series *The Flintstones*' (p.169, p. 173). Notice how similar the tenor of these criticisms are to the criticism made of Pinker's tastes. But in this case, Carroll claims he has criteria.

What will these criteria look like? Carroll says they will not be realism as such – that there is room to entertain fantastic scenarios: 'literal accuracy in depiction is not a *decisive* factor in the quality of fiction' (p. 185). Instead, he is after something closer to a *psychological* realism, hence: 'seriousness and honesty in the conception of a subject, in contrast, are decisive factors, and for those factors, Darwinian social science can provide us with important conceptual measures' (p.185). The 'conceptual measures' he has in mind are metrics calibrated against the expectations of evolutionary psychology. Hence: 'One measure is internal organismic coherence of a represented subject' (that is, fidelity to biological possibility) 'and another is the integration of an organism and its environment' (p. 185) – a snug adaptive fit between the character and setting. The test for *literary merit* is thus consonance with *biological feasibility* as per the expectations of evolutionary

---

27 So although they are nominally fictions, Carroll has much at stake in their content – particularly as one of the fictions has an additional motive: Kurtén is himself a paleoanthropologist seeking via a fictional representation to buttress the plausibility of one account of human evolution over the others. For more on this style of argument see Jane Gregory's essay on Fred Hoyle (Gregory, 'The Popularization and Excommunication of Fred Hoyle's "Life From Space" Theory', *Public Understanding of Science*, 12 (2003), 25-46.

28 Carroll, *Literary Darwinism*, p.177.

psychology. Consequently, Carroll is more or less compelled to assess the three novels as he does: 'When I assess Kurtén's novel as a bad work of art, I am making a judgement based on the criteria I have outlined above. The representation of character in *Dance of the Tiger* is not true to human nature. The motives and feelings of its characters are not plausible' (p.167). Where Kurtén and Auel fail, Golding succeeds: *The Inheritors*' (purported) fit with the actual world scores an aesthetic point.[29] After this, the disclaimer that follows is strangely unconvincing: 'We cannot claim any absolute, transcendent source for aesthetic values [...]. What we can claim is that certain specific qualities satisfy our own cognitive needs' (p. 185) – where '*cognitive needs*' are understood as being prescribed by Darwinian fitness.

The conversations Kurtén imagines between the Cro-Magnons and Neanderthals are unrealistic for both physiological and psychological reasons: not only has Kurtén failed to properly 'take account of anatomical limitations in Neanderthal vocal tracts' (p.170) but his characters are both too articulate and too polite. Carroll is adamant that the brutal Palaeolithic world would have no place for 'civilities,' and 'could not conceivably produce or sustain the manner of mincing politeness that characterises the speech attributed to the Neanderthals' (p.169). The negative comparison Carroll then makes between Kurtén's work and Jane Austen's ('Austen's characters operate in the high tension of a ruthless social world focused on status and mate selection. They are civilised, but they are not saccharine, and they are not fatuous,' p.129) is intended to underline that Austen's work is superior to the extent that Austen's work is closer to how evolutionary psychology thinks humans are.

Carroll reads this failure in realism (at least a failure against the account of prehistoric hominids he is more sympathetic to) as an attempt by Kurtén to 'fudge the evidence' (p.171). Kurtén's picture of prehistoric life is not merely inaccurate, it is fundamentally dishonest, and that Stephen Jay Gould was 'enthusiastically responsive to the bad art of Kurtén's novel' amounts to

---

[29] Note here how close this is to the 'movie physics' analyses – a game among physicists which involves calculating the feasibility of stunts in movies: this entails working out if the bus-jump in *Speed* would have been possible, given the available acceleration, forces acting on the bus, etc (answer: no). It's good fun, if you like calculating forces, and probably provides a nice distraction from what must be fairly crushing work. Of course, those Gradgrind physicists who point out that the bus never could have made the jump are greeted with much eye-rolling: it's an *at best* humorous exercise. There is more to fiction – even realistic fiction – than operating within the parameters of actual possibility. Fiction involves a cognitive contract to the effect that we allow certain transgressions – of the laws of physics, of the laws of probability (like they'd *ever* meet again), and indeed of the laws of strict psychological realism (inasmuch as these are even *in principle* 'strict').

a 'vindication of dishonesty' (p.171). Carroll is not surprised – this sort of dishonesty is to be expected: 'Gould himself does not object to the badness of Kurtén's art, but then, he does not object to bad logic in science theory, either' (p.173). It almost looks as if Gould's choice of fiction might be indicative of wider character flaws – that poor taste in art would reflect poor judgements generally. Carroll seems to believe just this: 'Integrity in art and integrity in science are closely related characteristics' (p.171).

Quality and Normality

In ranking the novels as he does, Carroll has found by way of a Darwinian analysis a remarkably close agreement with his existing tastes. But – as hinted at already in Carroll's attitude to Gould – perhaps the strangest aspect of Carroll's approach is his eagerness to read backward from the 'quality' (by biological standards) of the work to the quality of the 'temperament' of the author (by the same standards): 'The coherence of artistic organization – or its absence – provides clues about an author's motives and the quality of his or her mind' (p.166). So not only can a work's artistic merit be measured by its fit with the actual world (as described by evolutionary psychology), but secondly, the 'quality of the art' constitutes a direct reflection of the character of author.

It's difficult to read this as being anything other than a moral judgement – and this is very worrying. Given that 'quality' here is apparently calibrated against biological norms, Carroll appears to have locked himself into an account which measures the *quality* of an individual's mind by its *biological* normality. Thus a deviant mind would likely produce deviant thoughts – where 'deviance' was understood in (naturalistic) terms of 'deviation from Darwinian fitness'. Homosexuality, for example, would be 'deviant'. *Surely* Carroll wouldn't want to argue this?

But, astonishingly, he does:

> Obviously not all and probably not even most poststructuralists are homosexuals, but many are, especially among the dominant generative theorists, and one can plausibly speculate that the theory lends itself to certain elementary functions within the psychological economy of homosexual life.[30]

The 'psychological economy of homosexual life' is different because it operates with different 'normative criteria'. As with his assessment of the Palaeolithic novels, Carroll is at pains to stress that these normative standards simply offer guidance: 'The appeal to normative criteria of intellectual judgement enforces no necessary evaluative judgement, but they

30 Carroll, *Evolution and Literary Theory*, p.166.

can provide a framework within which we can articulate our evaluative judgements.'[31] The argumentative structure is identical to that employed in the evaluation of the three Palaeolithic novels. And thus, when it comes, the 'articulation of the normative judgment' locates nothing less than an isomorphism between Foucault's deviation from 'normative criteria' and his deviation from biological norms:

> The recent biographical revelations about Foucault's sadomasochistic amusements and his bathhouse promiscuity – activities that apparently continued after he knew about the danger of AIDS and even had reason to suspect that he himself had already contracted AIDS – are wholly commensurate with the quality of temperament displayed in his writings. Foucault's temper is malignant, and his philosophy is primarily destructive in intent. […] one could draw a parallel between the dissemination of his writings and his apparently knowing participation in practices that contribute to the spread of an infectious, fatal disease.[32]

One needn't be sympathetic to Foucault's intellectual agenda to find this style of argument disturbing. That it is so deeply rooted in the conceptual foundations of Darwinian Literary Criticism is equally worrying. Here, then, is the other, seedier side of the evolutionary psychology program, unpleasantly reminiscent of the worst parts of Social Darwinism: the apparent attempt to derive a normative 'ought' from a naturalistic 'is,' of which the more minor aim to derive literary merit from consonance with biological expectations is only a facet.

It is very important to stress that such commission of the naturalistic fallacy is *by no means* universal among proponents of evolutionary psychology – it doesn't, for example, seem to occur anywhere in Pinker. But it is undeniably present in Carroll, and ultimately because, unlike Pinker, he is seeking to secure and preserve a special place for high art in a system that has no criteria on which to base such a distinction.[33] Although it seemed like

---

31 Carroll, *Evolution and Literary Theory*, p.171.

32 Ibid., p.427.

33 How pervasive this stance is among the Darwinian Literary Critics is unclear. They are mainly members of literature departments, and their adoption of evolutionary psychology in the first place is in many ways an instrumental move designed to exclude theorists such as Foucault, yet whether they would want to defend Carroll's equating the quality of art with the quality of character is a different matter. Meanwhile, perhaps the judgements proceeding from Foucault's lifestyle to Foucault's philosophy (and back again) might be excused as Carroll's early attempts to rough out his critical position; or an expression of his frustration at Foucault's ascendancy. Yet the retention of the same argumentative structure a decade later – in 'Human Universals and Literary Meaning' (Carroll, *Literary Darwinism*, pp.129-45) – militates strongly against this interpretation.

such a small quarrel, that minor difference between inclusive and exclusive definitions of art has yielded some striking differences.

Legitimate and Illegitimate Uses of Evolutionary Psychology

Note that Pinker would have no problem at all with explaining the success of Auel's *Clan of the Cave Bear* – the mass appeal of soap operas is amply accounted for by evolutionary psychology's claims. Indeed, evolutionary psychology generally offers a much better account of popular, low-brow productions than it does of high-art. (Pinker himself expresses precisely this reservation about the biases in Carroll's work: 'Darwin may be more important in explaining the ambivalent appeal of wealthy suitors in *Sex and the City* than in *Pride and Prejudice*. The question of whether evolutionary lit-crit is better suited to low culture or high culture is rarely mentioned in *The Literary Animal*.')[34] What's difficult to defend – what the application of scientific methods to literary study have always sought (and thus far failed) to achieve is a rigorous account of what Jakobson called 'Literariness' – the quintessence of quality: the mark that makes *Foucault's Pendulum* better than *The Da Vinci Code*; that makes *Madame Bovary* more than Mills and Boon.[35] Studying narrative fiction disinterestedly from a sociological or anthropological perspective has a long and distinguished history, and was never something in need of a defence. The difficulty has always been explaining the ringfencing: why are these texts canonical and those not? And that sort of demarcation will remain indefensible until criteria can be established. Until they can, canon formation is to evolutionary psychology what ley-lines are to geography.[36]

---

34 Pinker, 'Toward a Consilient Study of Literature', p.167.

35 Roman Jakobson,. 'Modern Russian Poetry: Velimir Khlebnikov. [Excerpts]', in *Major Soviet Writers: Essays in Criticism*, trans. and ed. by Edward J. Brown. (London: Oxford University Press, 1973), pp.58-82 (p.62).

36 One needn't deny that a canon could still be formed, but a canon formed using the criteria of evolutionary psychology would be a very different cut. Indeed, Carroll seems alert to this – noting that poststructuralist critics, with their 'automatically hostile stance toward all forms of normative order' typically run into problems with Austen. This is because Austen's 'own ideology is conservative,' and so at odds with the 'programmatically radical' poststructuralist critics' attempts to 'radicalize and problematize all literary texts'. Austen's work thus 'exercises a strong constraining force on the distorting pull of postmodern ideology' (Carroll, *Literary Darwinism*, p.135). Of course, the flip side of this criticism seems to be that Austen works especially well with modes of criticism which are similarly conservative. Postmodernist critics will produce successful readings (and thus tend to privilege) such figures as John Barth, Robert Coover, Thomas Pynchon; while the Darwinian Literary Critics will find Austen, Dickens, and Homer especially appealing. What

The evolutionary psychologists (as distinct from the Darwinian Literary Critics) know this as well as Jung did when he said that those works of most interest to the psychologist were apt to be lowbrow. Pinker's broad range of examples is a deliberate attempt to capture as wide an audience as possible – even if that is to the exclusion of critics such as Carroll. Pinker is trying to persuade as many of his readers as he can that they are already in agreement with him. To make the case for evolutionary psychology's descriptive accuracy, he needs to have as few exceptions as possible. Indeed, if anything, the distinctive feature of much high literature is not consonance but *dissonance* with evolutionary psychology's expected criteria. Despairing of the Darwinian Literary Critics' fixation on those 'forms of art that appeal to 1% of the population (people like themselves),' Pinker makes much the same point: 'since highbrow and avant-garde genres often define themselves in defiant opposition to low- and middlebrow culture, and to the high culture of a previous period, they are bound to refute just about any generalization of the nature of art that anyone will ever make.'[37] Whatever evolutionary psychology is bringing to the table, it isn't a *sortative* criteria. Assuming that evolutionary psychology offers even a broadly correct account of human psychology, establishing that canonical literature fits the pattern gives us precisely this conclusion: it was written by a human.

Conclusion

Attempting to maintain a place for a value gradient and to justify that value gradient by appeals to criteria of biological normality is a pernicious and illegitimate use of evolutionary theory: pernicious because it seeks to calibrate moral-aesthetic value against biological fitness; and illegitimate because evolutionary psychology (inasmuch as it is a scientific enterprise) is descriptive not prescriptive.

Pinker's decision to employ an inclusive and loose definition of art, and to restrict his analysis to content and not value was no accident. Inasmuch as it has any claims to be a science, evolutionary psychology offers no legitimate grounds on which to evaluate texts. Ultimately, the aims of the evolutionary psychologists and the aims of the Darwinian Literary Critics are divergent – as Carroll's quarrel with Pinker, and Pinker's reply, underlines. They are two separate projects. As Pinker has it: 'The goal of explaining the human taste for fiction (a problem in evolutionary psychology) sits uneasily

---

Carroll seems to be tacitly admitting is that different schools of criticism generate different canons.

37 Pinker, 'Toward a Consilient Study of Literature', p.176.

with the goal of improving the analysis and criticism of specific works of fiction (a problem in departments of English and other literatures).'[38]

To argue in this fashion is not to deny that value judgements are possible, simply that evolutionary criteria are a poor base on which to ground them. Nor does such an argument entail the claim that evolutionary psychology is a false theory of mind, simply that it provides no basis for discriminating between texts. And in the absence of that capacity, it's not at all clear what it can offer to literary study.

38 Pinker, 'Toward a Consilient Study of Literature', p.175.

The Book of Nature: Biosemiotics and the Evolution of Literature

Wendy Wheeler

Abstract: This essay argues that recent developments in evolutionary biology require us to reformulate the Darwinian Synthesis which has dominated evolutionary understandings from the 1930s to Neo-Darwinism and Evolutionary Psychology in the present. Introducing the new interdiscipline of biosemiotics, which understands all living things – from cells to organisms to ecosystems – as communicative makers of meanings, the essay argues that we can understand cultural and aesthetic life both as emergent from natural biosemiotic life and also as rearticulating nature's patterns at a new symbolic level in humans. Drawing on the semiotic philosophy of Charles Sanders Peirce and on Michael Polanyi's understanding of tacit embodied knowledge, the essay suggests that the power of literature lies in its capacity to remind us of the generative power and creative evolution of semiosis, both in nature and in culture.

There were once, in Christendom, two books: the Book of Nature and the Book of God's Word. Both were God's work: complementary, to be read together. Neither was superficial; both involved many levels of understood development and interpretation.[1] Ever since, as Hegel says, the Greeks recognised the shape of our own reasoning mind in the human body, and exalted it in sculpture, the origin and meaning of that body-shaped mind has perplexed us.[2] In seeing our own embodied selves in the collective mind bodied forth in human art, we have searched the mind's models in all our ways of knowing: in aesthetics, in ethics, and in science. But the earliest art, what Hegel calls 'symbolic art,' was recognition of pattern itself.[3] Before the dull simplicities of a dully materialistic age, we looked for what Gregory

---

1 'Medieval exegesis considered the Bible to have 'a *cortex*, or level of surface meaning, covering a *nucleus* of truth' which was to be revealed by the exegete. The hermeneutic tradition distinguishes three stage of scriptural analysis: The first stage called *littera*, was concerned with grammatical structure. At the second stage, called *sensus*, the obvious literal or narrative meanings had to be understood. *Littera* and *sensus* form the outer cortex of the text. At the same time, *sensus* comprises the first of four levels of spiritual meanings. As such it is called literal or historical meaning (*sensus litteralis, sensus historicus*). The more profound meaning concerning the scriptural nucleus is revealed only at a third stage of exegesis, called *sententia*.' Winfried Nöth, *Handbook of Semiotics* (Bloomington: Indiana University Press, 1990), p.335.

2 G. W. F. Hegel, ch. 5 'Division of the Subject', section CVI, *Introductory Lectures on Aesthetics*, trans. by Bernard Bosanquet, ed. by Michael Inwood (London: Penguin, 2004), p.84.

3 See, for example, the Blombos Cave, South Africa, ochre etchings which are reliably dated to between 70,000 and 75,000 years BCE. The pattern here consists of repetition (diagonal lines repeated) and difference (diagonal lines repeated in a contrasted direction, i.e. cross-hatched). Illustration available at <http://cogweb.ucla.edu/ep/Art/BlombosOchre.jpg> [accessed 14 May 2009].

Bateson called 'the pattern which connects' all living things (and nature *and* mind) across theology, the arts and natural philosophy.[4] Due to the religious doctrine of human exceptionalism, we rarely looked *in nature* for the pattern (of semiosis) which seems so remarkable in us; for historical reasons, we have temporarily, disastrously, ceased properly both to look and to *read* our meanings in both occult books.[5] With the advent of modernity, old evolved wisdoms have been scorned, and the book of nature has been turned into commodities, or what Heidegger calls 'standing reserve,' and exploited.[6] We have buried the mystery of making sense (the interpretation of signs in nature and in language) in a fantasy of positivistic realism which confuses the experience of a real world (which every organism has) with all the world might be. Prospero has drowned his grimoire, and the world itself has become disenchanted.

From as early as Epicurus (341-270 BCE), evolutionary possibilities were recognised, but it was not until the nineteenth century that Charles Darwin brought forth a decisive account in *Origin of Species* in 1859.[7] Although Darwin's theory of evolution via variation and natural selection was not mechanical, it was brought forth in what Carlyle in 1829 had typified as 'the Mechanical Age'.[8] Despite Kant's arguing that there would never be a Newton of a blade of grass, western science has been a science dominated by physics and the idea of physical mechanisms and determinism.[9]

Since then, Darwinism has both remained at the centre of biologists' understanding of evolution and has also had a contested evolutionary life of

---

4 Gregory Bateson, *Mind and Nature: A Necessary Unity* (Cresskill, NJ: Hampton Press, 2002), p.7.

5 The exceptions resisting the growth of mechanistic views in their own times were the German romantics, Friedrich Schelling in particular. See Robert J. Richards, *The Romantic Conception of Life: Science and Philosophy in the Age of Goethe* (Chicago: University of Chicago Press, 2002); Stephen Toulmin, *Cosmopolis: The Hidden Agenda of Modernity* (Chicago: Chicago University Press, 1990).

6 Martin Heidegger, 'The Question Concerning Technology', *Basic Writings*, ed. by D. F. Krell (London: Routledge, 1993).

7 'If only the geologists would let me alone, I could do very well, but those dreadful hammers! I hear the clink of them at the end of every cadence of the Bible verses.' John Ruskin, letter to Henry Acland of 24 May 1851, in *The Works of John Ruskin*, ed. by E. T. Cook and Alexander Wedderburn, 39 vols (London: George Allen, 1903-12), XXXVI, p.115.

8 Thomas Carlyle, 'Signs of the Times', *Selected Writings*, ed. by Alan Shelston (London: Penguin, 1986), pp.61-85 (p.64).

9 'It is utterly impossible for human reason [...] to hope to understand the generation even of a blade of grass from mere mechanical causes.' Immanuel Kant, section 16 (77) in Div. II, 'Dialectic of Teleological Judgement', *Critique of Judgement*, trans. by J. C. Meredith (Oxford: Oxford University Press, 1952), p.66.

its own. From the Modern Evolutionary Synthesis of the 1930s and 40s, still dominated by a mechanistic, linear and reductionist point of view, new variations of the evolutionary theme have developed. Influenced by the development of cybernetic systems understandings in the nineteen-forties and fifties, these essentially ecological and non-linear views of biological and cultural evolution fall under a family of approaches often named as biological systems theory, evolutionary developmental theory, and living systems theory. What they share is a 'whole systems' and 'information' approach to understanding life.[10] The idea of 'information,' though, remains caught in a relatively predictive model. Evolutionary theory teaches that nothing comes from nothing; everything that evolves is a reworking of earlier forms. But what may be thought of as 'information' in abiotic feedback systems is not simply 'information' in us. The more accurate term for human communication is 'semiosis' – a much more nuanced account of life in signs. For the new interdiscipline of biosemiotics, the emergence of articulate semiosis in *Homo sapiens* is part of a much longer semiotic natural history which belongs to all living things.

In many ways, the developing insights of biological systems theory look set to change Darwinism quite radically. The Modern Synthesis was dominated by population geneticists, but the astounding developments in molecular biology in recent years have begun to alter understandings of the supposed centrality of the gene. The gene (more accurately, gene sequence) is a set of instructions, encoded in DNA, for the development of an organism; it is digitally encoded memory capable of passing on these instructions faithfully down the generations. On its own, though, DNA is inert. It makes nothing. As Richard Lewontin has written, 'First, DNA is not self-reproducing, second it makes nothing, and third, organisms are not determined by it.'[11] What actually happens, both in embryological development *and* postnatally, is environmentally contextualised.[12] How the

---

10 See, for example, Denis Noble, *The Music of Life: Biology Beyond Genes* (Oxford: Oxford University Press, 2008); Eva Jablonka and Marion J. Lamb, *Evolution in Four Dimensions: Genetic, Epigenetic, Behavioural, and Symbolic Variation in the History of Life* (Cambridge, MA: MIT Press, 2000); Sean B. Carroll, *Endless Forms Most Beautiful: The New Science of Evo Devo and the Making of the Animal Kingdom* (London: Phoenix, 2007).

11 Quoted in Jesper Hoffmeyer, *Biosemiotics: An Examination into the Signs of Life and the Life of Signs*, trans. by J. Hoffmeyer and Donald Favareau (Scranton: Scranton University Press, 2008), p.31.

12 The genetic code is interpreted in the cell's production of proteins – in a context, in other words. But this cellular activity takes place within the larger context of the body's systems in which the embryo is growing; and this body itself has a natural (and in some cases social and cultural) context. Just as words only have meanings in contexts, so with biosemiosis. The idea of 'nested systems' producing meanings may

DNA is interpreted by proteins depends on the cell's internal and external environment within the organism, and even on the organism's environment postnatally. In the first place it depends on the cell's interface, via its membrane (the skin which is essential to any 'self' at all), with other cells in the organism. This environmental responsiveness via membrane feelers and receptors causes protein activity within the cell, and within the cell's multicellular environment. Lastly, the entire organistic self-organisation responds also to activities within its own wider environment. As Jesper Hoffmeyer writes:

> If any agency in the body deserves to be called directive or controlling, it would not be DNA but instead the membranes which permeate the body. Yet it would be more accurate to say that the biological orchestration of an organism is created by a well-tuned symphony of biosemiotic relationships across the membranes, which in each instant of an organism's life controls and coordinates the biochemical, physiological, and even cognitive processes that together constitute life. [...]
>
> Where there is DNA, there is always also the cell with its membranes, whose patterned organization is autonomously determined by the continuity of cellular life through cell division. Living cells, through their membranes, use DNA to construct the organism, not vice versa. It is the active functioning of these membranes as well as the membrane-connected proteins that direct life's activity, not the passive and inanimate DNA. It is, in other words, in the semiotic functioning of the cellular membranes that we shall seek what can be called life's agency, its inherent future-directedness, its survival project.[13]

Biosemiotics: Culture is Natural and Evolutionary

It is a remarkable feature of writing by biologists who understand that life is organised by the causality of sign relations and that the biosphere is coextensive with the semiosphere, that the aesthetic realm, music especially, is so often invoked in trying to explain what are now called biosemiotic insights. There are good reasons for this, to do with Bateson's 'the pattern which connects,' and these reasons will be discussed after considering what Neo-Darwinism, and its attempted use by literary critics, *didn't* do. Late nineteenth- and early twentieth-century attempts to apply evolutionary theory to human societies led to eugenics and, eventually, in a perverse mix of (unacknowledged) Protestant theology with romanticist socialist holism,

---

be helpful here. See Wendy Wheeler, 'The Biosemiotic Turn, or, The Nature of Creative Reason in Nature and Culture', in *Ecocritical Theory: New European Approaches*, ed. by Axel Goodbody and Kate Rigby (Charlottesville: University of Virginia Press), forthcoming. For research demonstrating postnatal effects of maternal behaviour on biological processes, see Devin Powell, 'Mum's Behaviour May Make Rats More Butch', *New Scientist*, 2690, 9 January 2009.

13 Hoffmeyer, *Biosemiotics*, pp.31-32.

to the realised fantasy of purifying state populations by selective and mass culling. This made biology, as a way of discussing humans and their motivations, a dangerous topic until it was revived in the nineteen-seventies first as Sociobiology, then later (in the nineteen-eighties and nineties) as Evolutionary Psychology. The results, both in biology and in the literary criticism which critics such as Joseph Carroll tried to develop from Neo-Darwinism, were uninspiring.[14] The physics-dominated reductionism of these enterprises meant that they had nothing interesting at all to say about the complex motivations of political, ethical and aesthetic human life. The development of evolutionary biological systems science, in which living systems are understood as complex, non-linear (i.e. recursive), and emergent, mounted a significant challenge to the idea that biological systems must be understood only by final reduction to the (relatively) deterministic laws of physics. Rather, the relationship between physics, chemistry, biology, and finally culture itself came to be understood as evolutionary: each subsequent level is understood as, itself, involving greater extents of complexity and emergence. With the development of cybernetics, then bio-cybernetics (ecological systems as feedback systems), biological systems were understood first in terms of 'information,' and then, with the development of biosemiotics, as semiosic.

The difference between cybernetic 'information' in physical systems and bio-cybernetic semiosis in biological systems can be found in the word 'communication'. Purely physical 'information' systems do not communicate. Both kinds of systems might be thought of as involving resonance (feedback loops which are either amplifying via positive feedback or dampening via negative feedback), but in living systems such feedback is neither mechanical nor probabilistic. On the contrary, the resonances (feelings, intuitions) living things respond to as signs are *interpreted* (as possible meanings) according to the *self-maintaining purposes which only belong to living things*. The difference in living systems is that the latter are teleological. Life is purposeful; it wishes to endure, and its endurance takes the form of living organisms with membranes ('skins') defining inside from outside (and 'self' from 'not-self'), and, necessarily, of communication between this 'self' and its others which, thereby, constitute and shape it.

It is here – with an understanding of the evolution of life as the play of identity and difference (communication), and the necessary plasticity of interpretation involved in purposeful self-maintenance – that biosemiotics can contribute to dissolving the divide between the sciences and the

---

14 Joseph Carroll, *Literary Darwinism: Evolution, Human Nature, and Literature* (New York: Routledge, 2004).

humanities. Semiosis (and interpretation) does not belong to humans alone. As Peirce averred, 'all this universe is perfused with signs.'[15] The 'pattern which connects' the aesthetic life of an individual text, both within itself and to its historical context, and which gives it its meanings, is the same pattern which connects the ethical life of an individual to the collective life upon which it depends and which gives it its meanings. Both are ruled by the play of identity and difference in contexts; and this same play is at work in biological life too. In human mind, language, and creative making we call this 'pattern which connects' metaphor. But it is clear that the resonances of this form of sign relation and creative play are at work in organisms and evolution too.[16] Similarly, 'downward causation' in living things, may remind us of T. S. Eliot's evolutionary insight in 'Tradition and the Individual Talent' that, like biologically evolved forms, new aesthetic forms both depend upon a tradition and also reorganise it.[17] Observation of 'the pattern which connects' means that we can also 'read backwards,' from human aesthetics and ethics, for example, to biology and other animals.[18] From a biosemiotic point of view, the insights of readers (interpreters) of cultural artefacts in the semiosphere offer valuable insights to the understanding of what 'interpretation' might mean in the biosphere. The ways in which literary language resonates for a reader – via the iconic play of signs in metaphor, and indexical signs in metonymy – is merely a more complexly

---

15 Charles Sanders Peirce, 'The Basis of Pragmaticism in the Normative Sciences', in *The Essential Peirce: Selected Philosophical Writings*, ed. by the Peirce Edition Project, 2 vols (Bloomington: Indiana University Press, 1998), II: 1893-1913, p.394.

16 Jesper Hoffmeyer, 'Biosemiotics: Towards a New Synthesis in Biology', *European Journal for Semiotic Studies*, 9.2 (1997), 355-76. http://home.comcast.net/~sharov/biosem/hoffmeyr.html> [accessed 1 May 2009].

17 'Downward causation' refers to the capacity, within living organisms, for evolutionarily 'later' emergent processes to effect changes to evolutionarily 'earlier' ones. This is an ecological commonplace which merely emphasises the interrelatedness of all living things; but it is, of course, anathema to a view of living things as most fully comprehended when reduced to separable elements (things) rather than as processes and relations. See Claus Emmeche, Simo Køppe, Frederik Stjernfelt, 'Levels, Emergence, and Three Versions of Downward Causation', in *Downward Causation: Minds, Bodies and Matter*, ed. by P. B. Andersen, C. Emmeche, N. O. Finnemann and P. V. Christiansen (Aarhus: Aarhus University Press, 2000), pp.13-34. <http://www.nbi.dk/~emmeche/coPubl/2000d.le3DC.v4b.html> [Accessed May 2009]; and D. Noble, *The Music of Life*, pp.42-54. T. S. Eliot, 'Tradition and the Individual Talent', *The Sacred Wood* (London: Faber, 1997), pp.47-59 (p.41).

18 Mark Bekoff and Jessica Pierce, *Wild Justice: The Moral Lives of Animals* (Chicago: Chicago University Press, 2009).

articulated (i.e. in symbolic language) form of the resonances felt by organisms, and even cells.

To grasp the insight offered by the development of biosemiotics is to understand that communication/semiosis is synonymous with life, and that the evolution of biological complexity is at the same time the evolution of semiosic complexity. In this way we can easily see that human culture continues the evolution of nature by increasing communicational, semiosic, complexity. Physics, then chemistry, are the necessary grounds upon which biological life can come into being, but at each stage we are dealing with emergent forms of organisation (physical, chemical, biological, cultural) which are less constrained than the levels which precede them. In biological evolutionary forms, the evolution of greater biological complexity always involves the emergence of greater extents of semiotic, communicational, freedom.[19] Later forms are *constrained* by antecedent forms, *but they are not reducible to them.* The existence of antecedent constraints (the present and future always emerge from antecedent forms of semiotic organisation: all complex systems have histories, in other words) teaches us another lesson. Both natural *and* cultural evolution are teleological. However, the Aristotelian Final Cause does not lie outside the system; it is autopoietically immanent in living things (both as systems and as individuals as systems within systems) as the creative force by which living things evolve and grow, according to the constraints of their kinds, and in creative, adaptive, response to the environments (natural and cultural feedback systems) in which they find themselves. This is true of systems (and cultures) and of the individual lives within them. In human cultures the name we give to these processes is 'learning'.

Cultural processes (individual and collective) are, thus, not random. Each system (including you and me) is both constrained by the past (individual and collective) and (at best) is creative in the present in making the future. Once understood as the evolution of greater extents of semiotic/communicational freedom via 'learning,' it becomes clear that human progress will always and everywhere involve the development of 'tools' (from languages to lenses to laptops) which extend our embodied semiosic capacities and reach. But not only is 'learning' always partial, it is also directional and cumulative. It involves a modelling of the world (just as animals must have 'models' of their *Umwelten* in their minds) which, in humans, tends to be developmentally progressive (and to speed up over time); but it also (as in natural evolution) always carries the past (antecedent

19 Jesper Hoffmeyer, *Signs of Meaning in the Universe* (Bloomington: Indiana University Press, 1996).

forms) within it. Skills, in other words, develop exponentially. In other words, myth, religious organisation, art, are not displaced by science; rather (and although generally, and sometimes damagingly, unacknowledged), historically later forms always contain within them what are effectively transforms of the past. As paleobiologist Neil Shubin has written in *Your Inner Fish*:

> Looking back through billions of years of change, everything innovative or apparently unique in the history of life is really just old stuff that has been recycled, recombined, repurposed, or otherwise modified for new uses. This is the story of every part of us, from our sense organs to our heads, indeed our entire body plan.[20]

Newness is real, but it evolves from, is a transform of, what has been; in human cultural life it is also, more strangely, an abductive 'guess' at the future based on a sense, an intuition, of a deepening coherence in our understanding of reality. As Michael Polanyi writes on Meno's paradox concerning the search for knowledge:

> This is how I would resolve the paradox of the *Meno*: we can pursue scientific discovery without knowing what we are looking for, because the gradient of deepening coherence tells us where to start and which way to turn, and eventually brings us to the point where we may stop and claim a discovery.[21]

The Evolution of Literature

In *The Music of Life: Biology Beyond Genes*, systems biologist Denis Nobles writes:

> Genetic and cultural forms of evolution share this messiness, or, to use a less derogatory term, inventiveness. For it is through a complicated series of bodges that nature has arrived at the huge diversity of life as we know it. Tangled intricacy is the mother of nature's invention.
>
> The idea of metaphor is important here, too. On that basis, we can say that, as the genome has developed, nature has switched from one metaphor to another. It has plundered the treasure chest of old DNA modules to form new combinations and to give old genes new functions.[22]

---

20 Neil Shubin, *Your Inner Fish: A Journey into the 3.5 Billion-year History of the Human Body* (London: Allen Lane, 2008), p.201.

21 Michael Polanyi, 'The Creative Imagination', in *The Concept of Creativity in Science and Art*, ed. by D. Dutton and M. Krausz (London: Nijhoff, 1981), p.98.

22 Denis Noble, *The Music of Life: Biology Beyond Genes* (Oxford: Oxford University Press, 2006), pp.103-04.

The implication of this is that everything that is found in highly complex semiotic form in humans is also found in less complex, evolutionarily antecedent, biological forms. Signs, too, have an evolutionary, and emergent, history. The creative evolution which Noble describes lies in nature's capacity to play with what is to hand in order to adapt old forms to new uses. Play, as a number of people over the years have realised, is at the heart of all creative being.[23] Whether artist or artisan, or writer or philosopher, or scientist or engineer, when people are being creative what they are doing is playing – with things and ideas. The play-state is neither wholly conscious nor wholly rational; it feels intoxicating, and is characterised by intuitive moves (what Peirce called 'abduction' and what Michael Polanyi called 'foreknowledge'), and it seems to have a life of its own – which in some sense, a teleological one, it does.[24] Its sources seem to lie in tacit knowledges which both build and grow over time and are also directly related to reality.[25] As Polanyi writes:

> A scientist's originality lies in seeing a problem where others see none and finding a way to its pursuit where others lose their bearings. These acts of his mind are strictly personal, attributable to him and only to him. But they derive their power and receive their guidance from an aim that is impersonal. For the scientist's quest presupposes the existence of an external reality. Research is conducted on these terms from the start and goes on then groping for a hidden truth towards which our clues are pointing; and when discovery terminates the pursuit, its validity is sustained by a vision of reality pointing still further beyond it.[26]

In the human mind, this same play of identity and difference, now abstracted out of embodied experience, takes the form of metaphor. As Paul Ricœur writes in *The Rule of Metaphor*, this very basic form of thinking involves 'the fusion of "sense" and "sensa", understood [...] as an iconic

---

23 Kant in *Critique of Judgement*, Friedrich Schiller in *On the Aesthetic Education of Humankind in a Series of Letters*, and Peirce himself (in 'the play of musement' in 'A Neglected Argument for the Reality of God' in *The Essential Peirce*, vol II, pp.434-50.) offer eighteenth- and nineteenth-century examples. In the twentieth century Johan Huizinga in *Homo Ludens*, D. W. Winnicott in *Playing and Reality*, and Hans-Georg Gadamer in *Truth and Method*, are all concerned with the importance of play. In the twenty-first century, Stuart Brown's *Play: How it Shapes the Brain, Opens the Imagination, and Invigorates the Soul*, and Sergio and Vivien Pellis's *The Playful Brain: Venturing to the Limits of Neuroscience* offer accounts of scientific development in this field.

24 Mihaly Csikszentmihalyi, *Flow: The Psychology of Optimal Experience* (New York: Harper and Row, 1990).

25 Michael Polanyi, *The Tacit Dimension* (Chicago: University of Chicago Press, 1966) and, with a new Foreword by Amartya Sen, University of Chicago Press, 2009.

26 Polanyi, 'Creative Imagination', p.106.

unfurling of sense in imagery'.[27] As in all art, metaphors are iconically generated, as Ricoeur later adds, by improvisations on a theme. These themes, telling from the point of view of evolutionary biology, are of the same sort by which we recognise (in the usual, as also in the Wittgensteinian, sense) 'family resemblances'. Ricoeur writes: 'A family resemblance first brings individuals together before the rule of a logical class dominates them. Metaphor, a figure of speech, presents in an *open* fashion, by means of a conflict *between* identity and difference, the process that, in a *covert* manner, generates semantic grids by fusion of differences *into* identity.'[28] This ability, to play with similarities and differences, and to carry meanings across (or away – abduct) from one schema to another in new fruitful ways, and to develop the implications of this bearing across (Greek: *metapherein*) from one place to another, is the basis of all creative thinking. Peirce (who advises such 'play of musement' in facilitating productive thought) describes it as the logic of abduction – antecedent to deduction and induction – as the *only* source of new ideas:[29]

> Abduction is the process of forming an explanatory hypothesis. It is the only logical operation which introduces any new idea; for induction does nothing but determine a value, and deduction merely evolves the necessary consequences of a pure hypothesis. Deduction proves that something must be; Induction shows that something actually is operative; Abduction merely suggests that something may be.[30]

Peirce notes how strange it is that abductions, essentially informed guesses, are so very often correct. Both Peirce and Michael Polanyi, who expressed precisely the same idea in the concepts of tacit knowledge and 'foreknowledge,' have understood the teleological nature of all living things, including human ideas in every sphere where thoughtful practice inheres. Life and thought are structured practices. It is possible to go down blind alleys, but very often skilful practitioners have intuitions about the most fruitful directions, and these are not random. In 'The Creative Imagination,' Polanyi writes:

> But there exists also a more intensely pointed knowledge of hidden coherence: the kind of foreknowledge we call a problem. And we know that the scientist produces problems, has hunches, and elated by these anticipations, pursues the

---

27 Paul Ricœur, *The Rule of Metaphor: The Creation of Meaning in Language*, trans. by Robert Czerny with Kathleen McLaughlin and John Costello (London: Routledge, 2003), p.250.

28 Ibid., p.234.

29 Peirce, 'A Neglected Argument', *Essential Peirce*, II, pp.434-50.

30 Peirce, 'The Nature of Meaning', *Essential Peirce*, II, pp.208-255 (p.261).

> quest that should fulfil these anticipations. This quest is guided throughout by feelings of deepening coherence and these feelings have a fair chance of proving right. We may recognise here the powers of dynamic intuition.[31]

We can see that the thinker, whether scientist or writer or philosopher or artist, is pursuing a project. The papers, poems, novels, books, paintings produced and communicated are way-stations, markers of a journey: an intense dialogic relation their maker shares, both with past forms and knowledge, with those he or she addresses, and thus, always, with the future into which he or she leans.

When Kant, in the *Critique of Judgement*, wrote about aesthetic judgement as the disinterested (i.e. non-ends driven, non-instrumental) free play of the understanding and the imagination, and Schiller linked this to his own, mediatory notion of the *Spieltrieb* (play-drive), he was articulating the purposefulness of creative life in all human beings as, precisely, non-determinate in the same way that nature's products are.[32] Like all living things, art is process and free play. A work of art is both the marker of a way-station in the creative artist's telic discovery of 'a host of yet hidden implications' and also a 'question' and an invitation to further semiotic reciprocity and play.[33] Thus Thierry de Duve argues that the purposefulness in Duchamp's 'Urinal' is that it asks us, in the historical moment in which art appears to offer itself fully in commodity form, 'what *is* art?,' and answers that it is the possibility of a relationship.[34]

In considering culture as evolutionary and natural, and works of art as the opening, or offer, of communicative reciprocity and semiotic relationship in play and process in time, we can see that works and ideas are like natural organisms. Like nature itself in modernity, art occupies two parallel modes of communication and exchange. In one it is reified as commodity and, like labour and land (both belonging to nature), is ordered as exchange value in terms of money; in the other it offers itself as non-instrumental use in the service of relationship as the creative opening to the other as another order. In *this* order, 'profit' is not laid up in the accumulation of material goods glistering with the promise of salvation like dead gold in banks, but lies instead in the promise of living relationship.

---

31 Polanyi, 'Creative Imagination', p.116.

32 On play, see Kant, *Critique of Judgement*, section 54 in Book II, Analytic of the Sublime, also note 10 above.

33 Michael Polanyi, *Personal Knowledge: Towards a Post-Critical Philosophy* (London: Routledge and Kegan Paul, 1962), p.64. 'The work of art [...] is essentially a question, an address to the responsive heart, an appeal to affections and to minds.' Hegel, *Introductory Lectures on Aesthetics*, p.78.

34 T. de Duve, *Kant after Duchamp* (Cambridge, MA: MIT Press, 1996).

Like friendship and love between lively human beings, the works of art we love invite ongoing dialogue, communication, interpretation, and change.

Literature: Caliban's Curse and the Art of Not Losing

In order to think properly about literature itself from a biosemiotic point of view, we must ask ourselves what evolutionary pressure its emergence was responding to. Once it is fully understood that culture is evolutionary, the many things which in our currently conceptually impoverished public life are thought unimportant will come to be re-evaluated; nothing comes into being and survives which is not adaptively useful. Thomas Sebeok has argued that the evolutionary pressure driving the exaptive evolution of articulate speech from mimetic language was not communication *simplex*, but the need to better model the human world.[35] All organisms' *Umwelten* are models organised in embodied experience to make coherent sense of (and a model of reference to) the signs the organism apprehends as the condition of its existence and survival. It is easy to imagine that bipedalism led to manual dexterity, and manual dexterity is closely tied to imaginative dexterity, play, the capacity to solve problems, and language.[36] To be able to play with signs, first iconic, then indexical, and finally symbolically in articulate language, is a tremendous gain from an evolutionary perspective. It leads not only to better tools for shaping the human *Umwelt* physically, but also to better tools-as-ideas for imagining answers to the great questions which an increasingly agile human mind, achieving greater self-consciousness, must ask.

Michael Polanyi spent a lifetime considering such tacit knowledge and its extraordinary structured capacity to intuit the solution to problems. Himself a successful scientist, Polanyi was particularly interested in this question in relation to the strange processes of scientific discovery. In his many books and essays, he argued that the structure of foreknowledge consists in progressive tacitisation of knowledge whereby, faced with a problem, the scientist (true of the artist also) disattends from what he or she knows (whether consciously known or merely internalised as never conscious practice) which henceforth act as 'signs' in order to lean into the future host of as yet unknown things in the direction he or she intuits that the answer lies.

---

35 'Language at its inception was not used for exterior communication, but only as an interior modelling device – a modelling device or system being a tool wherewith an organism analyses its surroundings.' T. Sebeok, 'The Evolution of Semiosis' in T. Sebeok, *Global Semiotics* (Bloomington: Indiana University Press, 2001), p.28.

36 F. R. Wilson, *The Hand: How Its Use Shapes the Brain, Language, and Human Culture* (New York, Vintage Books, 1999); Brown, *Play*.

Perhaps one of the most striking things about humans lies in the determination with which they normalise what is, in fact, extraordinary. For Hegel, it is in aesthetic experience pre-eminently that the struggle to give form to such disturbing self-reflective consciousness takes place.[37] Cultures which resist such visionary experience do so because of its powers to unsettle unreflective life. Scientists, as much as artists, have cleaved to their strongly intuited 'visions' of a deepening reality even when ratification by other means lies far ahead and even beyond their own lifetimes. (Copernicus's vision finally ratified by Newton, and Einstein's childhood intuition that Newton was wrong about absolute inertia, are examples Polanyi offers.) And yet we do not think this extraordinary intuition of a more coherent reality strange in itself, even while we might marvel at such astonishingly accurate foresight. All great, visionary science in fact proceeds by such means.[38]

The conjuration of ideas in words is a similar case. We normalise language, and forget its vast and evolving generative capacity. And it is here that we can see the evolutionary power of literary language. Its enchantment, and adaptive power in the persistence of our kind, lies in the fact that it reminds us, in the midst of the normal disattentions of the prose of the world, that language is the conjuring power to *make* (i.e. to model) worlds anew. The old etymological roots of 'grammar' (and 'glamour') in 'grimoire' tell us that this spellbinding connection with magic was once closer to the fronts of less belaboured imaginations. Prospero's drowning of his book of spells at the end of *The Tempest* indicates, perhaps, Shakespeare's foreknowledge of disenchantment to come. In leaving 'the airy music' of the isle, in the end, to unobservant Caliban – whose 'only profit' from language was that it taught him 'how to curse,' and upon whom nearly everything truly meaningful was lost – perhaps we hear the poet's farsighted judgement on the fate of language in an increasingly disenchanted and mercantile world. 'All that glisters is not gold'; cultural prosperity, as *The Merchant of Venice* reminds us, is not measured in money and earthly wealth alone. It is justly measured both in love and also in the power of what measured words, brought to court and carefully interpreted, can do. Portia makes new meanings from dead legal metaphors in an exuberance of language which reminds us of its power to transform. Literary art is the art of not losing the 'sea change' power of language in its 'rich and strange' transformative making of worlds. In literature we not only exercise afresh the primal power of imagination which still drives the shaping of worlds in words, but we also

37 Hegel, *Introductory Lectures on Aesthetics*, p.83.

38 Polanyi, 'The Creative Imagination', p.107.

pay close attention to the details of that poiesis as its resonances teach us, in the words of Henry James, to try to be those observant and self-reflective kinds of people 'on whom nothing is lost'.[39]

[39] Henry James, 'The Art of Fiction', *The Portable Henry James*, ed. by J. Auchard (London: Penguin, 2004), pp.426-47 (p.435).

Men, Monkeys, Monsters and Evolution in Fiction from the *Fin-de-siècle* to the Present

Christopher Lloyd

Abstract: Fascination with creatures that challenge boundaries between humans and other species long predates Darwin's theory of evolution by natural selection. Narratives have evolved from the fabulous to the more scientifically grounded, and tend to follow either the regressive or relativistic model, established respectively by Swift and Voltaire: men as apes, debased, stinking brutes; or apes as men, as potential rivals and substitutes. My aim here is to sample and compare a select corpus of fictional texts in French and English ranging from the late nineteenth century to the present, which deal with monstrosity or the incursion into the human domain of threatening, rival primate species, in the context of human evolution, whether individual, social or biological. Stevenson, Zola, Maupassant and Hugo all present variations of the self dispossessed by some monstrous other, whether through physical deformity caused by a malicious external agent, or more subtly through some internal disturbance of ontological equilibrium. This unleashes creatures that devour their human host, but at the cost of their own extinction. H. G. Wells and Octave Mirbeau, on the other hand, adopt the perspective of the external observer who poses as a scientific witness of attempts to modify human and animal evolution. In conclusion, I discuss two twentieth-century novels which rewrite the whole course of primate evolution, Boulle's *La Planète des singes* (1963) and Self's *Great Apes* (1997).

Fascination with creatures that challenge boundaries between humans and other species long predates Darwin's theory of evolution by natural selection and man's descent from 'ancient, lower and extinct forms'.[1] Narratives have evolved from the fabulous to more scientifically grounded. Thus Edward Tyson produced an anthropological study comparing pygmies, apes, monkeys and men in 1699 (concluding that pygmies were a type of ape).[2] Subsequent literary accounts tend to follow either the regressive or relativistic model established respectively by Swift and Voltaire: men as apes, debased, stinking brutes like the Yahoos encountered by Gulliver in part four of his *Travels* (1726), 'the most unteachable of all Animals [...] insolent, abject, and cruel'.[3] Or apes as men, as potential rivals and substitutes: in chapter sixteen of *Candide* (1759), the naïve traveller in South America

---

1 Charles Darwin, *The Descent of Man and Selection in Relation to Sex*, (London: Murray, 1909), p.4.

2 See Paulo Medeiros, 'Simian Narratives at the Intersection of Science and Literature', *Modern Language Studies*, 23, 2 (1993), 59-73. Steve Jones reports the exhibition of an African pygmy in the same cage as an orang-utan at the Bronx zoo in 1906 (he was released after he started shooting arrows at mocking spectators). See *The Language of the Genes* (London: Flamingo, 1994), p.245.

3 Jonathan Swift, *A Selection of His Works*, ed. Philip Pinkins (Toronto: Macmillan, 1965), p.256.

shoots two apes intent on ravishing two naked girls, only to discover that the beasts were their accredited lovers, for monkeys 'sont des quarts d'homme' ('are a quarter men'). While their justification that 'le droit naturel nous enseigne à tuer notre prochain' ('natural law teaches us to kill our neighbour') exposes Candide and his companion to the wrath of the bereaved women's tribe, it is by confessing to having also murdered a Jesuit (a colonialist oppressor loathed by all) that they escape execution.[4] Gulliver too is the object of unwelcome sexual approaches from a female Yahoo (to the amusement of his equine masters, who see little difference between them). Species identity, gender relations, sexual permutations and 'fantasies of miscegenation' are recurrent themes in simian narratives, whether literary or anthropological, from the folkloric tradition of the *1001 Nights* (originating around the tenth century but popularised in western Europe by Galland's translation from 1704), through Flaubert's 'Quidquid volueris' (1837) and a film like *Max mon amour* (Oshima, 1986), to popularising studies of primates and primatologists by Desmond Morris and Carole Jahme.[5] And scientific or pseudo-scientific discourse, or a naturalistic setting in the case of fictional narratives, are often used as ideological vehicles for allegorical or satirical reflexions about human evolution.

If apes and monkeys are used to challenge humans' primacy among primates, this realignment of species boundaries does not necessarily threaten the natural order: this is the more radical function of monsters. Evelleen Richards asserts that monsters disrupt 'the boundaries of human identity. Typically, they denote physically or morally deviant states of nature'.[6] However, Darwin himself notes that the central notion of evolution, that 'all the members of the vertebrate kingdom are derived from some fishlike animal,' is likely to seem monstrously implausible to sceptics.[7] Elsewhere, he uses the term in a narrower, technical sense to refer to aberrant effects in reproduction. Thus: 'unnatural treatment of the embryo causes monstrosities; and monstrosities cannot be separated by any clear line of distinction from mere variations.'[8] While some of Darwin's

4 Voltaire, *Romans et contes* (Paris: Garnier-Flammarion, 1966), pp.212-13.

5 Medeiros, p.68; Desmond Morris, *The Naked Ape* (London: Jonathan Cape, 1967); Carole Jahme, *Beauty and the Beasts: Woman, Ape and Evolution* (London: Vintage, 2002). See also Jared Diamond's 1991 *The Rise and Fall of the Third Chimpanzee* (London: Vintage 2002) for a more searching and less sensationalist study.

6 Evelleen Richards, 'A Political Anatomy of Monsters, Hopeful and Otherwise: Teratogeny, Transcendentalism, and Evolutionary Theorizing', *Isis*, 85 (1994), 377-411 (p.377).

7 Darwin, p.244.

8 Charles Darwin, *On the Origin of Species by Means of Natural Selection, or the Preservation of Favoured Races in the Struggle for Life* (Harmondsworth: Penguin, 1968), p.72.

contemporaries (like Richard Owen) claimed that 'evolutionary change occurs by preordained monstrous deviations,' Darwin concluded that monstrosities were sudden, chance peculiarities, maintained artificially, unlike adaptive variations caused by external conditions; monstrous change therefore could not cause or explain species change.[9]

Whereas Darwin marginalised monsters (or sudden deviations, known as radical saltation) in favour of gradual, steady evolution, fictional monsters continue to thrive and engender a vast critical literature. Perhaps this is because in a broader cultural sense, Darwin's revolution not only demoted the human species but also promoted biology as 'the most authoritative medium for thinking about the human condition': imagining what biology seems to promise or legitimise thus becomes a kind of existential enterprise.[10] My aim here is to sample and compare a select corpus of fictional texts in French and English ranging from the late nineteenth century to the present, which deal with monstrosity or the incursion into the human domain of threatening, rival primate species, in the context of human evolution, whether individual, social or biological.

Taking an extreme relativistic position, Marian Scholtmeijer affirms that 'in a post-Darwinian world, *all* stories are stories about apes told by other apes.'[11] This overlooks the fact that humans may well be animals, but animals are not humans and don't tell stories, except in fables written by humans. To put in another way, all stories about apes or monstrous beings invite interpretation as being allegories about humans, or humans' changing relationship to the natural order; the most memorable or successful narratives, however, tend to be those that allow space for the literal and naturalistic, that are informed by close and subtle observation of a plausible social and psychic reality (or by an informed reading of anthropological and evolutionary texts). As John Glendening observes in his recent lucid account of *The Evolutionary Imagination in Late-Victorian Novels*, the 'universal acid' of Darwinism does not dissolve pre-existing binary distinctions (such as that between human and animal), but leaves them 'as traces and sources of

---

9 See Richards, pp.399, 408-10. In his recent book *Mutants* (London: Harper Perennial, 2005), Armand Marie Leroi discusses the 'querelle des monstres', the debate among eighteenth-century French anatomists as to whether deformity was due to divine design or accident, while pointing out that genetically most 'monstrous' people have mutations, that is deficiencies in particular genes caused when DNA is copied or repaired.

10 Janet Browne, 'Constructing Darwinism in Literary Culture', in *Unmapped Countries: Biological Visions in Nineteenth-Century Literature and Culture*, ed. by Anne-Julia Zwierlein (London: Anthem, 2005), pp.55-69 (p.69).

11 Cited by Susan Bridget McHugh, 'Horses in Blackface: Visualizing Races as Species Difference in *Planet of the Apes*', *South Atlantic Review*, 65.2 (2000), 40-72 (p.40).

confusion'.[12] In their attempt to come to terms with Darwin's disruption of a man-centred universe, novelists work with five oppositions: human/animal; modern/primitive; male/female; progress/degeneration; nature/culture. In the context of the monstrous, one could add two more: similarity/difference; self/other.

In his lengthy treatise on the monster in western art, Gilbert Lascault argues that the monstrous was perceived as fundamentally alien by classical and early modern authors. Thus 'Rabelais liait le monstre à une anti-nature, à une production enfreignant les lois d'un cosmos organisé' ('linked the monster to an anti-nature, a product breaking the laws of an organised cosmos').[13] Since we now are uncertain even about what 95% of the cosmos is made of, our perception of nature's laws and boundaries is somewhat more fluid and provisional. What seems unnatural or supernatural may simply reflect our limited understanding of the diversity and emergence of different species. The man-made monsters sampled here offend human ethics because of the way they are created and treated; the simian societies which supplant humanity show how evolution could have followed an alternative path. The remainder of this discussion will develop these points by examining and comparing specific literary texts.

Victor Hugo's heroic clown Gwynplaine in *L'Homme qui rit* (1869) has been deliberately disfigured as a child to turn him into a viable fairground freak, in a cruel act of 'chirurgie à rebours' ('reverse surgery'), as part of the commercial 'production artificielle de cas tératologiques' ('artificial production of teratological cases'); morally, this is 'un essai de ramener l'homme au singe' ('an attempt to reduce man to monkey').[14] In an impassioned speech to the House of Lords, he proclaims that his deformity is analogous to that inflicted on all humankind by civilisation: 'L'homme est un mutilé. Ce qu'on m'a fait, on l'a fait au genre humain' ('Man is a mutilated being. What was done to me, has been done to the human race').[15] While Hugo extends his melodramatic plot contrivances and metaphysical bombast over hundreds of pages, Maupassant deals with a similar topic more tellingly in six pages. In 'La Mère aux monstres' (1883), the narrator is taken to see a local curiosity, a woman known as 'la Diable' ('the Devil'), for her main occupation is producing 'chaque année, volontairement, des enfants difformes, hideux, effrayants, des monstres enfin' ('deliberately, every year, children who are deformed, hideous, frightful, monsters in other

---

12 John Glendening, *The Evolutionary Imagination in Late-Victorian Novels* (Aldershot: Ashgate, 2007), p.17.

13 Gilbert Lascault, *Le Monstre dans l'art occidental* (Paris: Klincksieck, 1973), p.31.

14 *L'Homme qui rit*, 2 vols (Paris: Garnier-Flammarion, 1982), I, pp.69-71.

15 *L'Homme qui rit*, II, p.293.

words'). She is 'le vrai type de la paysanne robuste, demi-brute et demi-femme [...] pareille à une bête féroce' ('a typical case of a robust peasant, half brute and half woman [...] akin to a wild beast'). Having deformed her first child by attempting to conceal an unwanted pregnancy with a corset, she then does a deal with freak-show organisers to produce a whole series of deformed infants, 'pour se faire des rentes comme une bourgeoise' ('to get a regular income like a bourgeois woman'). 'Elle en eut de longs et de courts, les uns pareils à des crabes, les autres semblables à des lézards' ('she had long ones and short ones, some like crabs, others like lizards').[16] The narrator and his companion hear her latest creation mewling pathetically in another room before the woman throws them out of her house. In a curious sententious postscript, he warns his readers that middle-class women whose vanity makes them wear corsets while pregnant are also likely to give birth to monsters.

Whereas Hugo and other writers like Jules Vallès depict the monster as a social outcast and victim of human malignity, Maupassant, despite his references to peasant venality and bourgeois vanity, and to decidedly unnatural treatment of the foetus, ultimately sees nature as women's true enemy, 'car elle nous ramène sans cesse à l'animal' ('for nature constantly reduces us to the animal'). Maternity is a repugnant bestial condition. God is imagined as 'un monstrueux organe créateur inconnu de nous' ('a monstrous creative organ unknown to us') mindlessly spawning worlds like a fish ('L'Inutile Beauté').[17] In 'Le Docteur Héraclius Gloss' (written in 1875), a more burlesque tale, which he left unpublished, Maupassant portrays the misadventures of an eccentric savant whose efforts to understand the chaotic diversity of creation lead him to become obsessed by metempsychosis, the belief that souls undergo a series of transmigrations as they are reincarnated in lower beings to atone for sins committed in a previous existence as a human. Dr Gloss himself is a strange hybrid: 'Il y avait en lui du rat, de la fouine et du basset' ('He was part rat, part marten and part basset hound'; he also resembles a paper knife). As 'un homme privé de la parole' ('a man deprived of speech'), the ape is the highest-ranking beast below man.[18] Héraclius acquires an ape from a fairground barker, mistakes it for his double, and lets it become master of his household. Metempsychosis leads to psychosis: he is interned in an asylum, where his quarrels with a rival cause him to abandon his beliefs. Returning home, he kills the ape and massacres all the other creatures to which he had given refuge; as a consequence, he is sent back to the asylum.

---

16 Guy de Maupassant, *Contes et nouvelles* (Paris: Gallimard, 1974), I, pp.842-46.

17 *Contes et nouvelles*, 2 vols (Paris: Gallimard, 1979), II, pp.1216-17.

18 *Contes et nouvelles*, I, pp.10, 21.

While these tales show the predictably disastrous effects of either degrading humans or humanising animals, Maupassant's better-known story 'Le Horla' (final version 1887) is about the evolution of a monstrous being which has less to do with human immorality and folly. Admittedly, the obsessive narrator can be seen as callous and insane: believing he has been possessed by an invisible, vampirical being, which is draining his vital forces while he sleeps, 'sa bouche sur la mienne, [buvant] ma vie entre mes lèvres' ('its mouth on mine, draining my life between my lips'), he is driven to burn down his house in a futile attempt to destroy his enemy. His unfortunate servants perish in the inferno, and the narrator is left contemplating suicide. A great deal of textual evidence suggests however that he is not suffering from a delusion: the Horla shows physical traces of its presence, infects other people, appears to originate in Brazil, and it is predicted that this 'rôdeur d'une race surnaturelle' ('harbinger of a supernatural race') will enslave mankind.[19] What makes it truly monstrous is its unknowable quality, its failure to belong to any identifiable species or to conform to any recognisable law.[20]

Maupassant's grasping, brutish peasant woman in 'La Mère aux monstres' belongs to that category of bestialised humanity most frequently associated with Zola's degenerate Rougon-Macquart clan, whose members are constantly assailed by the 'griffes du monstre héréditaire' ('claws of the hereditary monster'), what David Baguley calls 'the elemental self itself, reaching back into the distant reaches of the supposed origins of the human species and out into those marginal states in which human identity breaks down'.[21] Robert Louis Stevenson's *Strange Case of Dr Jekyll and Mr Hyde* (1886) embodies this elemental self in the antipathetic Edward Hyde, a figure overloaded with undesirable attributes. He is a violent, satanic midget, more troglodytic or ape-like than modern humans, an atavistic incarnation of inexplicable evil. Only in the penultimate chapter is it revealed that Hyde is actually Jekyll's double and alter ego, and that he has usurped Jekyll's identity and body as the transformation becomes irreversible.

Stevenson, Zola, Maupassant and Hugo all present variations of the self dispossessed by some monstrous other, whether through physical deformity caused by a malicious external agent, or more subtly through some internal

---

19 *Contes et nouvelles*, II, pp.919, 930.

20 The Horla thus conforms to Darwin's definition of a parasite as 'An animal or plant living upon or in, and at the expense of, another organism' (*Origin*, p.473), as indeed do most vampires.

21 Émile Zola, *Œuvres complètes*, 6: *Le Docteur Pascal* (Paris: Cercle du livre précieux, 1967), p.1254; David Baguley, *Naturalist Fiction: The Entropic Vision* (Cambridge: Cambridge University Press, 1990), p.215.

disturbance of ontological equilibrium. This unleashes creatures that devour their human host, but at the cost of their own extinction. H.G. Wells and Octave Mirbeau, on the other hand, adopt the perspective of the external observer who poses as a scientific witness of attempts to modify human and animal evolution. Such observers may avoid personally enduring dehumanising physiological catastrophes, but they too ultimately suffer social and spiritual estrangement. Wells's *The Island of Doctor Moreau* (1896) is narrated by Edward Prendick, a somewhat effete voyager who is shipwrecked somewhere in the Eastern Pacific and reluctantly rescued by Dr Moreau, his assistant Montgomery and their strangely deformed islander servants. Montgomery is 'an outcast from civilisation' after an unexplained misdemeanour in London, while Moreau has been barred from working in England after practising vivisection.[22] As he encounters the uncanny inhabitants of the remote island on which Moreau has established his 'biological station,' 'a kind of Bluebeard's Chamber' (pp.84, 87), Prendick becomes aware of their diabolical, misshapen and bestial features, their furtive aggressiveness (which he contrasts with the frank curiosity normally shown by uncultured savages). When Moreau sets about torturing a puma ('It was as if all the pain in the world had found a voice,' p.94), Prendick flees to the forest, fearing he is likely to be the next victim. Moreau reveals however that his larger creatures are 'humanised animals – triumphs of vivisection' (p.132); he calls them 'show-monsters' and cites Hugo's novel as a source of inspiration. Inflicting pain is meant to induce rationality in his subjects. The reader who wonders how these creatures acquired speech and brains able to support language has to make do with Moreau's glib explanation that, 'The mental structure is even less determinate than the bodily' (p.136). In any case, the creatures always revert to animality (p.147), although prohibitions implanted in their minds prevent them from attacking Moreau and vegetarianism keeps them from tasting blood. Their offspring usually die and do not inherit acquired human characteristics. Moreau's 'painful disorder' (p.168) collapses when the meat-eating Leopard Man defies him and the puma escapes. As the Beasts revolt and abandon the laws which make them human simulacra, Moreau and Montgomery are killed, leaving Prendick once again as the sole survivor. On his return to civilisation, he sees all humans as versions of the Beast People (like Swift's Gulliver).

---

22 H.G. Wells, *The Island of Doctor Moreau*, ed. by Leon Stover (Jefferson: McFarland, 1996), p.75.

The Island of Doctor Moreau is categorically different from the preceding texts discussed in that it is explicitly about evolution, the transformation of one species into another, with pseudo-scientific discourse replacing the supernatural agency invoked by Stevenson or Maupassant. Glendening calls it 'a hazy evolutionary allegory,' arguing that 'Moreau's project is Lamarckian in his effort not only to bring about rapid evolution, but to create evolutionary order in place of Darwinian randomness'.[23] The problem with this view is that Moreau's humanised beasts do not, as Lamarckian evolutionary theory would require, pass on their acquired characteristics, whether physiological or cultural, to their offspring. How could they? – since their transformation is caused not by genetic mutation through breeding, but by Moreau's surgical interventions, which seem mainly to involve sculpting adult animals into human body forms. Their rapid regression also shows how poorly adapted they are to their immediate environment. The reader is tempted to conclude, like Prendick, that this is a monstrous failure and that Moreau's 'sadistic assertion of semi-divine agency' is insanely hubristic. [24] (Both 1895's *The Time Machine* and 1901's *The First Men in the Moon* offer a more plausible version of evolutionary change and its impact on advanced cultures.) Leon Stover argues, however, that Wells is more sympathetic towards Moreau than more squeamish readers might suppose. As the concluding echo of Swift's Yahoos suggests, Wells's preoccupation 'is not the humanizing of animals but the deanimalizing of man,' and 'Directed evolution [...] is the essence of Wells's religion of progress,' for 'social man is a manufactured product of which the natural man is the raw nucleus'. [25] Hence Wells's view that 'In the utopian modern state, social surgery replaces nature's law of wanton murder'.[26]

Wells's totalitarian, eugenic fantasising probably is more disturbing for twenty-first century readers than for his contemporaries, who were spared its historical manifestations in the twentieth century. His dreams of 'social surgery' create a convenient link with Octave Mirbeau, whose novel *Le Jardin des supplices* (1899) offers some striking (and rarely noticed) parallels with *The Island of Doctor Moreau*. Both texts are about voyages to exotic locales recounted by first-person narrators supposedly engaged in scientific

---

23 Glendening, pp.49, 50. Peter Morton argues in *The Vital Science: Biology and the Literary Imagination, 1860-1900* (London: Allen & Unwin, 1984), that Moreau's beasts are 'transformed surgically into travesties of men' and that their humanity derives from their rudimentary culture and legal-theological system; this is an anthropological rather than a biological fable (p.218).

24 Glendening, p.58.

25 Stover, *Dr Moreau*, pp.14, 12.

26 Wells, cited in *Dr Moreau*, pp.21, 22.

projects, who witness scenes of horror, presented in graphic detail (bodily mutilation is described with unpleasant relish, particularly in Mirbeau's case). Both invite interpretation as allegories about the clash of nature and civilisation. The prologue of *Le Jardin des supplices* involves a discussion about nature, human nature and society conducted by a group of male intellectuals, who claim that murder is an 'instinct vital' akin to the 'instinct génésique' ('reproductive instinct'), that 'le besoin de tuer naît chez l'homme avec le besoin de manger, et se confond avec lui' ('the urge to kill is born in man with the urge to eat, and they are indistinguishable').[27] The urge to kill is the source of human actions and the basis of society, which uses its laws somewhat arbitrarily to contain the homicidal impulse within acceptable practices such as violent pastimes. Mirbeau treats his speakers with a certain derision, so we are not necessarily meant to lend credence to their crude social Darwinism, with its caricatural emphasis on conflict, injustice and hypocrisy and omission of more constructive instincts like cooperation and nurturing.[28] The ravaged man who will narrate his voyage to the Chinese torture garden suggests that woman, when released from sentimentalised social conventions, can exemplify nature in its most destructive form, and presents his narrative as an illustration of this thesis.

Unlike Wells, Mirbeau is an anti-authoritarian who views those who aspire to direct and educate mankind with contempt and suspicion, although his humanitarian posture is undercut by his blatant delight in spectacles of suffering and torture inflicted by those he is ostensibly attacking. The narrator's purported mission is to seek 'la cellule primordiale' ('primordial cell') in the Far East, although he is a shabby charlatan with no scientific training. In fact Darwinism and embryology are dismissed as 'une immense blague' ('a great joke') by the narrator and his sponsor.[29] A British embryologist, Sir Oscar Terwick (Mirbeau following Voltaire, Hugo and Maupassant in giving foreigners absurd names to emphasise their ludicrous pretensions), encountered in Ceylon, talks only puerile nonsense. In a later novel, *Dingo* (1913), another British naturalist, Sir Edward Herpett (author of *La Dentition des grands singes*), summarises Darwinian evolution as follows: 'Rien n'est immuable. Tel qui était poisson jadis est devenu oiseau; tel qui fut singe est aujourd'hui pape, roi, ministre, général ou philosophe'

27 Octave Mirbeau, *Le Jardin des supplices*, ed. by Michel Delon (Paris: Gallimard, Folio, 1988), pp.44, 52.

28 For further discussion, see Louise Lyle, 'Charles Darwin dans *Le Jardin des supplices*', *Cahiers Octave Mirbeau*, 14 (2007), 47-64, and Claire Margat, 'Sade avec Darwin. A propos du roman d'Octave Mirbeau *Le Jardin des supplices* (1899)', *Analyse Freudienne Presse*, 6 (2003), 47-64.

29 *Le Jardin des supplices*, pp.97, 98.

('Nothing is immutable. What was once a fish has become a bird; what was once an ape is now a pope, king, minister, general or philosopher'). Short-circuited in this fashion, the mutability of species is reduced to an interchangeable process.[30] One of the most disturbing scenes in *Le Jardin des supplices* is the narrator's encounter with an obese, verbose *bourreau* (torturer/executioner), who like Dr Moreau takes a surgeon's aesthetic delight in his craft, although the narrator considers him an obscene monster. Thus the torturer boasts of his prowess in working human flesh, carving a woman from a man and vice versa, while complaining that his art is being destroyed by colonialists, bureaucrats and technological progress.[31]

Although Mirbeau treats scientific pontificating and investigation with derision, *Le Jardin des supplices* can still be seen, like *Dr Moreau*, as a somewhat confused allegory about nature and culture. It is confused, in the sense that the more literal interpretation provided by the narrator (i.e. rejection of what he witnesses as monstrously inhumane) is not necessarily supported by the narrative economy, which exposes the gaps and failings in conventional morality. Playing on the duality of similarity and difference, Michel Delon suggests that Mirbeau makes China both the antithesis of and a metaphor for Europe: its exotic, erotic atrocities make it ripe for the civilising intrusion of Western powers, but these excesses are mirrored in the violence inherent in all societies.[32] Hence Clara's assertion to the narrator that 'Ce que tu appelles des monstres ce sont des formes supérieures ou en dehors, simplement, de ta conception' ('what you call monsters are superiors forms or simply beyond your conception'); monsters live 'au-dessus des mensonges sociaux, dans la divine et resplendissante immoralité des choses' ('above social lies, in the divine and resplendent immorality of things').[33] Her eventual breakdown demonstrates, however, that she is no more capable than the narrator of escaping the shackles of human guilt and sinfulness. On a more mundane, observational level, Mirbeau turns the naturalists' obsession with the beast in man into a teratological enterprise, endlessly cataloguing the physical and behavioural deformities of his human specimens in this and other texts.

---

30 Octave Mirbeau, *Dingo* (Paris: Éditions de Maule, 1987), p.19. Yet this parodic version is actually quite close to Leroi's account of genetic evolution: 'the homeotic genes were merely the first indication that many of the molecular devices that make our bodies are ancient. Over the last ten years it has become plain that we are, in many ways, merely worms writ large' (p.99).

31 *Le Jardin des supplices*, pp.203-06.

32 Ibid., p.17.

33 Ibid., p.225.

Wells's and Mirbeau's voyagers return to human societies whose supposed superiority over more primitive, natural cultures and repression of the beast have been exposed as a pretence. A more radical solution is to annihilate human civilisation entirely. In conclusion, I want to discuss two twentieth-century novels which rewrite the whole course of primate evolution. Pierre Boulle (1912-94) is rarely recognised as the original creator of the *Planet of the Apes* film franchise, with his philosophical novel *La Planète des singes* (1963). He also wrote *Le Pont de la rivière Kwaï* (1952), often perceived as an archetypal British war film thanks to David's Lean's 1957 adaptation, although the original novel is evidently an allegory of collaboration. Lawrence M. Porter has noted that the Japanese are often compared to apes in Boulle's novel, despite their dominance and humiliation of their captives.[34] Inspired perhaps by his own experience of the Resistance in the Second World War and by the threat of mass extermination posed by the cold war, Boulle transposes the wartime narrative of *Kwaï* into science fiction in *La Planète des singes.*

A narrative found in a bottle by two space vacationers recounts the voyage of three human astronauts and a chimp to a planet which they name Soror, in the vicinity of Betelgeuse. This is 425 light years distant from Earth, so the fact their journey lasts only three years suggests a certain disregard for the practicality of space travel (or alternatively, definitive proof of Einstein's special theory of relativity); in the film versions, the astronauts arrive at an alternative version of Earth, where everyone still conveniently speaks American English, irrespective of species. On landing, they are captured by a band of naked humans, who murder the chimp and destroy all their manufactured objects, including their shuttle craft. These creatures do not have language, see smiling as threatening, are unable to manipulate objects, and live in temporary nests like great apes. The band is hunted down and massacred by a force of gorillas and chimps, who possess language, fire arms, clothes, and motor-driven vehicles. The narrator, symbolically named Ulysse, finds himself a prisoner with the other survivors in a medical research centre where he begins to regress to the despondent, animal behaviour of his fellow captives. Finally, he is able to establish communication by demonstrating his understanding of geometry to the female chimpanzee scientist Zira, to learn the apes' language, achieve his release, become a celebrity and escape from the planet. But on returning to Earth (seven hundred years after his departure, thanks to the effect of

34 Lawrence M. Porter, 'Text of Anxiety, Text of Desire: Boulle's *Planète des singes* as Popular Culture', *French Review*, 68 (1995), 704-14.

reaching the speed of light), he discovers that apes have taken over there too; even the readers of his story prove to be chimps.

La Planète des singes clearly belongs to the long tradition of philosophical tales where a naïve traveller is used to offer a contrast between a familiar and an exotic society and to subvert the reader's assumptions about hierarchies and norms. At first sight, Boulle seems just to have reversed humans and simians: apes are smart, humans are dumb brutes, much like Yahoos (only sexier: Ulysse acquires a beautiful mate and child). This raises certain objections, if one expects a reasonable degree of plausibility and coherence within this parallel world. Boulle shows very little knowledge about the actual social behaviour of apes and monkeys: most of the time, his apes behave like humans, and his humans like mindless automata. For instance, he dodges the evolutionary conundrum of why and when humans lost their bodily hair (why do humans go naked, and apes need to wear clothes?). Boulle has none of the savage wit of Swift, Voltaire (or Will Self) and apparently little of their desire to satirise the ideological absurdities of his own society through its counterfactual equivalent.

Nevertheless, Boulle's novel does slyly subvert certain preconceptions about human superiority on the evolutionary scale. Humanity (i.e. civilised and rational behaviour) is shown to be a fragile construction; after a few months' confinement in a zoo, Ulysse's companion Professor Antelle has regressed to a state of speechless abjection. The chimps' explanation of their evolutionary superiority offers a mocking counter-example to human conceit:

> Le fait que nous soyons quadrumanes est un des facteurs les plus importants de notre évolution spirituelle. Cela nous a servi d'abord à nous élever dans les arbres, à concevoir ainsi les trois dimensions de l'espace.
>
> (The fact we are quadrumanes is one of the most important factors in our spiritual evolution. This allowed us initially to raise ourselves into the trees, and thus to conceive the three dimensions of space.)[35]

This is followed up by a satirical account of the *Bourse* (stock exchange), where the agitated dealers swarm over the walls and ceiling, as though in a three-dimensional madhouse. Boulle ignores (or is unaware of) the anatomical differences between humans and other primates regarding the capacity for speech, that is that apes are physiologically incapable of producing articulated sounds resembling human speech. His apes speak because they have acquired the will to do so. In a surprising plot twist, it turns out that humans have lost their former ability to speak and that apes may have inherited or replaced a lost human civilisation. However,

---

[35] Pierre Boulle, *La Planète des singes*, in *Étrange planète* (Paris: Omnibus, 1998), p.287.

experiments on the brains of human subjects reveal that not only can they recover speech but indeed the whole collective memory of their past dominance, thousands of years before. At this point, Ulysse ceases to be a freakish novelty and becomes a revolutionary threat to simian society.

Boulle's novel thus is as much a fable about decadence, the collapse of civilisation, as about inter-species or interracial rivalry (the three species of dominant apes in his tale enjoy an uneasy alliance).[36] Will Self's *Great Apes* (1997) can be seen as a splendidly inventive sequel to Boulle, who is however an unacknowledged ancestor (although Self references *The Planet of the Apes* films, as *Planet of the Humans*, the source novel is forgotten). Self retains Boulle's human-chimp inversion, while dispensing with the summary trappings of science fiction.[37] Waking from a 'night of routine debauchery' (cover blurb), the artist Simon Dykes discovers that his girlfriend Sarah and the rest of humanity have become chimpanzees. So has he, although for much of the novel he is reluctant to accept the fact that as far as everyone else is concerned he and they have always been chimpanzees. What makes the book a delight is the naturalistic skill with which Self recreates a chimpanzee society in contemporary England, where chimpanzees have language, culture and intelligence but continue otherwise to behave like real apes. Self's chimps communicate partly through phatic vocalisations, but mainly through sign language, a form of visual and tactile communication which is cleverly related to their social behaviour. In fact Self creates a great deal of provocative comedy by emphasising the significant differences between chimp and human behaviour, particularly concerning family and sexual relations. More than mischievous satire of contemporary human social mores and politically correct taboos (such as child abuse and incest) is involved, since humanity as a dominant species ultimately is shown to have become redundant.

Self's amalgamation of human and simian characteristics in his evolved chimps does seem a monstrous aberration to his deluded protagonist, whereas for other characters and the reader this redrawing of boundaries is entirely plausible. This sample of texts inspired by post-Darwinian understanding of and anxiety about human evolution thus shows a continuing shift from the reductive to the relativistic model: humans are neither unique nor bestial, but therefore might readily be supplanted by rival species. These literary works take the facts and consequent ideological implications of evolution as unavoidable and incontestable; what they offer us is speculation embodied in imagined characters and settings rather than

36 For a fuller discussion, see McHugh.

37 It is worth noting that several late and also forgotten Victorian writers had already used this 'satirical trope': see Browne.

scientific or philosophical enquiry, although Hugo and all the fin-de-siècle authors cited here also wrote extensive essays on literature, science and human nature. How then might one sum up their contribution to the evolution of literature, in an aesthetic or formal sense?

In 1891, the journalist Jules Huret published his *Enquête sur l'évolution littéraire*. Based on interviews with 64 leading French writers of the day, this study offers a taxonomy of the literary scene and reveals as much about writers' rivalries and struggle to survive as their literary theories. Zola's rivals and erstwhile disciples are eager to proclaim the imminent extinction of naturalism, 'surtout parce qu'[il] a été l'expression de tout un mouvement positiviste et matérialiste qui ne répond plus aux besoins actuels' ('especially because it was the expression of a whole positivist, materialist movement that no longer responds to present needs'), as the Swiss novelist Édouard Rod put it.[38] If however naturalism is taken to be a mode of writing posited on a Darwinian worldview, which blends documentary realism with imaginary speculation, which situates human nature and the subjective in an objectively realised natural world, then the proclamation has clearly proved to be premature, since all the authors discussed here practise it in varying fashions. Hence Joseph Carroll's assertion that Maupassant exemplifies a 'biotic orientation,' what Henry James disapprovingly called 'the physiological explanation of things'.[39] I would add that this does not preclude Maupassant from inventing unknown beings like the Horla or from investing their existence with a form of spiritual anxiety.

By way of a postscript, in conclusion it is interesting to consider what effect developments in molecular genetics in the past half century have had on fiction. Does the shift from the anatomical to the biochemical, from tracing evidence of natural selection and extinction in the fossil record, to genetic engineering of improved or new species, merely extend the apocalyptic possibilities for imagining the erosion of human singularity and mass annihilation? In practice, the most imaginative texts often combine knowledge both of scientific developments and a literary tradition which blends fable and adventure story (thus Michael Crichton rewrites Arthur Conan Doyle in *The Lost World* (1995), while Paul McCauley updates Mary Shelley, Wells and Conrad in *White Devils* (2004), re-imagining the heart of darkness of a mid-twenty-first century Africa surrendered to plague, corporate greed and biotechnology). The enduring vitality and popularity of such texts demonstrate what Joseph Carroll calls the 'comprehensive force of the Darwinian worldview,' our continuing need to re-situate ourselves

---

[38] Jules Huret, *Enquête sur l'évolution littéraire* (Paris: Charpentier, 1891), p.14.

[39] Quoted by Joseph Carroll, *Evolution and Literary Theory* (Columbia: University of Missouri Press, 1995), p.259.

through imagined narratives in a natural world and environment over which our control seems increasingly fragile and illusory.[40]

[40] Joseph Carroll, *Literary Darwinism: Evolution, Human Nature, and Literature* (New York: Routledge, 2004), p.151.

Zola and Darwin: A Reassessment

David Baguley[1]

Abstract: It is widely assumed that the French 'naturalist' novelist was largely indebted to Darwin. But if we take into account the evidence of his early novels, criticism and sources in the context of the resistance to Darwin's ideas and the prestige of Lamarckism in France at the time, this assumption must be seriously questioned. However, the later novels of the *Rougon-Macquart* series discreetly embody a remarkable range of Social Darwinist themes and there is an even more discreet tribute to the English naturalist in Zola's utopian novel, *Travail* (1901).

The names of Britain's most influential naturalist and France's most famous so-called naturalist writer have long been almost inextricably linked, but, under close scrutiny, the match is less harmonious than is commonly supposed. To lend a certain credence to the link, critics have frequently invoked a somewhat startling outburst from the young French novelist, supposedly directed to the Goncourt brothers in December 1868 and reported by one of Zola's first modern biographers, Armand Lanoux: 'Les caractères de nos personnages sont déterminés par les organes génitaux. C'est de Darwin! La littérature, c'est ça!' ('The actions of the characters we write about are determined by their genital organs. That's what Darwin says, and that's what literature is!').[2] Fortunately for Zola's reputation, it is highly unlikely that he ever made such a statement since there is no record of it in the brothers' *Journal*. Lanoux seems to have indulged here, as elsewhere, in the licence of the *biographie romancée*. More significant, however, is Lanoux's further claim that Zola read Darwin 'greedily' in Clémence Royer's translation of *The Origin of Species*.[3] Zola critics, including the most eminent amongst them and with only rare exceptions, seem to have taken it for granted that Zola read Darwin's works. [4] Yet there is no extant evidence to justify this assumption.

It could be argued, of course, that there is no proof that Zola did not read Darwin. But there is plenty of circumstantial evidence to suggest that this was the case, for there is not a single direct reference to Darwin in the eleven-volume edition of Zola's correspondence and his name only appears

---

1 The author wishes to thank the Master and Fellows of Fitzwilliam College, Cambridge, for their hospitality during the preparation of this study. Unless otherwise noted, translations of quotations in French are my own.

2 *Zola*, trans. by Mary Glasgow (London: Staples, 1955), p.87. From *Bonjour, Monsieur Zola* (Paris: Amiot-Dumont, 1954), p.99.

3 Lanoux, p.88.

4 Notably Roger Ripoll, *Réalité et mythe chez Zola* (Paris: Champion, 1981), pp.159-60, and Geoff Woollen, 'Le darwinisme chez Zola: réflexe ou réflexion?', *Cahiers de l'U.E.R. Froissart*, 5 (1980), 27-36.

three times in passing references in the whole of the twenty-volume series of the *Rougon-Macquart* novels and, so it seems, only once in the novelist's substantial body of criticism. Given Zola's propensity for flaunting his scientific sources – disastrously, of course, in the case of his essay on the 'experimental novel' – it is highly likely that, if he had had a thorough and direct acquaintance with Darwin's texts (as opposed to some familiarity with his ideas), he would unfailingly have made it known to his readers.

Zola's case seems to confirm Gillian Beer's observation that reading Darwin is a question with 'softer edges than might at first appear' and even that 'ideas pass more rapidly into the state of assumptions when they are *unread*'.[5] Indeed, such a hypothesis can liberate us from the peremptory exclusiveness and illusory completeness of direct influence studies. We are led into a more indeterminate, but, paradoxically, more convincing zone of interaction, into a plexus of analogies, of facile invocations or profounder visionary perspectives, situated in this case in an indeterminate zone between the vaguely defined scientific climate of French intellectual life in the 1860s and the supposed seismic effect of the publication of *The Origin of Species*.

This much looser linkage between Darwin and Zola can to some considerable degree be confirmed by the circumstances of the reception of the English naturalist's works in France, which was far from enthusiastic and even outright hostile in the scientific community. In her comprehensive study of this topic, *L'Introduction du darwinisme en France au XIXe siècle*, Yvette Conry seriously puts in doubt the widespread assumption that, by the end of the nineteenth century, Darwinism had been thoroughly assimilated in France by studying the degree of compatibility between Darwin's ideas and the prevailing scientific discourses in France at the time. [6] She defines four phases. Initially, between the original date of the publication of *The Origin of Species* (1859) and the appearance of Clémence Royer's translation (1862), there was a period of indifference, of 'scorn' and, one suspects, of ignorance. Between 1862 and 1878 ensued a period of polemical resistance to Darwin's ideas amongst zoologists, botanists, geologists, palaeontologists and anthropologists alike, prompted by national pride and entrenched prejudice but fostered also by misunderstandings attributable to the problematical nature of the translation itself.[7] Clémence Royer, a sort of social Darwinist *avant la lettre*, used the occasion to air her trenchant

5 *Darwin's Plots: Evolutionary Narrative in Darwin, George Eliot and Nineteenth-Century Fiction* (London: Routledge & Kegan Paul, 1983), p.6.

6 Yvette Conry, *L'introduction du darwinisme en France au XIXe siècle* (Paris: Vrin, 1974).

7 Notably Pierre-Marie Flourens's outbursts in his *Examen du livre de M. Darwin sur l'origine des espèces* (Paris: Garnier frères, 1864; 2nd edn 1881). See Conry, p.30.

anticlerical, atheistic, feminist, and laissez-faire economic views in a controversial preface that she added to her translation. But there were also, along with misleading footnotes, serious errors of translation, notably 'concurrence vitale' for 'struggle for existence,' which pleased the liberal economists, and 'élection naturelle' for 'natural selection,' which implied an intentionality in nature alien to Darwin's views. A third phase, in Conry's scheme, was a period of 'consideration,' between 1878 and 1890, which witnessed more widespread discussion of Darwin's theories in academic works, but, significantly also, led to a resurgence of Lamarckism to gauge and counteract the foreign influence. Lamarck's belief in the existence of an immanent perfecting principle in the living being, subject to environmental modification, was more attuned to the prevailing French Enlightenment tradition than Darwin's emphasis on natural selection, just as the weight that the French theorist attached to the doctrine of inherited characteristics recommended itself to future naturalist novelists such as Zola, as we shall see, in providing them with a convenient, scientifically based theory of motivation to explain the actions of their characters. Finally, between 1890 and 1900, there was a period of so-called popularisation, not, however, in scientific circles, where there was only a grudging acceptance, but mainly in the political, sociological and economic fields in the guise of 'Social Darwinism'. According to certain commentators, this resistance to Darwin has even continued to more recent times. Little wonder then that, with the scientific establishment of mid-century France so resolutely conservative and hostile to Darwin's 'transformism,' Zola had no choice, as Pierre-Henri Guyon observes, but to incorporate Lamarckian principles of heredity into his novels.[8]

Furthermore, Conry makes the valid point that there are different modes and degrees in which a foreign doctrine is introduced into another national culture, ranging from, at one extreme, a limited level of integration through mere invocation ('interpolation') to, at the other extreme, a thorough assimilation, which involves a conceptual and theoretical reorganisation.[9] Clearly, in Zola's case, in the few examples available, direct references to Darwin belong to the former strain whereby the *name* of Darwin is called upon as a rhetorical invocation of the authority of the great naturalist to produce a kind of 'effect of scientificness,' as Barthes might have put it. More telling, however, is the absence of any reference to the English naturalist amongst the scientific sources of Zola's novels during the 1860s,

---

8 See 'François Jacob: éloge du darwinisme', *Magazine littéraire*, 374 (1999), 18-22 (p.22). See also *Le Nouvel Observateur* (20 December 2007-2 January 2008, 42-45) on the political dimension of the French resistance to Darwin.

9 Conry, p.22.

when, inspired by the example of Balzac and by the ideas of Taine and of Émile Deschanel, he first became convinced that the novel should have a scientific purpose. Soon after the publication of his first comprehensively naturalist novel, *Thérèse Raquin* (1867), Zola embarked upon the preliminary scientific research for his *Rougon-Macquart* series of (eventually) twenty novels. Perpetuating the myth, Maurice Larkin, for example, claims that 'the project revealed the influence of Darwin as well as Taine, Zola having read *On the Origin of Species* just three years before the appearance of the first volume of the series, *La Fortune de Rougon* (1871).'[10] Fortunately, we have the complete set of the novelist's preparatory notes and plans for the series, in which Darwin is conspicuous by his absence.[11] In fact, the contents of Zola's reading list of eighteen scientific texts on physiology are all pre-Darwinian and, since, unusually, he left no notes on these works, he may well not have read them at all. There are, however, detailed notes on four other texts that provided the novelist with the familiar traits of madness, hysteria and atavistic criminality that run through the genealogy of his famous fictional family. From these works, particularly Charles Letourneau's *Physiologie des passions* (1868) and Prosper Lucas's *Traité philosophique et physiologique de l'hérédité* (1847 and 1850), and from others as the series progressed, Zola acquired a wealth of anecdotal details suitable for dramatic fictionalisation, a confirmation of his scientific ideology and physiological determinism, the organisational principles of his fictional world and a credible basis, mainly through the laws of heredity, for the mythology of fatality, atavism and decadence that served his fictional purposes. In any case, Zola was quite cavalier in his use of his scientific sources. As he frankly and revealingly admitted, for instance, in the same preparatory notes for the series: 'Prendre avant tout une tendance philosophique, non pour l'étaler, mais pour donner une unité à mes livres. La meilleure serait peut-être le matérialisme, je veux dire la croyance en des forces sur lesquelles je n'aurai jamais le besoin de m'expliquer. Le mot force ne compromet pas' ('Above all, I should take a philosophical tendency, not to flaunt it, but to give unity to my books. The best will perhaps be materialism, that is to say the belief in forces about which I shall never need to provide an explanation. The word

---

10 Maurice Larkin, *Man and Society in Nineteenth-Century Realism. Determinism and Literature* (London: Macmillan, 1977), p.129.

11 Reproduced in *Les Rougon-Macquart*, 5 vols, ed. by Henri Mitterand (Paris: Fasquelle et Gallimard, 1967), v,1674-6. Unless otherwise stated, references to and quotations from the *Rougon-Macquart*, as well as quotations from Zola's working notes, are taken from this 5-volume edition (1960-67); volume and page numbers are included in parentheses in the text.

force is noncommittal.')[12] Such statements suggest that Zola was ultimately less wedded to a belief in scientific truth, at least as far as his methodology is concerned, than his public pronouncements pretended and that he was principally intent on establishing a convincing system of 'motivation' and of intelligibility for his fiction, more attentive to verisimilitude than to veracity. Thus, whereas, in Darwin's case, the literariness of his texts served his scientific purpose, for Zola the science was resolutely in the service of his fiction.

Emphasising in this way the lack of direct influence and the separateness of their projects does not by any means preclude, in a broader frame of reference, the possibility of significant shared assumptions that allow us to relate Zola to the corpus of evolutionary concepts and naturalist propositions commonly referred to as 'Darwinism,' that is to a common temper, to a shared cluster of 'dangerous ideas' (for the age). The familiar image of the tree may usefully illustrate their differences and affinities: Zola's genealogical tree extending, of course, over a much narrower time span than Darwin's tree of life, but both revealing a preoccupation with ancestral relations, inherited characteristics, the conviction that life is a process of both common descent and transmutation, and the view, as Darwin wrote, that 'we may all be netted together.' There is, furthermore, at least in Zola's early naturalist works, a perceived 'Darwinian' tendency to shear life of any moral direction or purpose and to subject living creatures, including human beings, to their deterministic, biological impulses. Zola's (early) Rougon-Macquart characters are, in a sense, almost invariably reduced to their instinctive behaviour, to their 'appetites,' with the predaceous Rougons prevailing in the struggle for survival, the Mourets virtually condemned to extinction by their inherited 'taints' ('tares') and the Macquarts similarly undone by their biological dispositions. In the human ontology of Zola's naturalist fiction, there is no transcendence. The individual is powerless before the deterministic forces that regulate his behaviour, before the irrepressible drives of 'the human beast'. Heroism becomes an irrelevant concept and love is decidedly a de-sentimentalised affair. Man is displaced, if not, as in Darwin's work, from his central role in the universe, at least as the primary agent of his destiny. Zola's emphasis on the environment as a primary determining factor is no doubt more Lamarckian than Darwinian, as is the prominence that he gives to the transmission of directly acquired traits in his characters. But Alain de Lattre, for example, makes a strong case for the Darwinist analogy by arguing, with some justification, that Zola's (naturalist) novels are characterised by a

---

12 From the novelist's 'Notes générales sur la nature de l'œuvre', f° 12 ; V, p.1744.

recurrent pattern in which the integrity of the individual character is constantly undermined by instinctive, somatic forces and impulses that bring about its disintegration.[13] Thus, in the typically naturalist manner, Zola bases his theory of characters, who are composites of the traits of their parents and ancestors, on relationships rather than on essences, thereby stripping them of their identity and their substance and presenting them as a kind of meeting point of an interplay of forces and subjecting them to an otherness inscribed within the individual.[14] The process occurs, of course, within the confined parameters of the novel with all the artfulness of the dramatic plot, and it is invariably pathological, leading to a decidedly jaundiced literary version of the transformist model. Nevertheless, if we look a little more closely at Zola's fiction, further links with Darwin's thought emerge, albeit in a rather oblique way.

In the initial volume of the *Rougon-Macquart* series, *La Fortune des Rougon*, which is set in deepest Provence and which was originally to bear the Darwinian title *Les Origines*, there appears an intriguing character, Pascal Rougon, who, as a case of 'inneity' (in Lucas's terminology), is conveniently immune to the laws of heredity and bears none of the physical or moral characteristics of his family. A practising doctor, but also an amateur natural historian and the author of several learned submissions to the Académie des Sciences de Paris, he is frequently seen by the locals, unaware of his distant fame, with his botanist's samples box around his neck and his geologist's hammer in his hand, heading for the hills. This Darwinian figure, with his passion for comparative natural history, will play a far more significant role, as we shall see, in a later novel, but, already here, he represents a second self of the novelist, a scientific observer of the family and of the 'menagerie' of bizarre characters of this highly political novel. However, in Zola's preliminary notes for the series, there is a highly significant statement of the novelist's views: 'Je ne nie pas la grandeur de l'élan moderne, je ne nie pas que nous puissions aller plus ou moins à la liberté, à la justice. Seulement ma croyance est que les hommes seront toujours des hommes, des animaux bons ou mauvais selon les circonstances' (V, 1739) ('I do not deny the grandeur of life's dramatic advances, I do not deny that we can evolve more or less to a state of liberty and of justice. And yet my belief is that men will always be men, animals that are good or bad according to their circumstances'). In this somewhat paralogistic pronouncement, there is an evident incompatibility between Zola's ideology and his ontology, between

---

13 *Le Réalisme selon Zola: Archéologie d'une intelligence* (Paris: Presses universitaires de France, 1975), p.137.

14 See Muriel Louâpre, 'Une fiction non euclidienne. Vérité et modélisation en régime naturaliste', *Romantisme*, 138 (2007) 89-105 (p.99).

his belief in progress and his naturalism. Yet these seemingly inconsonant views do represent two fundamental directions of his works, which are often in a delicate balance in certain texts, notably in *Germinal*, as the ambiguous title suggests. They also explain the shift in emphasis and tone taken by the second half of the series, with *Au Bonheur des Dames* (1883) as the pivotal text, the novel which Zola called (in his notes) the 'poem of modern activity,' entailing 'a complete change of philosophy' on his part and the resolution to move with his times, to express his age, 'an age of action and conquest, of effort in all directions.'[15] He is no longer content to recount the plight of characters condemned to their biological fate. But how could he, henceforth, invest history with direction and purpose and remain faithful to his bio-centric vision? The solution was to embrace certain tenets of what would later be referred to as Social Darwinism, a solution which had the distinct advantage of allowing the novelist to combat, within the framework of his series of novels, the emergent strain of thoroughly pessimistic naturalist fiction inspired by Schopenhauer, of which he heartily disapproved. It is generally agreed that there are two fundamental models of naturalist texts: a dynamic 'Darwinian' type in which all is struggle and strife and which ends, like *L'Assommoir*, in depersonalised biological misfortune; a repetitive Schopenhauerian or Flaubertian model according to which all is grief and boredom and ends in futility and weary resignation. Zola emancipated himself from the former all the better to oppose the latter.

As is well known, the appropriation of Darwin's theories by conflicting ideologies, frequently with a Spencerian edge, either to justify competitive capitalism by reference to natural selection or to extend the struggle for life of species to the class conflict with a belief in progressive change, led to vigorous debate at this time. Zola's own modest but not insignificant contribution to these debates in his fiction has been duly noted.[16] Elsewhere, Gérard Molina summarizes academic studies of Social Darwinism as consisting of two distinct phases: firstly, a rationalisation that justified liberal individualism, capitalist competition and the triumph of the most able, followed by, around 1900, a collective construal in which competition involves classes, nations and races, endorsing colonialism, imperialism and eugenics.[17] Interestingly, Zola's own development seems to

---

15 See III, 1679.

16 See Linda L. Clark, *Social Darwinism in France* (Alabama: University of Alabama Press, 1984), pp.113-15 and Jean-Marc Bernardini, *Le Darwinisme social en France (1859-1918)* (Paris: CNRS, 1997), pp.204-08, as well as Geoff Woollen's article previously cited (note 4) and Robert J. Niess, 'Zola et le capitalisme: le darwinisme social', *Cahiers naturalistes* 54 (1980), 57-67.

17 'Darwin au piège des idéologies', *Le Magazine littéraire*, 374 (1999), 38-41 (p.40).

follow a similar scheme. Given the rarity of direct statements by the novelist and the fact that such views are usually attributed to characters in his fiction, caution is required, though the insights provided by the working notes and the fact that the novelist's later texts are unashamedly thesis novels usually leaves little doubt about his views.

A brief survey of some of the most significant examples from the *Rougon-Macquart* series clearly shows the nature and extent of the novelist's Social Darwinism. In *Au Bonheur des Dames* (1883), the ladies fashion store run by the entrepreneurial Octave Mouret progressively kills off the competition from the small businesses in the neighbourhood. In his notes the novelist is alarmingly flint-hearted on the theme: 'Mais je ne pleurerai pas sur eux, *au contraire*: car je veux montrer le triomphe de l'activité moderne; ils ne sont plus de leur temps, tant pis! ils sont écrasés par le colosse' (III, 1680). ('But I shan't shed any tears over them, *on the contrary*: for I want to show the triumph of modern activity; they are no longer of their age, too bad! They are crushed by the colossus.')[18] Such insouciance is offset in the novel by the tender-hearted heroine, Denise Baudu, though even she comes to accept the inexorable Darwinian 'law of struggle'. In *Germinal* (1885), the novel of industrial strife and class conflict set in 1866-67, Zola introduces a brief and blatantly anachronistic debate on Darwin's views between his autodidactic hero, Étienne Lantier, who has come across fragments of Darwin's works and sees them as a vindication of revolutionary action, and an even more anachronistic Russian anarchist, Souvarine, who cuts the discussion short by denouncing 'la bêtise des socialistes qui acceptent Darwin, cet apôtre de l'inégalité scientifique, dont la fameuse sélection n'était bonne que pour des philosophes aristocrates' (III, 1524) ('the stupidity of the socialists who accept Darwin, that apostle of scientific inequality, whose famous natural selection was only fit for aristocratic philosophers.'[19] But Étienne is sufficiently convinced of his own view for it to feed into the famous inner soliloquy of the final paragraphs of the novel predicting apocalyptic change:

> Darwin avait-il donc raison, le monde ne serait-il qu'une bataille, les forts mangeant les faibles, pour la beauté et la continuité de l'espèce ? [...] Mais une idée dissipa ses doutes, l'enchanta, celle de reprendre son explication ancienne de la théorie, la première fois qu'il parlerait. S'il fallait qu'une classe fût mangée, n'était-ce pas le peuple, vivace, neuf encore, qui mangerait la bourgeoisie épuisée de jouissance? (III, 1589).

---

18 Zola neatly sums up the theme and tone in his general plan: '*The struggle for life*. Gaie' (III, 1695).

19 *Germinal*, trans. by Peter Collier, ed. by Robert Lethbridge (Oxford: Oxford University Press, 1993), p.450.

> [Was Darwin right, then, was the world nothing but a battlefield where the strong ate the weak, for the beauty and the survival of the species? [...] But one idea in particular dissipated his doubts and enchanted him, that of using his old explanation of the theory, as soon as he could make his first speech. If one class had to go under, wouldn't the people, who were still fresh and vital, trample all over the bourgeoisie, who were debilitated by their endless pleasure-seeking.][20]

As for *L'Argent* (1891), Zola's novel on the stock exchange presents the view of the liberal economic order taking its ruthless course which is supposed to have inspired Darwin's notion of natural selection through his readings of Adam Smith and Malthus and which later provided a natural explanation and justification for the capitalist system. There is certainly a pitiless process of struggle for survival depicted in the novel through the titanic combat between two rival banks on the battlefield of the stock exchange which leads to terrible casualties amongst the investors on the losing side. In a reprise of the model of *Au Bonheur des Dames*, the uncompromising speculator, Saccard, justifies such conflict as 'the very allurement of life,' whilst his more compassionate mistress, Mme Caroline, is resigned to the view that it is 'in nature's plan' (V, 135-36).

From the battlefields of commerce, industrial politics and finance, Zola turns to the real battlefields of the Franco-Prussian War in *La Débâcle* (1892). In this novel his Social Darwinism is the most explicitly articulated, as he sought to provide an overarching explanation of France's defeat. In fact, the correlation between war and natural selection appears in three related Zola texts: the preliminary sketch of the novel, an article on 'Sedan' in *Le Figaro* (1 September 1891) on the occasion of the twenty-first anniversary of the battle, and in the novel itself. But only in the private manuscript notes for the novel does Darwin's name appear, in the following phrase: 'la vision vraie de la guerre, abominable, la nécessité de la lutte vitale, toute l'idée haute et navrante de Darwin dominant le pauvre petit, un insecte écrasé dans la nécessité de l'énorme et sombre nature' (V, 1377) ('the true vision of war, appalling, the necessity of the struggle for life, the whole lofty and distressing idea of Darwin dominating the poor little individual, an insect crushed in the inevitability of the vast and sombre workings of nature.'). The lack of any direct reference to Darwin may well again be due to the fact that the novelist preferred to incorporate his interpretations into the fabric of the novel itself rather than presenting it as the narrator's commentary (except for tell-tale 'perhaps'), astutely using for dramatic effect the hallucination of a Bonapartist lieutenant, wrapped in the French flag, as he dies in a hail of bullets, 'voyant peut-être monter à l'horizon la vision vraie de la guerre,

[20] *Germinal*, p.521.

l'atroce lutte vitale qu'il ne faut accepter que d'un cœur résigné et grave, ainsi qu'une loi' (V, 704) ('perhaps seeing rise on the horizon the true vision of war, the terrible struggle for life to which we can only gravely resign ourselves as a law of life').[21]

This cycle of texts is quite remarkable for the way in which it both displays and veils its Darwinist tendencies and for the manner in which it presents a veritable anthology of Darwinist themes and options, demonstrating the potential for multivalent extrapolation in Darwin's own works. But in *Le Docteur Pascal*, the final novel of the *Rougon-Macquart* series and in the following two series of novels, the tenuous links to Darwin's original theories become more strained. In *Le Docteur Pascal* our young, retiring Darwinian figure of *La Fortune des Rougon*, now twenty-one years older, takes centre stage, becoming not only a fictional figure but far more overtly Zola's spokesman on scientific matters, as well as the novelist's *alter ego* with his files of dossiers on members of his family. He has become the arch-naturalist, studying the infinite proliferation of life in its origins and in its propagation in the sea, dissecting corpses in the local hospital, probing the mysteries of embryos, and bringing the reader up to date on the laws of heredity supposedly underpinning the series. For this purpose, Zola consulted and recycled a set of notes provided by Georges Pouchet, a doctor and professor of anatomy, along with a study by the neurologist, Jules Déjerine, *L'Hérédité dans les maladies des systèmes nerveux* (1886), a text which not only provided the novelist with some convincingly scientific details to apply to his mentally unstable characters but also with a brief, handy survey of more recent theories of heredity that create further 'effects of scientificness' and inform Pascal's own researches. Darwin's 'gemmules' and 'pangenesis' receive a brief mention, as do Haeckel's 'perigenesis' and Galton's 'stirps'.[22] The narrator even suggests, as Zola is more alert this time to the danger of anachronism, that Pascal intuitively anticipates August Weismann's theory of 'germ plasma,' a key stage in the development of modern genetics and the discovery of DNA.[23] But *Le Docteur Pascal* is not simply an endorsement of the benefits of the medical and natural sciences. In fact, Pascal, whose ambitions are boundless, is led to give up his experiments after his failed attempts to cure his patients by injections of a

21 See also, in the final phase of the novel, the equally internalized reaction of the German officer, Otto Gunther, as he watches Paris burn during the Commune, meditating on the renewed occurrence in which the 'Germains' had come to punish a perverse nation and sweep away the vestiges of Latin corruption (V, 887).

22 See V, pp.1576-79, on Zola's sources relating to heredity.

23 See V, p.946. See also Henri Mitterand's summary of Darwin's and Haeckel's theories (V, p.946) and Ripoll on Zola's sources, pp.178-79.

concoction made up of the substance from the nerves of sheep's brains and distilled water. In addition to his difficulties in meeting the butcher's bills, Pascal fails to provide more than temporary pain relief and does not succeed in fulfilling his aim of discovering a universal panacea. But the outcome of this dabbling in eugenics, which brings him face to face with the limits of science, contributes to the character's 'credo,' which the novelist, with the ponderous rhetoric that will characterise his later works, presents as the philosophical conclusion of the whole *Rougon-Maquart* series, combining a faith in the relative benefits of science and an exaltation of life itself: 'La vie où nous baignons, la vie aux courants infinis et contraires, toujours mouvante et immense, comme une mer sans bornes' (V, 1219). ('Life in which we are immersed, life with its infinite and contrary currents, always moving and immense, like a boundless sea.') This emancipation from the rigorous determinism of his earlier naturalist phase tips the balance of the novelist's own earlier 'credo' even further in the direction of his optimistic belief in human progress, aided by the practical benefits of science, countering more effectively the pessimistic stoicism of his former Naturalist *confrères*, with a devastating effect on the readability of his fiction in the next two novel series.

In these works, *Les Trois Villes* (*Lourdes*, 1894; *Rome*, 1896; *Paris*, 1898) and *Les Quatre Évangiles* (*Fécondité*, 1899; *Travail*, 1901; *Vérité*, 1903, *Justice*, unfinished), religious symbolism and redemptive narrative will abound, replacing the pseudo-certainties of the earlier scientific phase; his protagonists will acquire a heroic status and an unshakeable faith in the future. Paradoxically, it is precisely when Zola drifts away from his bio-centric articles of conviction to promulgate this broader, prophetic *Weltanschauung* that he comes closest to Darwin's own vision. As a young Léon Blum wrote on reading Zola's novel *Paris*: 'Jamais M. Zola n'avait développé avec plus de lucidité et de force sa vision optimiste de l'humanité en marche, ce panthéisme naturaliste qui est celui de Darwin et de Fourier, mais aussi des Stoïciens et de Goethe.' ('Never had Zola developed with more lucidity and conviction his optimistic vision of the progress of humanity, the naturalist pantheism that he shared with Darwin and Fourier, but also with the Stoics and Goethe.')[24] This may well be, to some degree, an overstatement of the case, but there is certainly in Zola's late works, notably in *Fécondité*, the same sense of wonder and exultation before the splendour of nature's prolificity that we find in the famous final paragraph

---

24 Quoted by René Ternois, p.672, from Blum's article in *La Revue blanche*, 1 April 1898, 551-54.

of *The Origin of Species*.[25] This perhaps explains why, even though, as we have seen, there is little to no evidence of any direct influence of the thinker on the writer, and even though, in their exploration of nature, their paths rarely and barely crossed, the connection is far from specious, for Zola wrote very much in the spirit of Darwin's heritage, never a disciple but undoubtedly a 'darwinisant'.

There is, as a rather quaint coda to this piece, one final and rather intriguing connection between the two figures to report. In the second of Zola's *Évangiles*, *Travail*, the charismatic evangelist, Luc Froment, is inspired by Fourier's ideas and, aided by the scientist, Martial Jordan, builds a utopia out of an industrial hell, redeeming the working classes in the process. Martial Jordan is a physicist and inventor, who leads a secluded life, watched over and cared for by his sister, Sœurette. He is the one who invents the machines that bring about 'truth, justice and final happiness' in the community and liberate the workers. Though he lives to be over ninety, he is in deplorably bad health, only capable of working for three hours a day. Clearly the character is modelled on Darwin, as a note in the preparatory dossier of the novel confirms: 'Prendre ce qu'on dit de Darwin: maladif et faible, ne pouvant donner que peu de vie, utilisant si bien ses forces pour son œuvre, qu'il construit des mondes.' ('Take what is said about Darwin: sickly and weak, only able to give a little to life, but using his strength so well for his work that he builds worlds.')[26] Thus, Jordan et Sœurette mirror Darwin and Emma, even in the similarity of some of the symptoms of the men's debilitating illness,[27] though the female servants who attend him in padded slippers to avoid making any noise are doubtless Zola's own invention. Significantly, in Zola's oblique tribute, Martial Jordan, whose name is a virtual anagram of Zola and Darwin combined, enjoys a serene and euphemistic death in his laboratory, though it is perhaps equally significant that this paragon of scientific endeavour is totally marginalised politically and socially, leaving the leading role to the reforming zeal of Luc Froment and his promise of radiance and fructification.

---

25 The main theme in *Fécondité*, of course, is the very survival of the French nation in an age of depopulation; Bernardini (pp. 207-08) sees in the 'politique hygiéniste et réformiste' of the novel the Social Darwinist assumption that a nation should be demographically strong and healthy to develop and survive in the struggle for existence among nations.

26 Quoted by Frederick Ivor Case in *La Cité idéale dans 'Travail' d'Émile Zola* (Toronto: University of Toronto Press, 1974), p.48, from Zola's notes on his characters (ms. 10334, f°s 1-187).

27 See *Émile Zola, Œuvres complètes,* 21 vols, ed. by Henri Mitterand (Paris: Cercle du Livre Précieux, 1968), VIII, pp.633-4.

On the Evolution of Humanity and the Oppressions of Darwinism in French Postwar Fiction

Louise Lyle

Abstract: Evolutionary motifs feature strongly in Vercors's novel, *Les Animaux dénaturés* of 1952 and Romain Gary's *Les Racines du ciel* of 1956. The former explores the moral and legal problems posed by the discovery of living examples of a 'missing link' between the higher primates and man. The latter, adapted for the cinema by John Huston in 1958, recounts the campaign to save the African elephant from extinction initiated by the protagonist, Morel, a survivor of the Nazi concentration camps. These works explore the theme of the evolution of humanity, revealing the authors' preoccupation with the major political and moral issues of the postwar period: coming to terms with the Holocaust, decolonisation, racism and the nature of civilisation itself. This chapter argues that Vercors and Gary are not only the inheritors of certain characteristically French modes of evolutionary thought, but are also strongly influenced by contemporary developments in the popular understanding of the life sciences. Drawing as heavily on the *transformiste* theories of Jean-Baptiste Lamarck as on a deeply ingrained resistance to social Darwinism, their works are also marked by contemporary thinkers such as the notorious Soviet agronomist, Trofim Denisovich Lysenko and the celebrated priest-palaeontologist, Pierre Teilhard de Chardin.

This chapter will examine the legacies of Darwin in French literary culture of the immediate post-World War II period, focussing on selected works of two important French authors, namely Vercors and Romain Gary.[1] While scholarship on the influence of evolutionism on French literature is already limited in comparison to that on English-language writings, attention, such as there has been, has been concentrated mainly on the period between 1880 and 1914 during which Darwinism attained mass popularisation status and

---

1 Jean Bruller (1902-91) was an active *résistant* during the war, adopting the pseudonym 'Vercors' from the Resistance unit based in the mountainous region of that name in Eastern France. Co-founding the resistance publishing house, Les Éditions de Minuit, with Pierre Lescure and Yvonne Paraf in 1941, he was the author of the house's first book, *Le Silence de la mer* (1942). Although he continued publishing into the 1970s, he is best remembered for his wartime activities. Romain Gary (1914-80) was born Romain Kacew into a Jewish family in Vilnius. Aged fourteen, he moved to Nice with his mother, subsequently becoming a naturalised French citizen and adopting French as the language in which he would write to great acclaim. Gary is the only author to have won the prestigious Prix Goncourt twice (albeit in contravention of the rules), firstly for *Les Racines du Ciel* in 1956, then for *La Vie devant soi*, published under the pseudonym 'Émile Ajar', in 1975. A pilot for De Gaulle's Free French, he entered the diplomatic service after the war. Postings in Los Angeles gave him access to Hollywood, and John Huston's 1958 adaptation of *Les Racines du Ciel* as *The Roots of Heaven*, starring Trevor Howard and Errol Flynn, was his first screenwriting success. He committed suicide just months after his ex-wife, actress Jean Seberg, in 1980. See Dominique Bona, *Romain Gary* (Paris: Gallimard, 1987).

consequently enjoyed its greatest notoriety internationally. In considering the significance of Darwinian themes for these authors, who, like so many of their contemporaries were driven by a desire to explore questions of human nature and responsibility, I will discuss the extent to which they were not only influenced by the major political and intellectual developments of their era, but were also the inheritors of certain characteristically French modes of evolutionary thought originating in the late nineteenth century. After outlining French responses to Darwin's ideas and to the alternative *transformiste* theories of the Enlightenment philosopher Jean-Baptiste Lamarck in the fin-de-siècle period, I will show how these attitudes re-emerge, subtly transformed by contemporary influences, in the writings of Vercors and Gary half a century later. In their literary reflections on the aftermath of the Holocaust and the on-going decolonisation process, the authors are simultaneously confronted by the bewildering challenges to evolutionary science presented by the Lysenko Affair in the USSR and the ideas of the internationally famous French Jesuit priest and palaeontologist, Pierre Teilhard de Chardin.

While the theory of natural selection provoked controversy everywhere, the French were particularly hostile towards it. The highly centralised scientific community of late nineteenth-century France disdained Darwin's theory on the grounds that it could not be verified experimentally, that it depended too heavily on random chance in respect of the mutations which are the raw material of natural selection, and that it did not *necessitate* progress, in the sense of being directed towards the production of organisms of increasing biological sophistication.[2] The French lay-public was strongly influenced by the social Darwinist themes, widely seen as the epitome of Anglo-Saxon brutality, foregrounded by the self-educated Herbert Spencer devotee Clémence Royer in the first French translation of *The Origin of Species* in 1862. Her translation was a source of consternation to Darwin, and her unauthorised preface contained provocative assertions such as the following:[3]

---

2 In 1982, no less a figure than Jean-Pierre Chevènement, the French Minister of Research and Industry, acknowledged the extent of French resistance towards Darwin's theories in his opening remarks at a conference commemorating the centenary of Darwin's death. See J.-P. Chevènement, 'Allocution d'ouverture au colloque organisé par l'Université de Picardie à l'occasion du centenaire de la mort de Charles Darwin', in *De Darwin au darwinisme*, ed. by Yvette Conry (Paris: Vrin, 1983), pp.9-16 (pp.10-11).

3 Although Darwin was surprised by Royer's extrapolations of his ideas into the social, moral and religious domains, he was particularly irritated by her misrepresentations in successive French introductions to *The Origin of Species* of certain of his other ideas,

> Que résulte-t-il de cette protection inintelligente accordée exclusivement aux faibles, aux infirmes, aux incurables, aux méchants eux-mêmes, enfin à tous les disgraciés de la nature?C'est que les maux dont ils sont atteints tendent à se perpétuer indéfiniment; c'est que le mal augmente au lieu de diminuer, et qu'il s'accroît de plus en plus aux dépens du bien.
>
> [What is the result of this pointless protection granted exclusively to the weak, to the infirm, to the incurable, to the wicked, indeed to all of nature's failures? It is that the ills with which they are afflicted tend to be infinitely self-perpetuating; it is that this scourge is on the rise instead of diminishing, growing ever stronger to the detriment of the greater good.][4]

Royer's ideas were neither typical of any political or social group, nor popular with the wider public. High-profile French exponents of social Darwinism were rare indeed and, beyond Royer, only Georges Vacher de Lapouge and Gustave Le Bon stand out.[5] As Linda Clark notes: 'Many French writers discussed the social import of Darwinism but few labelled themselves "social Darwinists." Self-proclaimed enemies of social Darwinism were more in evidence.'[6] This commonly held aversion led numerous late-nineteenth-century French writers, concerned with a variety of social and political questions, to discern a certain utility in social Darwinism, since accusations of espousing such ideas could usefully serve as a damning indictment of one's adversaries, whoever they might be. By the mid eighteen-eighties, ideological authors from across the political spectrum were making liberal use of the rhetoric of *le struggleforlife* [*sic*] to condemn their respective enemies. Émile Zola, the leading literary naturalist and a passionate exponent of so-called scientific republicanism, viewed capitalism as economic social Darwinism. Right-wing reactionaries Maurice Barrès and Paul Bourget found unexpected common ground with the anarchist sympathiser Octave Mirbeau in their respective attacks on the early Third Republic's meritocratic public education system and its colonial ambitions,

---

notably the theory of pangenesis. See J.-M. Bernardini, *Le darwinisme social en France (1859-1918)* (Paris: CNRS, 1997), p.74 n.4.

4 Clémence Royer, 'Preface to the First Edition', in Charles Darwin, *De l'origine des espèces*, trans. by C. Royer, 2nd ed. (Paris: Victor Masson et fils, 1866), pp.i-lix (pp.liii-liv).

5 George Vacher de Lapouge (1854-1936), a librarian trained in law, published widely on social Darwinist and eugenic theories. While his theories aroused enthusiasm in Germany and America, his influence in France was marginal. Gustave Le Bon, a social psychologist with medical training, wrote extensively on a range of scientific topics, but is best remembered for his internationally influential treatise on group psychology, *The Crowd* (1895). See Bernardini, pp.163-76.

6 Linda L. Clark, *Social Darwinism in France* (Alabama: University of Alabama Press, 1984), p.7.

viewed as examples of social Darwinism in practice.[7] Pronatalist, anti-feminist authors such as Colette Yver and Marcel Prévost meanwhile worried that by offering young women educational access to modern scientific ideas such as Darwin's theory, the French State was encouraging them to abandon their traditional child-bearing role in order to participate actively in an economic 'struggle for existence'. Still others, notably Gustave Le Rouge and Capitaine Danrit, authors of popular pro-militarist serialised fiction aimed at a young male readership, feared that British colonial arrogance and the USA's growing economic supremacy amounted to nothing less than an Anglo-Saxon espousal of the 'survival of the fittest' at international level.[8] The popular success enjoyed by these works helped to conflate the theory of natural selection with the social Darwinist themes accentuated by Royer in the collective consciousness of non-specialist French readers.

The *transformiste* theories of the French Enlightenment thinker Jean-Baptiste Lamarck (1744-1829) were meanwhile enjoying renewed popularity in fin-de-siècle France. With, as Stuart Persell notes, 'the idea of the inherited transformation of acquired characters serv[ing] as a biological justification for [...] theoretical writings in politics, economics, criminology and a host of other fields,' neo-Lamarckism offered biological justification for the notion that the French citizens of the future could be formed through socio-political conditioning, and was central to the development of both the educational and colonial enterprises of the early French Third Republic (1870-1914).[9] Positing *transformisme* as a teleological process leading to the emergence of increasingly sophisticated organisms of which mankind was the ultimate product, neo-Lamarckism justified humanity's superiority over other species and was also the basis of a useful natural analogy for the development of increasingly 'progressive' and advanced societies with 'governmental organisations or ethical belief systems'.[10] Persell highlights the notion that: 'Contests for survival were essentially British [...], while doctrines of reciprocal aid and biological tolerance were part of the French genius. Worse, in German society such ideas had become a dangerous creed of force and immorality.'[11] A chauvinistic adherence to a French scientific

---

7 See Louise Lyle, 'Charles Darwin dans *Le Jardin des supplices*', *Cahiers Octave Mirbeau*, 14 (2007), 47-64 and '*Le Struggleforlife*: Contesting Balzac through Darwin in Zola, Barrès and Bourget', *Nineteenth-Century French Studies*, 36 (2008), 305-19.

8 See Louise Lyle, *Le Struggleforlife: Social Darwinism in the French Ideological Novel, 1880-1920*, forthcoming with Rodopi, Amsterdam.

9 Stuart Persell, *Neo-Lamarckism and the Evolution Controversy in France, 1870-1920* (Lewiston, NY: Edwin Mellen Press, 1999), p.178.

10 Persell, p.187.

11 Persell, p.186.

paradigm, serving as the basis for a set of progressive social and political values deemed also to be characteristically French, coupled with a strong dose of xenophobia towards the nation's major military and economic rivals, thus underpinned a widespread preference for Lamarck over Darwin in early Third Republic France. Darwinian ideas (as known through highly tendentious social Darwinist interpretations thereof) were popularly deemed unscientific, anti-progressive, brutally conflictual and, worst of all, foreign; the home-grown *transformiste* theories of Lamarck suggested a model which was at once adaptive, progressive, non-conflictual and, therefore, much more in keeping with the fraternal and egalitarian values of French Republican society.

While the vogue for ideological authors to use social Darwinist rhetoric as a kind of big stick with which to beat their adversaries had all but disappeared by 1914, it is my contention that the connotations of Darwinian and Lamarckian evolutionary theories described above remained ingrained in French intellectual consciousness for a long time thereafter. The remainder of this chapter will therefore consider the extent to which these evolutionary preconceptions inform certain French texts concerned with the major political developments of the post-World War II period, namely Vercors's 1952 novel, *Les Animaux dénaturés* (translated as *You Shall Know Them* in 1953) and Romain Gary's 1956 novel, *Les Racines du ciel* (translated as *The Roots of Heaven* in 1958).

The first of these historical influences is, unsurprisingly, the Holocaust, understood as the realisation of Nazi philosophy developed by Adolf Hitler primarily in *Mein Kampf* (1925-26). Hitler's appropriation of social Darwinism as one of the bases of his own ideas is well known. In his analysis of the role of academic knowledge in *Mein Kampf*, Werner Maser suggests that religious dogma was 'outweighed in the formation of [Hitler's] ideas by biological theory (primarily pseudo-Darwinism)'.[12] This view is readily verified in respect of the numerous examples of social Darwinist rhetoric in *Mein Kampf*, in which Hitler recounts his personal 'struggle for existence' during the hard times in Vienna, for example, before considering a somewhat anthropomorphised view of nature's role in 'select[ing] the best from an abundance of single elements and stamp[ing] them as fit to live and carry on the conservation of the species,' and reflecting on the international struggles foreseen thus: 'Nobody can doubt that this world will one day be the scene of dreadful struggles for existence on the part of mankind. In the end the instinct of self-preservation alone will triumph.' He later describes

12 Werner Maser, *Hitler's 'Mein Kampf': An Analysis*, trans. by R.H. Barry (London: Faber, 1970), p.67.

the emergence of a new racialised strain of nationalism using the language of biological speciation: 'The instinct for the preservation of one's own species is the primary cause that leads to the formation of human communities. Hence the state is a racial organism and not an economic organisation.'[13]

That social Darwinism had served as one of the foundations of Nazism, an ideology directly associated in the French collective consciousness with the terrible memories of occupation, collaboration and the Holocaust, could hardly have served to improve Darwin's already impoverished image in France. Social Darwinist ideas had also, however, long been used as a standard justification for the imperialist enterprises of Great Britain, France and the other colonial powers prior to World War II. As Elleke Boehmer notes: 'Even before Darwin, colonisation was represented as a survival of the fittest, in De Quincey's phrase "winnowing the merits of races". The struggle for survival dictated that the strong, or those best at imposing their power, were deserving of hegemony.'[14] With the struggle between colonised nations and their imperialist masters taking on a new impetus in the postwar period, the rhetoric of social Darwinism acquired a further layer of negative associations in France, linked not only with Nazi ambition, but also with the values of an outdated world order whose lengthy and violent undoing was to confront the French public first in Indochina, then in Algeria.

The Lysenko Affair of 1948 only served to reinforce these pejorative connotations. In an evolutionary theory which owed much to the ideas of Jean-Baptiste Lamarck, the Soviet agronomist Trofim Denisovich Lysenko conceptualised heredity as a function of the organism's relationship with its environment, reasoning that environmental modification could transform living organisms and that new species could even be created out of old ones. According to Soviet Lysenkoists, any attempt to defend the rapidly advancing Mendelian genetic science of the West, shortly to yield new insights on the very fabric of life through the discovery of the double-helix structure of the DNA molecule, was tantamount to supporting Western bourgeois imperialist values derived from a social Darwinism whose theoretical utility extended from the ruthless competition of American capitalism to the racial and social hierarchies of *Mein Kampf* and the horrors of the Final Solution beyond. Astute and opportunistic, Lysenko presented his theories as the realisation of certain ideas of Marx, Engels and, crucially, Stalin himself, gaining influence over the Soviet leader and communist party hierarchy. His ideas were actively propagated around communist circles in

---

13 Adolf Hitler, *Mein Kampf*, trans. by R. Manheim (Boston: Houghton Mifflin, 1943), pp.80, 124, 128, 140.

14 Elleke Boehmer, *Colonial and Postcolonial Literature: Migrant Metaphors*, 2nd edn (Oxford: Oxford University Press, 2005), p.77.

Europe, and were briefly influential in respect of certain French authors with left-wing sympathies, including Vercors, as I have argued elsewhere.[15]

Given the continuing bad press to which natural selection had, through the constant reiteration of its social Darwinist interpretations, been subject in France, it is hardly surprising that Vercors and Gary, both clearly interested in evolutionism and in the question of what 'humanity' might actually mean in a world torn apart first by war, then by the ideological and racial divisions imposed by Cold War politics and decolonisation, should evince a measure of suspicion towards Darwin. Despite important differences in their attitudes towards man's relationship with nature, their writings considered here reveal a shared sentiment that 'humanity' is no simple physiological or psychological quality conferred though the combined effects of heredity and environment. It is rather a state of existence privileging the rights and responsibilities of the individual and the collective, towards which mankind may, or may not, by means of a conscious and reflective evolution of moral and social values, progress. Both reveal their concern with the biological *and* moral aspects of the evolution of humanity in the specific context of their own time, as we shall consider.

Vercors's *Les Animaux dénaturés* tells the story of Doug Templemore, an English journalist who accompanies an expedition to Papua New Guinea during which living examples of a 'missing link' between man and the higher primates are found. If the 'Tropis' are classified as animals, they stand to be exploited as slave labour by Australian corporate interests aiming to compete with the British textile industry. Legal recognition of the Tropis' humanity will therefore provide the key to their freedom, while securing British economic interests. Determined to force the issue for pure humanitarian reasons, Templemore kills the offspring he has sired by artificial insemination with a Tropi mother and subsequently has himself charged with murder. The question of his guilt, which, in keeping with the times, would entail capital punishment, rests on whether the hybrid offspring is deemed in an English court of law to have been human or animal. After the first trial's failure to produce a verdict, a compromise is reached in the retrial when the judge has Templemore acquitted while acknowledging the full human rights recently conferred upon the Tropis by an Act of Parliament.

The novel is a whimsical piece which draws on the French tradition of the *conte philosophique* in order to address contemporary scientific and political

15 See Louise Lyle, 'Science Bourgeoise et Science Prolétarienne: French Literary Responses to the Lysenko Affair', *New Readings / New Screenings* e-journal, 9 (2008),<http://www.cf.ac.uk/euros/subsites/newreadings/volume9/abstracts/lyle-abstract.html> [accessed 9 September 2010].

issues of international relevance. Its central theme is the difficulty, at once of a scientific, legal and ethical nature, entailed in providing a precise delineation between man and his closest evolutionary relatives, the higher primates. This, for Vercors, is no abstract philosophical debate. It is rather a framework within which to uncover and condemn man's inhumanity towards those whom he is willing to acknowledge only as lesser versions of himself. Recent examples of this phenomenon included not only the Holocaust, but also racist responses to the decolonisation question, upon which Templemore reflects:

> So here we have staring us in the face again [...] the grimacing ghost of racial discrimination, already to rise again, with its hellish attendants. And what discrimination [...]! A racialism in whose name whole populations can tomorrow be stripped of their human status and the rights that go with it, and be sold, in turn, like livestock [...]! Where would the boundary be fixed [...]? *Where it pleases the strongest*! Imagine what will happen to the natives in the colonies, to the Negroes in the states where segregation is in force! And, generally speaking, to any ethnical [*sic*] minority![16]

This sentiment is even more explicit in *Zoo, ou l'Assassin philanthrope*, Vercors's 1964 theatrical adaptation of the novel, in which an expedition member cynically seeks to justify the white man's historical dominance, noting: 'Que le racisme, c'est la loi du plus fort, rien d'autre. Et que le jour où les peoples d'Asie ou d'Afrique le deviendront, les plus forts, ils pourront tout aussi bien nous rendre la pareille.' ['That racism is the law of the survival of the fittest, nothing more. And that the day that the peoples of Asia and Africa become the strongest, then they'll be able to do the same to us.'][17]

With biological similarity sufficiently close to permit inter-species breeding, the close resemblance between Tropi and human repeatedly proves troubling to Vercors's protagonists. While Templemore is, for instance, relieved to be able to transport the veiled female Tropi named Derry who will bear his offspring back to England on a passenger-plane without incident, his new wife, Frances, is 'queerly moved' to note that Derry's carefully manicured hands are more beautiful than her own.[18] Once safely installed in their new home, the couple are soon embarrassed by Derry's carefree nudity and by her 'really too sensual appreciation' of the

---

16 Vercors, *You Shall Know Them*, trans.by R. Barisse (Boston: Little, Brown, 1958), p.102.

17 Vercors, *Zoo, ou l'Assassin philanthrope* (Paris: Magnard, 2003), p.93.

18 Vercors, *You Shall Know Them*, pp.111-13.

most innocent touch.[19] Provocatively holding up the social and moral difficulties posed by our unwillingness to accept our resemblance with our closest animal relatives, occupying a 'lower' rung on the evolutionary ladder than ourselves, Vercors aims to challenge the white imperialist's disinclination to view the colonised subject as anything other than an inferior in intellectual, moral and even biological terms. Yet in seeking to highlight the inequalities between white and black arising as the social and political consequences of an unjust world order through a perhaps rather naïve analogy between humans and primates, Vercors himself became the butt of accusations of racism. His implied comparison of the opposition between white and black with that between human and subhuman was particularly strongly criticised by his adversaries in the Eastern bloc.[20] This clumsy, though well-meaning misjudgement, revealing perhaps of the author's inability to abandon deeply ingrained preconceptions about racial hierarchies even in a text intended to combat racism, is not the only paradox of Vercors's work. For while the title of the novel's 1953 English translation, *You Shall Know Them*, evokes the central themes of resemblance and recognition, it fails to capture the author's assertion that, as the 'animals removed from nature' of his original title ('*Les Animaux dénaturés*'), our humanity resides precisely in our capacity to eschew brutal and instinctive 'animal' behaviours, including the Darwinian 'struggle for existence' thought to pit individual against individual and species against species.

Another denunciation of social Darwinist justifications of the right of force is found in *The Roots of Heaven*, Romain Gary's account of the efforts of his concentration camp survivor hero, Morel, to halt the slaughter of elephants in Chad, still part of French Equatorial Africa at the time of publication in 1956. The African location is significant not only in terms of the on-going decolonisation issues, but also because recent archaeological discoveries supported the idea of Africa's special status in mankind's evolutionary history: in the novel's opening paragraphs, for instance, a secondary character enquires whether, following recent excavations in Nyasaland (part of modern Malawi), Africa could be considered the 'true cradle of humanity'.[21] By virtue of the different meanings of 'humanity,'

19 Vercors, *You Shall Know Them*, p.118.

20 Vercors, who enjoyed a lukewarm relationship with communism at the best of times, defended himself against these allegations in his 1955 preface to the Hungarian and Polish editions of the novel. See Russell Barnes, 'Vercors: Morality and Commitment', *Modern and Contemporary France*, 50 (1992), 3-9, and Vercors, 'Préface pour les éditions hongroise et polonaise des *Animaux dénaturés*', *P.P.C.* (Paris: Michel, 1957), pp.212-24.

21 John Griffin's 1958 translation gives this as 'the cradle of the human race'. Gary's choice of expression in the original text is, however, 'le vrai berceau de l'humanité',

Africa is thus conceived not only as the historic birthplace of mankind but also as the site of a contemporary conflict destined to test its protagonists' capacity to act humanely and benevolently towards both the elephants and one another. The evolution of humanity is not, therefore, for Gary, simply the concern of science, but also a question of our ability to embrace moral standards appropriate to our position in relation to the rest of nature. In common with Vercors, he questions the extent to which the boundaries of speciation justify our disregard for animals in order to trace a parallel between our abuses of nature and of our fellow men. In contrast to Vercors, however, Gary proposes that salvation lies in the understanding that we ourselves are part of nature, bound within an all-encompassing ecosystem to the flora and fauna that we presume to dominate. His environmentalist-humanitarian hero, Morel, an idealistic and charismatic figure who sees the preservation of the elephants as the start of a broader enterprise to save mankind, even acquires a reputation in the international press as 'the man who changed species'.[22]

This proposed continuity between ecological morality and the acquisition of humanity is highlighted when a young French army officer, Laurençot, suggests that 'colonization has been based partly on the corpses of elephants; it was ivory-hunting that made it possible for the business men and colonists to meet the initial costs of establishment.' His superior officer responds that: 'There are men fighting and dying at this very moment, in the streets, on the barricades and in the prisons... One may be allowed still to prefer to take an interest in them,' to which Laurençot replies sadly: 'But the elephants are part of that fight. Men are dying to preserve a certain splendour of life. [...] They are dying to preserve a certain natural splendour.'[23] The price of such inhumanity, past, present and future, is evoked in the personage of Waïtari, the aspiring Marxist dictator of his country, who, we learn, 'was the magnificent product of a selection which it would be difficult to call "natural": for centuries, his forefathers had rid themselves of the weaker products of their race by delivering them into the hands of Arab or Portuguese slave traders.'[24] Thus characterised as the hateful result of the 'struggle for existence' at the heart of a slave trade that existed prior to the white man's arrival and which served as one of the foundations upon which colonialism was built, the French-educated Waïtari

---

which translates directly as 'the true cradle of humanity'. See Romain Gary, *The Roots of Heaven*, trans. by J. Griffin (London: Joseph, 1958), p.12 and Romain Gary, *Les Racines du Ciel* (Paris: Gallimard, 1980), p.17.

22 Gary, *Roots of Heaven*, p.69.

23 Ibid., pp.70-71.

24 Ibid., *Roots of Heaven*, p.121.

is less his countrymen's saviour from the European powers he contemptuously plots to overthrow than a collaborator in their exploitative enterprise.

Gary's message throughout the novel is that humanity is a condition to be attained, not through the suppression of any other human or animal, but rather through an alternative struggle towards the reconciliation of the contradictory and the resolution of conflict. In this connection, the importance of the character Father Tassin, a Jesuit priest and palaeontologist, is not limited to his function within the narrative as the near-silent listener to whom the tale constituting the novel is recounted. Indeed, this character's real significance derives from his uncanny resemblance not only to a Benedictine counterpart in Vercors's *You Shall Know Them* but also, primarily, to the real figure of Pierre Teilhard de Chardin, whom Gary came to know and admire during his time in New York. Teilhard (1881–1955) combined a scientific training with holy orders, undertaking research in the palaeontology laboratories of Paris before commencing field work across Africa and East Asia, where he was one of the discoverers of 'Peking Man' in 1929. While his research in palaeontology and geology was internationally respected, his attempts to reconcile religious faith with scientific knowledge brought him to the notice of a wider public. His philosophical ideas also brought him into conflict with the Catholic hierarchy, and his major works, notably *The Phenomenon of Man*, were only published after his death to great public acclaim, though, as Daniel Dennett notes, 'the scientific establishment [...] was just as resolute as the Church in rejecting [his ideas] as heretical.'[25]

Dennett explains, '[Teilhard] proposed a version of evolution that put humanity at the center of the universe, and discovered Christianity to be an expression of the goal – "the Omega point" – towards which all evolution is striving,' even making provision for Original Sin in his theory.[26] Given the eccentricity of his thinking, it is unlikely that Teilhard's appeal as a model for characters in the works of Gary and Vercors was based on an unqualified admiration for his philosophical theories. Indeed, Gary would later describe his feelings for Teilhard in the following terms:

> J'aimais ce grand capitaine et il m'arrive de rêver de lui, debout à la barre, avec son profil de boucanier, voguant vers l'horizon sur le pont d'un navire métaphysique dont il avait bâti à la fois la coque, la boussole et...qu'il me pardonne ! la destination. [...] Je lui ai emprunté [...] quelques idées que j'ai fortement

[25] Daniel Dennett, *Darwin's Dangerous Idea* (London: Simon & Schuster, 1996), p.320.

[26] Dennett, p.320.

> romancées, pour en faire le jésuite Tassin, dans *Les Racines du ciel*, que j'écrivais alors...
>
> [I liked this great captain and I sometimes dream of him, standing by the rail, with his buccaneer's profile, riding the waves towards the horizon on the bridge of a metaphysical ship of which the hull, the compass and...begging his forgiveness! the destination were all of his own fabrication. [...] I borrowed from him [...] certain ideas which, in a heavily romanticised form, became the basis of the characterization of the Jesuit Tassin in *The Roots of Heaven*, which I was writing at the time...][27]

Given these reservations, Teilhard's appeal might alternatively be explained in terms of the powerful theme of reconciliation which the paradoxical figure of the priest-palaeontologist, ready to incur the wrath of the Church in order to embrace an evolutionary philosophy, evokes. As a character in the novel tells Tassin, 'in your writings, you are giving our old Christian notion of salvation a scientific meaning of a biological mutation which will bring our species to a new spiritual horizon, and [...] you announce this salvation as coming "soon."'[28]

A further explanation for the presence of the Teilhard doubles in the works of these authors, may, however, derive from the celebrated priest's unwillingness to embrace views of the sort generally associated nowadays with Anglophone scholars such as Richard Dawkins and Daniel Dennett, i.e. that 'evolution is a mindless, purposeless, algorithmic process.'[29] The teleological aspect of Teilhard's evolutionary model, evoking the progressive aspects of Lamarckian *transformisme* which posit mankind as the ultimate product of evolution and which held sway in France longer than elsewhere, stands in opposition to the chaotic randomness of variation at the heart of the Darwinian model. The characterisation of Vercors's Benedictine, Father Dillighan, as a 'supporter of orthogenesis,' the 'theory that successive members of an evolutionary series become increasingly modified in a single undeviating direction [...] with variation in nature [...] directed toward fixed

27 Romain Gary, *La nuit sera calme* (Paris: Gallimard, 1974), pp.185-86.

28 Gary, *Roots of Heaven*, p.15.

29 Dennett, p.320. The views of Dennett, Dawkins, et al. are not, of course, universally upheld by English-speaking evolutionary thinkers, nor do there exist any clear national divisions governing modern interpretations of Darwinian thought. Comparison could, for instance, be made between Patrick Tort's notion of the '*effet réversif de l'évolution*', according to which the mechanisms of selection favour altruism , and John Dupré's argument that co-operation and ethical behaviours can emerge as products of Darwinian evolution. See Patrick Tort, *L'Effet Darwin: sélection naturelle et naissance de la civilisation* (Paris: Seuil, 2008) and John Dupré, *Darwin's Legacy: What Evolution Means Today* (Oxford: OUP, 2005).

goals and [...] species evolv[ing] in a predetermined direction irrespective of selection ,' upholds this interpretation. Dillighan and Tassin, the literary doubles of Teilhard in the respective works of Vercors and Gary, may finally, then, hold the key to the authors' shared evolutionary aspirations. [30] For if Teilhard is viewed not only as the exponent of a theory of directed evolution leading to the higher order of existence conceptualised in his 'Omega Point,' but also as the personification of the rapprochement of opposites in his attempts to reconcile science and religion, he could be understood as the antithesis of a Darwin whose ideas describe a Godless natural order derived from meaningless conflict and chaos. His attraction for Vercors and Gary, both educated, albeit in different ways, in the French philosophical tradition and in the failings of modern civilisation, as well as being the inheritors of a literary tradition which had, since at least the time of Émile Zola, sought to mobilise evolutionary biology in the name of ideology, may well reside in Dennett's assertion that: [31]

> The esteem in which Teilhard [...] is still held by non-scientists, the respectful tone in which his ideas are alluded to, is testimony to the depth of loathing of Darwin's dangerous idea, a loathing so great that it will excuse any illogicality and tolerate any opacity in what purports to be an argument, if its bottom line promises relief from the oppressions of Darwinism.[32]

30 Vercors, *You Shall Know Them*, p.35; 'Orthogenesis', *Encyclopaedia Britannica Online*, <http://www.britannica.com/EBchecked/topic/433448/orthogenesis> [accessed 13 December 2008].

31 In his 1842 'Avant Propos' to the *Comédie Humaine*, Honoré de Balzac posits the existence of human types as 'Social Species' whose quest for success and social mobility is a constant source of conflict between them, giving rise to the view that he had, to some extent, anticipated natural selection and the concept of the struggle for existence. See, for instance, André Wurmser, *La Comédie inhumaine* (Paris: Gallimard, 1964), p.264.

32 Dennett, p.321.

Houellebecq, Genetics and Evolutionary Psychology

Douglas Morrey

Abstract: The science fiction novels of Michel Houellebecq – *Les Particules élémentaires* (*Atomised*, 1998) and *La Possibilité d'une île* (*The Possibility of an Island*, 2005) – are true to the scientific demands of their genre in the informed use that they make of molecular biology and quantum mechanics in narrating humanity's arrival at its post-human future. But the predominant scientific influence on Houellebecq's writing surely remains evolutionary theory which casts a long shadow over the author's dispassionate, almost clinical descriptions of human behaviour, characteristically thrown into relief by references to other species. This frequent – if largely unspoken – appeal to evolutionary theory tends to saddle Houellebecq's novels with a rather determinist ideology, since economic and social inequalities in human life appear as the result of an atavistic heritage, while violence and aggression are explained away as adaptive behaviours. The most striking inequalities in Houellebecq's universe are, however, sexual and his constant reiteration of the ruthlessness of sexual competition tends to imply a reductive evolutionary psychology which suggests that individual and species reproduction remain the sole significant objectives of human life. This brutal sexual selection betrays a conception of evolution that remains rooted in the nineteenth-century origins of the theory, and that are criticised in this essay, using both nineteenth-century thought (Nietzsche) and more recent evolutionary theory. The theoretical inconsistencies in Houellebecq's presentation of evolution are analysed through a consideration of the paradoxical figure that stands at the centre of his science fiction: that of a humanity governed by base, atavistic impulses, but that nonetheless chooses and orchestrates its own overcoming.

The contemporary French novelist Michel Houellebecq makes frequent use in his writing of scientific discourse and reference to evolutionary theory, particularly, though not exclusively, in his science-fiction inflected novels *Les Particules élémentaires* (*Atomised*, 1998) and *La Possibilité d'une île* (*The Possibility of an Island*, 2005), both of which narrate the advent of a post-human successor to our species. In part, Houellebecq's explicit appeal to scientific ideas must be understood as a response to the perceived epistemological stranglehold enacted over the western world by scientific thought, almost to the exclusion of any other discourse, give or take a few harmless examples of New-Age nonsense. As Houellebecq has suggested in one of his published essays, the current era has witnessed the definitive completion of the Enlightenment project, with rational science replacing religion as a totalising means of understanding the universe.[1] This essay seeks to offer a preliminary investigation into the kinds of evolutionary discourse that appear in Houellebecq's writing, asking how and why they appear as they do. My approach will certainly not be that of Alan Sokal and Jean Bricmont

---

1 Michel Houellebecq, 'Lettre à Lakis Proguidis', in *Interventions* (Paris: Flammarion, 1998), pp.49-56 (pp.53-54).

in their *Intellectual Impostures*: it is not my intention – at least not exclusively – to challenge the accuracy of Houellebecq's scientific references, or his 'right' to adopt a scientific discourse.[2] Houellebecq himself has offered a critique of Sokal and Bricmont's book, pointing up the authors' apparent lack of either a sense of humour, or of any social graces, whilst Houellebecq's contemporary on the French literary scene Maurice Dantec has launched a searing attack on the 'Stalinist' pedantry of *Intellectual Impostures*.[3] My intention in this essay is rather to ask how Houellebecq's claims to a scientific discourse inform the ideological project of his work. I will not have room to discuss all of Houellebecq's references: his frequent allusions to quantum mechanics, for instance, will obviously fall largely outside the scope of this essay. Nor do I wish to be drawn into internecine debates about the relative merits of evolutionary psychology, human sociobiology and human behavioural ecology. Instead, I will interpret Houellebecq's novelistic narratives as a broadly evolutionary discourse asking – with the help of science, philosophy, and the philosophy of science – to what extent they fall into the trap of genetic determinism that tends to characterise bad evolutionary science. Much of this discussion will be focused around a close reading of Houellebecq's most recent novel, *La Possibilité d'une île*. The novel tells the story of Daniel, a successful French comedian of our era, as he relates his sexual misadventures and his casual, sceptical contact with a religious sect, the Elohimites, who advocate human cloning. Eventually this movement gains considerable momentum and a new breed of cloned, genetically enhanced humanoids populate the Earth, each living alone in a private compound. Meanwhile, the 'old' human race reverts, through war and epidemic, to a state of savagery. The first-person narrations of Daniel are interspersed with those of his cloned descendents, the relation between them only becoming clear as the novel progresses.

Houellebecq has a tendency to describe the social, and above all the sexual, activities of his contemporaries as though he were giving a lesson in natural history. Much of the comic effect of *Platforme* (*Platform*, 2001), for instance, is derived from incisive descriptions of the instinctive herding behaviour of tourists on package holidays. In *La Possibilité d'une île*, the sexual behaviour of the male members of the Elohimite sect is compared to that of a horde of apes dominated by an alpha male. The retrospective point of view conferred on such observations by the science-fiction narratives of *Les*

---

2 Alan Sokal and Jean Bricmont, *Intellectual Impostures: Postmodern Philosophers' Abuse of Science* (London: Profile Books, 1998).

3 Michel Houellebecq, 'Beau travail!', *Les Inrockuptibles*, *Hors série Houellebecq* (2005), 94-95; Maurice G. Dantec, *Le Théâtre des opérations: Journal métaphysique et polémique 1999* (Paris: Gallimard, 2000), pp.453-58.

Particules élémentaires and *La Possibilité d'une île* lends them a quasi-scientific authority. For instance, Daniel25 (i.e. the twenty-fifth generation clone of the original Daniel) observes the brutal sexual hierarchy and the merciless dispatching of weaker members of the tribe among the humans that have regressed to a stage of savagery and compares them to the descriptions of erotic competition and undignified ageing that he has found in the journals left by Daniel1, concluding that 'they were only the caricature-like residues of the worst tendencies of ordinary mankind, the kind that Daniel1 knew already.'[4] Daniel25's temporal and genealogical distance from the behaviour he is describing casts him in the position of the evolutionary historian and sanctions as a reasoned comparison what might otherwise appear rather preposterous. The gesture inverts the order of evolutionary psychology's deductions and thereby seems to confirm its central hypothesis: that the psychological mechanisms controlling the most basic of human behaviours – especially in the matter of sex – are adaptations selected for survival in our ancestors of the Pleistocene era, a primitive humanity to which Houellebecq's future people very quickly revert.

This naturalising discourse has a tendency to present social, financial and, once again, especially sexual inequality as the inevitable result of an atavistic heritage. Love relations are presented as a simple case of the *survival of the fittest*, with the stronger partner blithely destroying the weaker one. For the younger generation, in *La Possibilité d'une île*, inequalities such as economic disparity appear as a mere fact of life: 'to strike in protest of planned redundancies would have seemed to [Esther] as absurd as striking against the weather getting colder, or the invasion of North Africa by crickets.'[5] The novel also contains reflections on the moral heredity of violence, with Daniel25 suggesting that, if it is the strongest and cruellest of creatures that survive to breed, then the potential for aggression will naturally be transmitted from generation to generation. Such considerations actually run contrary to currently received evolutionary wisdom, the Seville Statement on Violence for instance insisting that 'it is scientifically incorrect to say that in the course of human evolution there has been a selection for

---

4 Michel Houellebecq, *The Possibility of an Island*, trans. by Gavin Bowd (London: Phoenix, 2005), p.402. 'Ils n'étaient que le résidu caricatural des pires tendances de l'humanité ordinaire, celle que connaissait déjà Daniel1', Houellebecq, *La Possibilité d'une île* (Paris: Fayard, 2005), p.462.

5 *The Possibility of an Island*, p.163. 'Une manifestation contre un plan de licenciements lui aurait paru aussi absurde qu'une manifestation contre le rafraîchissement du temps, ou l'invasion de l'Afrique du Nord par les criquets pèlerins', *La Possibilité d'une île*, pp.192-93.

aggressive behaviour more than for other kinds of behaviour.'[6] Evolutionary psychology, while recognising the aggressive competition for mates among higher primates like gorillas, has tended to focus instead on the development of a nurturing side to human males, with the evolution of greater parental investment helping to ensure the survival of human infants. Nonetheless Daniel1 opines, 'the root of all evil was biological, and independent of any imaginable social transformation.'[7] Such suggestions locate a broad biological *cause* for social and cultural phenomena, but to what extent can this discourse be regarded as scientific? Karl Popper, for instance, rejects as 'metaphysical' the principle of causality itself, which he defines as 'the assertion that any event whatsoever can be causally explained – that it can be deductively predicted,' because to believe as much is to imply an end to science, where hypotheses can no longer be refuted or superseded.[8] Meanwhile, Steven Rose reminds us of the need to distinguish between *enabling* and *causal* factors: certainly biological evolution has *enabled* all of the socio-cultural phenomena we see today, but 'this does not entitle one to say, for any observed behaviour, that it is *caused* by such processes.'[9] Particularly since, as Kevin Laland and Gillian Brown point out, 'a fundamental part of the causal pathway is missing,' namely a neurobiological theory describing how genes determine the growth and connection of neurons which would then influence psychological states and ultimately behaviour.[10]

Houellebecq's 'scientific' dissection of society, then, rests principally on a discourse of evolution, especially in the genealogical narratives of *Les Particules élémentaires* and *La Possibilité d'une île*. Houellebecq's latest novel repeatedly insists that sexual congress, with the ultimate – if unconscious – aim of the transmission of genetic information, remains the one and only goal of all animal life, including the human. Sex is 'the sole pleasure, the sole objective of human existence,' and 'when the animal is no longer good for

---

6 The Seville Statement on Violence (1986). <portal.unesco.org/education/en/ev.php-URL_ID=3247&URL_DO=DO_TOPIC&URL_SECTION=201.html> [accessed 27 April 2009].

7 *The Possibility of an Island*, p.134. 'La racine de tout mal était biologique et indépendante d'aucune transformation sociale imaginable', Houellebecq, *La Possibilité d'une île*, p.159.

8 Karl Popper, *The Logic of Scientific Discovery* (London: Routledge, 2002 [1959]), p.39.

9 Steven Rose, 'Escaping Evolutionary Psychology', in *Alas, Poor Darwin: Arguments Against Evolutionary Psychology*, ed. by Hilary Rose and Steven Rose (London: Vintage, 2001), pp.247-65 (p.251).

10 Kevin N. Laland and Gillian R. Brown, *Sense and Nonsense: Evolutionary Perspectives on Human Behaviour* (Oxford: Oxford University Press, 2002), p.185.

reproducing, it is absolutely no longer good for anything.'[11] Throughout the novel, the stress is on predation, cruelty and ruthless sexual selection as the *truth* of life. As such, Houellebecq's position bears some similarity to Richard Dawkins's theory of the 'selfish gene,' which has been criticised as reproducing a kind of Schopenhauerian metaphysics that sees life as characterised by a base inheritance, constant toil and conflict, and the possibility of salvation through the paradoxical achievement of altruism.[12] The view of evolution presented by Houellebecq is an ideologically loaded one. In fact, evolutionary theory argues that cooperation is as important as conflict and predation in determining the patterns of evolution, while the notion of sexual selection, in addition to having little biological foundation, is an evolutionary paradox, since the organs and behaviours developed to attract a mate often render an animal less fit for survival, in some cases even leading to extinction (which may be relevant to Houellebecq's argument).[13] In other words, the kind of evolution discussed by Houellebecq remains thoroughly rooted in its nineteenth-century origins of statistical mechanics, Malthusian population theory and Victorian sexual morality: as Friedrich Nietzsche put it, this theory 'exudes something like the stuffy air of English overpopulation, like the small people's smell of indigence and overcrowding'.[14] Nietzsche was already highly critical of the anthropomorphism involved in attributing 'laws' to nature and even more so of reading off an 'ethics' fit for man. Nietzsche attempted to recast Darwinian adaptation as a kind of 'survival of the weakest,' 'a largely negative feedback mechanism that encourages the physiologically weak and ill-constituted to gather together in herds in order to maximise their opportunities for self-preservation,' the theory of natural selection thus appearing as a conservative one that rests upon the 'value' of preservation.[15]

---

11 *The Possibility of an Island*, p.341; 'l'unique plaisir, l'unique objectif en vérité de l'existence humaine', Houellebecq, *La Possibilité d'une île*, p.392; *The Possibility of an Island*, p.189; 'lorsque l'animal n'est plus bon à se reproduire il n'est absolument plus bon à rien', *La Possibilité d'une île*, p.222.

12 See notably Keith Ansell Pearson, *Viroid Life: Perspectives on Nietzsche and the Transhuman Condition* (London: Routledge, 1997), p.13 n.3.

13 See for example Robert Wesson, *Beyond Natural Selection* (Cambridge, MA: MIT. Press, 1991), pp.157, 163; Rose, pp.250-51; Wesson, pp.167-69.

14 Friedrich Nietzsche, *The Gay Science*, trans. by Josefine Nauckhoff (Cambridge: Cambridge University Press, 2001 [1887]), p.208.

15 Ansell Pearson, p.88. Houellebecq, intriguingly, follows Nietzsche's lead when he suggests that love may be 'a fiction invented by the weak to make the strong feel guilty, to introduce limits to their natural freedom and ferocity' (*The Possibility of an Island*, p.294; 'une fiction inventée par les faibles pour culpabiliser les forts, pour introduire des limites à leur liberté et à leur férocité naturelles', *La Possibilité d'une île*, p.341). In this, Houellebecq obviously diverges from the official line of evolutionary

For Nietzsche, adaptation is merely a secondary process operating after forces have shaped the organism from within and tested themselves upon the environment – the very notion of the 'useful' only arises 'as an indirect result of this complex process'.[16] This is borne out by contemporary evolutionary theory which maintains that, if natural selection can account for the disappearance of species, it is helpless to explain their appearance which is due instead to internal processes of mutation and molecular drive.[17]

Houellebecq's vision of natural selection operates not only upon individuals but on whole societies, since *Les Particules élémentaires* and *La Possibilité d'une île* both present a moribund human race that has become unfit for survival and is shortly to be superseded by a species that is better adapted to the particular challenges facing higher forms of conscious life. This extension of the discourse of selection is valid to the extent that genes are distributed not only throughout an individual organism, but throughout a population or an ecosystem, making these, as much as the individual, the 'units' of evolution.[18] However, Keith Ansell Pearson, following Nietzsche and Deleuze and Guattari, argues that the species, just as much as the individual, is retrospectively constituted by the discourse of selection and anthropomorphises the 'flows, intensities, and pre-vital singularities of pre-stratified, non-organic life' that are the real material upon which evolution operates.[19] Houellebecq's view of social and species evolution is expressed in the language of thermodynamics, the 'laws of social entropy' stating that 'When a social system is destroyed, this destruction is definitive, and there can be no going back.'[20] His diagnosis of the current crisis is based on a simple energetics whereby the constant increase of sexual energy within the system, deprived of a suitable outlet within which to expend or transform itself, can only bring about the destruction of that system: 'To increase desires to an unbearable level whilst making the fulfilment of them more and more inaccessible: this was the single principle upon which Western

---

psychology, for which love is a psychological by-product of the 'sexual imprinting' that favours long-lasting pair bonds in humans, again thought to have adapted in order to ensure infant survival. See Lance Workman and Will Reader, *Evolutionary Psychology: An Introduction* (Cambridge: Cambridge University Press, 2004), p.90.

16 Ansell Pearson, p.97.

17 See Richard J. Bird, *Chaos and Life: Complexity and Order in Evolution and Thought* (New York: Columbia University Press, 2003), pp.37-38 and Wesson, pp.140-41.

18 Rose, p.259.

19 Ansell Pearson, p.130.

20 *The Possibility of an Island*, p.310. 'Lorsqu'un système social est détruit, cette destruction est définitive, et aucun retour en arrière n'est possible', *La Possibilité d'une île*, p.358.

society was based.'[21] This appeal to thermodynamics further reinforces the determinist slant of the discourse, Daniel25 suggesting that, if life were to reappear on earth at some future date, it would necessarily follow the same pattern: 'nothing new could be expected from it.'[22] But, once again, this is an anthropomorphic reification which transforms evolution into a directed teleology rather than a blind experimentation. As Ansell Pearson reminds us, thermodynamics is itself metaphysical, presenting a view of the universe that proceeds through the cancellation of difference on its linear trajectory towards a state of absolute heat death, which conceals an anthropomorphic *desire* for death conceived in terms of being.[23] In addition to ignoring the reality of the universe as a 'monster of energy' engaged in a directionless becoming without end, such a view also overlooks the central insight of complexity theory which suggests that life itself is a kind of managing of disorder, operating through a gradual and controlled increase of entropy. Meanwhile, since population development has been shown to display some of the statistical tendencies of chaotic systems, then the notion of sensitive dependence on initial conditions, central to chaos theory, would seem to make a mockery of the idea that life could ever develop and evolve again according to patterns already observed.

In the remainder of this essay, I wish to consider a series of problems, or paradoxes, that arise out of Houellebecq's use of science and that perhaps point to contradictions within his apparent ideological project. I will summarise these problems briefly here, before going on to discuss them in more detail below. First, we have the paradox of a species (humanity) ruled by a determinist heredity and a ruthless sexual selection that nonetheless chooses, apparently freely, to bring about its own demise. Second, humanity is therefore shown to be capable of organising its own overcoming, despite appearing, in all other instances, to be fixated on the immediate satisfaction of short-term goals (typically sexual). Third, a cloned future initially seems to offer the promise of a single consciousness existing across discrete incarnations but, in fact, Houellebecq's future effectively shows that this is only possible to the extent that the very notion of individual consciousness becomes meaningless and irrelevant to existence. Fourth, in a final reversal, however, the cloned Daniel strikes out alone, seemingly finding a need for individuality in a civilisation where such a concept should be redundant.

---

21 *The Possibility of an Island*, p.68. 'Augmenter les désirs jusqu'à l'insoutenable tout en rendant leur réalisation de plus en plus inaccessible, tel était le principe unique sur lequel reposait la société occidentale', *La Possibilité d'une île*, p.85.

22 *The Possibility of an Island*, p.420; 'rien de nouveau ne pouvait en être attendu', *La Possibilité d'une île*, p.482.

23 Ansell Pearson, pp.61-63.

Perhaps the most significant of these problems emerges from Houellebecq's evolutionary schema. On the one hand, as we have seen, the novels repeatedly describe human beings as the passive victims of atavistic forces, lacking any real freedom to change their biological fate. Yet on the other hand, both *Les Particules élémentaires* and *La Possibilité d'une île* envision an evolutionary leap beyond the human race that is *chosen and set in train by humanity itself.* The asexual, immortal clones of *Les Particules* and the genetically 'rectified' neo-humans of *La Possibilité* are both presented as the result of a conscious choice by their precursors. Humanity thus *wills its own demise.* Now, it is known that populations will sometimes deliberately limit their growth in order to maximise the resources of their habitat, and Houellebecq seems almost to present his science-fiction future as a logical extension of such Malthusian considerations when he describes the increasing popularity of a culture of childlessness.[24] Nonetheless, a collective decision to bring about the end of one's own species would be an evolutionary novelty, to say the least. Daniel24 describes the disappearance of the human race as a kind of 'mass suicide,' whereas the narrator of *Les Particules* notes 'the meekness, the resignation, perhaps even the relief of humans at their own passing away'.[25] In this way, humanity is gradually overcome by an idea that takes on a life of its own at the expense of the species responsible for the conception and transmission of that idea. Here, then, is an extreme demonstration of Richard Dawkins's theory of memetics – a kind of survival of the fittest for ideas, whereby concepts that are particularly attractive or replicable may thrive even at the evolutionary cost of their hosts.

Yet elsewhere, humans are shown to be short-sighted or impotent with regard to their own future: the neo-human Supreme Sister teaches of humanity's weakness in failing to take advantage of an ageing population in the west and of epidemics in the developing world in order to solve its crises of resources; she also points out that their ability to imagine a catastrophic future in science-fiction novels did not enable them to put in place any preventive measures. How could such a decadent and ill-constituted humanity thus have had the foresight to adopt and organise a programme for its own relief by a superior species? Not only does this seem improbable – since evolutionary history offers few, if any, examples of a successfully

---

24 Laland and Brown, p.73.

25 *The Possibility of an Island*, p.33; 'suicide collectif', *La Possibilité d'une île*, p.45; Houellebecq, *Atomised*, trans. by Frank Wynne (London: Vintage, 2000), p.378; 'avec quelle douceur, quelle résignation, et peut-être quel secret soulagement les humains ont consenti à leur propre disparition', *Les Particules élémentaires* (Paris: Flammarion, 1998), p.316.

adapted class of animals giving rise to its successor – but it appears as a further anthropomorphising of the very process of evolution: is Houellebecq not simply rationalising the projection of a willed direction to evolution by attributing it to humanity itself?[26] Ultimately, then, this view of evolution rejoins older metaphysical systems, which find their apogee in Hegelian idealism, and which project a teleological moral sense on to the process of evolution by describing an omni-directional ascent towards a pinnacle of consciousness. Oddly enough, for instance, Houellebecq's evolutionary vision is not so different from that of Pierre Teilhard de Chardin, object of an affectionately bemused portrait in *La Possibilité d'une île* where he is accused of a guileless inanity. When Teilhard de Chardin, in the profoundly Hegelian *Le Phénomène humain*, envisages the possibility of 'an organic crisis in evolution,' he might be describing the advent of neo-humans as postulated by Houellebecq: 'There is a danger that the elements of the world should refuse to serve the world – because they think; or more precisely that the world should refuse itself when perceiving itself through reflection. Under our modern disquiet, what is forming and growing is nothing less than an organic crisis in evolution.'[27]

In this sense, then, Houellebecq can be seen to be presenting another version of 'hi-tech Hegelianism' or 'antiseptic posthumanism' of the kind that has been so incisively attacked by Keith Ansell Pearson.[28] He argues that the notion of a teleological development evolving towards a sort of technological, disembodied consciousness belongs to an ascetic ideal that, rather than seeing the human animal as the fertile ground from which to launch its transhuman potential, presents the contingent human body as something to be negated, sublated or otherwise *escaped from*. Admittedly, Houellebecq's new breeds still have bodies, but the cleansed neo-human body of *La Possibilité d'une île*, freed from the shameful burden of those abject organs, the digestive tract and the anus, and released from the irksome bind of physical interaction, are one step closer to existing as pure thought. More to the point, in order for neo-human clones to exist at all, the memory of their human precursor has to be stored as data in order to be downloaded into the brain of the future clone, who emerges as a fully-

---

26 Wesson, p.159.

27 Pierre Teilhard de Chardin, *The Phenomenon of Man*, trans. by Bernard Wall (London: Collins, 1959), pp.253-54. 'Les elements du Monde refusant de servir le Monde parce qu'ils pensent. Plus exactement encore, le Monde se refusant lui-même en s'apercevant par Réflexion. Voilà le danger. Ce qui, sous l'inquiétude moderne, se forme et se grossit, ce n'est rien moins qu'une crise organique de l'Évolution', Teilhard de Chardin, *Le Phénomène humain* (Paris: Seuil, 1955), p.255.

28 Ansell Pearson, pp.33, 35.

formed adult, the species having dispensed with the messy business of childhood. The novel thus comes close to suggesting that personality can be reduced to a series of memories, themselves reducible to reinforced neural circuits which may be stored as data to be reproduced at a later date. This, the ultimate dream of Artificial Intelligence, has been challenged by Roger Penrose who suggests that consciousness may have a physical basis in processes taking place at a quantum level in the brain, the computer-like system of connection between neurons appearing as the mere shadow of a smaller-scale cytoskeletal activity operating in the brain cells.[29] Penrose has further offered a mathematical demonstration that human understanding relies on processes that are fundamentally non-computable, a demonstration to which Houellebecq seems to allude, when Daniel25 admits that the dream of up- and downloadable human memory was abandoned in favour of a provisional combination of biochemistry and behaviourism: 'The metaphor of the human brain as a fuzzy Turing machine was finally shown to be completely sterile.'[30]

With such considerations, *La Possibilité d'une île* revisits the old philosophical and theological problem of the persistence of consciousness across discrete physical incarnations. This problem is raised by Houellebecq's beloved Schopenhauer in his essay 'On the Indestructibility of Our Essential Being by Death:' Schopenhauer imagines an individual agreeing to be reincarnated with the same unique consciousness after a completely unconscious death-sleep of three months. But, he points out, if this sleep is completely unconscious, then it becomes entirely immaterial to the sleeper whether it lasts for three months or ten thousand years, or indeed whether he is ever woken at all.[31] In other words, the very notion of a unique consciousness only has meaning from an anthropocentric position that valorises individual identity as stable, eternal *being*. From the 'perspective' of evolution, this is a nonsense. Houellebecq's answer to this problem in *La Possibilité d'une île* is disconcertingly honest: as we read the commentaries of the successive Daniels, it becomes clear that these clones *do not* have a continuous consciousness, but that this doesn't matter because they know their lives to be rigorously *identical*. That is to say that the only way to overcome the anguish associated with the extinguishing of

---

29 Roger Penrose, *Shadows of the Mind: A Search for the Missing Science of Consciousness* (London: Vintage, 1994).

30 *The Possibility of an Island*, p.210. 'La métaphore du cerveau humain comme machine de Turing à câblage flou devait se révéler en fin de compte parfaitement stérile', *La Possibilité d'une île*, p.245.

31 Arthur Schopenhauer, *On the Suffering of the World*, trans. by R. J. Hollingdale (London: Penguin, 2004 [1850]), pp.48-49.

consciousness is to relinquish the narcissistic claim to individuality and reduce life to its *essential* characteristics. As Daniel25 puts it: 'the limited, respectable memories we keep of existences that have identical contours do not have any of the pregnancy that would be necessary for an individual fiction to take hold.'[32]

And yet this rejection of individuality proves untenable. The neo-human clones of *La Possibilité d'une île* display a certain nostalgia for individuality, for instance in the poetry written by Marie, another clone with whom the series of Daniels communicate, or in Daniel25's departure from his secure compound in search of an alternative community. The overcoming of humanity can only fail because the announcement of the 'death' of the human remains an all too human, anthropocentric gesture. Worse, the departure of the human marks the suspect return of Grand Narratives, which tend to disallow all practical political choices by pre-empting them. As Ansell Pearson argues, the notion that carbon life may be evolved out of existence by a higher level of silicone-based intelligence leaves us only with such empty options as 'ethical purism, futile Ludditism, or vacuous cyber-celebrationism'.[33] This, as Ansell Pearson points out in his trenchant critique of Jean-François Lyotard, is the danger of being seduced by the depoliticisation of capital, of confusing its monstrous development – which is nonetheless grounded in social relations of production – with a reified concept of *evolution itself*. (And we know that evolutionary theory has provided a justification for the most ruthless of capitalist business practices ever since Herbert Spencer first coined his simplifying slogan 'Survival of the fittest' in 1864.)

We might argue that Houellebecq falls into this trap, mistaking a certain *promotion* of sexuality by consumer capitalism with an organic evolution of humanity's reproductive competence, and hence of its capacity for survival. Houellebecq, as I suggested in my introduction, sees science as having replaced religion but, as *La Possibilité d'une île* demonstrates, in doing so it *ceases to be science*. Karl Popper reminds us that there is no *essential difference* in the nature of the knowledge offered to us by science, on the one hand, and metaphysics on the other. The difference lies rather in their *methodology*, with knowledge ceasing to be scientific when it resists the possibility of refutation.[34] *La Possibilité d'une île* makes this difference clear when it reveals

---

32 Houellebecq, *The Possibility of an Island*, p.371; 'les souvenirs limités, avouables, que nous gardons d'existences aux contours identiques, n'ont nullement la prégnance nécessaire pour que la fiction individuelle puisse y prendre appui', Houellebecq, *La Possibilité d'une île*, p.426.

33 Ansell Pearson, p.148.

34 Popper, pp.12-13.

that the Elohimite religion, despite its apparent scientific credentials, is actually based on a publicity stunt. The event that transforms Elohimism from a minority cult into a global religion – the supposed 'resurrection' of the Prophet – is in no sense a *proof* of the sect's scientific promise of immortality, but on the contrary a cheap, if neatly stage-managed *trick* enacted for the world's media. All of which suggests, perhaps, that to discuss serious science in relation to Houellebecq may be to miss the point entirely: for what his latest novel implies is that the deepest desires of today's humanity cannot be met by the rigour, the patience and the humility of the scientific method, but only by the clamour and cry of media and merchandise. What we want is not controlled experimentation, restricted results and improved understanding; what we want is a good story, preferably with pictures. What we want is something to believe in.

'Once in Human Nature, a Thing Cannot be Driven Out':[1] Evolutionary Aesthetics in Wilhelm Jensen's *The Legacy of Blood* (1869). An Early Response to Darwin

Nicholas Saul

Abstract: Whilst Darwinian evolutionary aesthetics in their more sophisticated versions are an interesting and increasingly plausible theory, they still tend to devalue both the autonomy and the cognitive power of literature and so lay themselves open to the charge of foundationalism. In this respect the modern debate unwittingly recapitulates German discussion in the late nineteenth century. The contributions to evolutionary aesthetics by Nordau and Bölsche in the eighteen-eighties on precisely these issues are reconstructed. It is then demonstrated that there is a still earlier, pioneering phase of reception, not by theoreticians but by German novelists of the eighteen-sixties, notably Raabe and Jensen. Jensen especially attempts to show as a convinced practitioner of evolutionary aesthetics how aesthetic experience (narrative) discloses cognitive horizons of personal development beyond Darwinist prescriptions of inherited behaviour. Darwinian evolutionary theory thus creatively generates literary form, and literature can be described as Darwinian without being reduced to a servant discourse.

As Darwin's year of 2009 approached, theories of evolutionary aesthetics came to be elaborated with ever more frequency and sophistication and been applied with increasing energy and conviction to the task of resolving one of modern literature's – or literary theory's – recurrent legitimation crises. Now this is not the place to attempt a representative overview, but perhaps one of the most striking features of this burgeoning movement is its confident reversal of the accepted order of things in the world of scholarly humanities discourse.[2] Since Kant – unless we count the most elementary forms of Marxian or Freudian criticism – the theory of the autonomy of the literary text has never seriously been challenged. Even in today's pluralised

---

1 'Was einmal im Menschen liegt, ist nicht heraus zu treiben', in Wilhelm Jensen, *Das Erbtheil des Bluthes.* Erzählung (Berlin: Expedition des Sonntags-Blattes, Duncker, 1869), p.1. Translations throughout are my own.

2 Fuller accounts can be found elsewhere in this volume, notably in the essay by Katja Mellmann. In the anglophone world – doubtless the dominant zone – suffice it to mention Joseph Carroll, *Literary Darwinism: Evolution, Human Nature, and Literature* (New York, London: Routledge, 2004); *The Literary Animal: Evolution and the Nature of Narrative*, ed. by Jonathan Gottschall and David Sloan Wilson (Evanston, IL: Northwestern University Press, 2005); and Brian Boyd, *On the Origin of Stories: Evolution, Cognition, and Fiction* (Cambridge, London: Harvard University Press, 2009). On the germanophone side, where evolutionary aesthetics are less prominent, the main work has been Karl Eibl, *Animal poeta: Bausteine der biologischen Kultur- und Literaturtheorie* (Paderborn: Mentis, 2004).

scene, from hermeneutics to new criticism and deconstruction to new historicism, literary autonomy continues axiomatically to serve as the (often unspoken) basis of most theories of literature. The Foucauldian tradition, to be sure, is more concerned with the deep-structural features of the *dispositif* than observations of surface form. But those who – to good effect – mediate between Foucauldian discourse theory and literary analysis take care to emphasise the unique ontological ability of the literary text to escape the determination of discourses in the presentation of reality, reflect critically on them, and offer a liberating counter-discourse.[3] Precisely this autonomy of the text and of cultural phenomena in general is challenged by the theoreticians of evolutionary aesthetics. The pragmatic philosopher Daniel Dennett, a committed Darwinian evolutionist, perhaps initiated the trend, when in *Darwin's Dangerous Idea* (1995) he insisted on the quasi-absolute foundational character of this strand of evolutionary theory. He thus insisted both its universal applicability (*qua* 'universal solvent') and issued a call to apply it in any given field of investigation.[4] That Dennett included literature in this scheme is indicated by the little song 'Tell Me Why,' which opens and closes his volume, in interpretations before and after the Darwinian treatment. Joseph Carroll, as one of the leading proponents of evolutionary criticism in Dennett's wake, makes the evolutionary position thoroughly explicit. For Carroll, the basic concept of 'cultural autonomy' in humanities and social sciences alike, is itself in crisis.[5] Modern literary theory of whatever flavour is dominated by a foundational discourse of constructivism, which by definition decouples all manifestations of culture from nature. Freeing itself from natural determinations, it by that very act also condemns itself to a private world of non-referential irrelevance.[6] But as evolutionary social and psychological theory now (rightly) dominates the life and social sciences, so too its time has come in literary theory. This

---

3 See *Diskurstheorien und Literaturwissenschaft*, ed. by Jürgen Fohrmann and Harro Müller (Frankfurt: Suhrkamp, 1988).

4 Daniel Dennett, *Darwin's Dangerous Idea. Evolution and the Meanings of Life* (Harmondsworth: Penguin, 1995), p.521.

5 Carroll 'Introduction', in *Literary Darwinism*, pp.vii-xxv (p.ix).

6 Carroll, 'Biology and Postructuralism', in *Literary Darwinism*, pp.15-27 (pp.15ff.). The doyenne of the English anglophone tradition, Gillian Beer, has, notably, maintained a distance from this line of theorising, and indeed could be seen as offering deconstructive investigations of the relations between Darwinism and literature. See Gillian Beer, *Darwin's Plots: Evolutionary Narrative in Darwin, George Eliot and Nineteenth-Century Fiction* (London: Routledge, 1983) and, especially, her essay 'Translation or Transformation? The Relations of Literature and Science', in *Open Fields: Science in Cultural Encounter* (Oxford: Clarendon, 1996), pp.173-95. Carroll's comment on this can be found in *Literary Darwinism*, p.27.

paradigm is admittedly still on the way to itself. But since by contrast to the constructivist paradigm the evolutionary explanation of human experience is, if not absolutely, at least 'relatively true,' then it 'necessarily provides the basis for any adequate account of culture and of literature'.[7]

What, then, does this plausible new evolutionary foundationalism – proof it seems against any anti-foundationalist critique à la Rorty – offer by way of new insights into the function and meaning of *texts*?[8] From an orthodox, not-yet-evolutionary perspective, one needs to look surprisingly hard. For the most part, it would seem, we find something like a sociobiological reading of familiar narrative texts such as *Pride and Prejudice*, in which parallels are sought between the structure of human motives and concerns (as derived from evolutionary psychology) and the organizing principle of literary representation (Carroll, p.108).[9] In short: the rules of the mating game can be discerned in Jane Austen, and literature of this kind can thus be said to serve an adaptive-educational function (p.116). Now of course there is more than one kind of reduction, adequate and inadequate.[10] There is, moreover, more (especially in the way of supporting argument) to Joseph Carroll's theory of literary Darwinism than my brief adversion indicates, and indeed more to the theory of evolutionary poetics in other, perhaps more differentiated accounts, than Carroll's. Karl Eibl, for example, frankly accepts the autonomy of culture and literature, but also seeks to mediate between natural and cultural domains through a dialectic based in the origin of natural language.[11] Eibl, moreover, follows Stephen Pinker in acknowledging one chief function of aesthetic experience as providing pleasure (Eibl, p.23), and is unique in recognising the particular illusionistic quality of aesthetic experience and the differentiating semantic power of literary form.[12] (This is perhaps the strength of operating in the received German tradition of aesthetic theory.) Yet one must search hard, even in Eibl, for an evolutionary theory of literature which recognises surely one basic and redemptive feature of aesthetic discourse, its – autonomous – *cognitive potential*, considered as stretching beyond the epistemological horizon circumscribed by the master discourse of Darwinian evolutionary

---

7 Carroll, 'Biology and Postructuralism', p.25.

8 See Richard Rorty, *Philosophy and the Mirror of Nature* (Princeton: Princeton University Press, 1980), especially pp.157ff. and (on Darwinism) p.366.

9 See Carroll, 'The Deep Structure of Literary Representations', in *Literary Darwinism*, pp.103-16.

10 See Eibl, *Animal poeta*, pp.29-31.

11 Eibl, *Animal poeta*, pp.14-23. On the origin of natural language see Joseph Emonds's essay in this volume.

12 See notably Brian Boyd's overview 'Evolutionary Theories of Art', in, *The Literary Animal*, pp.147-76; and Eibl, *Animal poeta*, pp.209-376.

psychology.[13] And yet, and yet: even this objection fails to negate entirely the seductive charisma of Dennett's initial proposition. If Darwin's theory and its derivations are the case – and there is no doubting their authority in other areas of scholarly endeavour – then in some way the human activity of producing culture *must* be relatable to evolutionary theory.

That said, another remarkable feature of this contemporary debate is the extent to which it recapitulates some of the major features of the debate which followed the initial publication of Darwin's major works in the nineteenth century, yet without recuperating that fact.[14] One key issue of this debate was precisely the foundational character of Darwinian evolutionary theory, which to his German readers at once (despite Darwin's careful rhetorical packaging) seemed to suggest the necessity of a new beginning across the scientific board. Nonetheless, it seems to have taken a little time for the aestheticians proper to take up the hints which Darwin himself (no sort of aesthetic theoretician) had sedimented in his works, notably of course in the theory of sexual selection.[15] The earliest example of an explicitly Darwinian aesthetic, so far as I can tell, is by Max Nordau (1849-1923). His well-known Lombrosian-Darwinian study *Entartung* (*Degeneration*, 1892-93), the touchstone of mass cultural anxiety in the Fin de siècle, was in fact preceded by a little essay 'Evolutionistische Ästhetik'

---

13 Ricœur's theory of the cognitive potential of poetic language vis-à-vis natural language and the philosophical proposition (as a referentiality shift of metaphor) would be an obvious source of this argument. See Paul Ricœur, *The Rule of Metaphor. Multi-disciplinary Studies of the Creation of Meaning in Languager* (Toronto: University of Toronto Press, 1977), especially pp.216-313.

14 The honourable exception is Karl Eibl, 'Darwin, Haeckel, Nietzsche. Der idealistisch gefilterte Darwin in der deutschen Dichtung und Poetologie des 19. Jahrhunderts. Mit einer Hypothese zum biologischen Ursprung der Kunst', in *Fritz Mauthner: Sprache, Literatur, Kritik. Festakt und Symposion zu seinem 150. Geburtstag*, ed. by Helmut Henne and Christine Kaiser (Tübingen: Niemeyer, 2000), pp.87-108. The standard works on German intellectual and cultural reception of Darwinism are Alfred Kelly, *The Descent of Darwin. The Popularisation of Darwinism in German 1860-1914* (Chapel Hill: University of North Carolina Press, 1981) and Paul Weindling, *Health, Race and German Politics between National Unification and Nazism, 1870-1945* (Cambridge: Cambridge University Press, 1989).

15 On Darwin, literature and taste see Janet Browne, *Charles Darwin: A Biography*, 2 vols (London: Jonathan Cape, 1995-2002), II: *The Power of Place* (2002), pp.68-71. Although Philip Ajouri, *Erzählen nach Darwin. Die Krise der Teleologie im literarischen Realismus: Friedrich Theodor Vischer und Gottfried Keller*, (Berlin: de Gruyter, 2007), pp.175-94, has made the case for Vischer's late revisions of his aesthetics in 1866 and 1873 as the transition between the Hegelian-idealist aesthetics of the poetic realist era and the Darwinian-Nietzschean materialist aesthetics of Naturalism.

('Evolutionist Aesthetics') of 1885.[16] Spencer's *Principles of Biology* (1864-67), as Nordau himself confesses, is his inspiration.[17] From Spencer's observation that most beauty in the organic and indeed the cultural world seems to derive from the process of sexual selection, and in declared total opposition to 'amtlich gelehrte [...] Ästhetik' (p.291; 'official scholarly aesthetics'), Nordau proposes to derive a 'natürliche[...] Schönheitswissenschaft' (p.291; 'a natural science of beauty'). The magic Darwinian word 'origin' is the key term here. As Kant's (notoriously underived) aprioristic categories of the understanding are in fact the acquired, now innate result of millennia of adaptive behavioural experiments, and as ethical behaviour is the outcome of the necessity for collective co-existence in the cause of survival, so our aesthetic experience has nothing to do with the traditional account of transcendent revelation (as Nordau puts it, p.293), but emerges from the learnt and reinforced responses of pleasure and unpleasure in repeated experience of that which fosters or threatens survival. It is the familiar feedback loop: the more something furthers our survival, the more pleasure we feel, the more we pursue the thing which gives us pleasure, the better the chances of survival and reproduction ... until the behaviour has become programmed into the mind and becomes inherited in successfully selected individuals. All aesthetic experience is explained by this evolutionary origin. Not all is sexual in origin. A sense of the sublime is occasioned in the individual by the pleasurable consciousness that the source of existential fear cannot really threaten us. But the experience of beauty is strictly sexual. It is primarily male (and internalised from the dominant male by the female), and consists in the 'organisch gewordene Gewohnheit' (p.305; 'organic habituation'), whereby the sight of a sexually mature and attractive woman immediately excites 'das höchste Geschlechtszentrum im Gehirn' (p.305; 'the highest sexual centre in the brain'). The familiar feedback loop operates, and results for example in figurations of female beauty exemplifying abstracts such as fatherland, fame, friendship, wisdom, etc. Art, concludes Nordau triumphantly, can succeed only if it follows in the path nature has already trodden (p.309). The filiations of this with Joseph Carroll's Austen interpretation (and others) are evident. Wisely, perhaps, Nordau (p.310) declined the chance to elaborate this theory systematically.

A much more powerful, yet equally problematic beginning is that by Wilhelm Bölsche (1861-1939). Like Nordau a committed Darwinian

---

16 Max Nordau, *Entartung*, 2 vols (Berlin NW: Duncker, 1896); Nordau, *Paradoxe* (Leipzig: Elischer, 1885), pp.290-310.

17 'Evolutionistische Ästhetik', pp.290-91. Compare Spencer, *The Principles of Biology*, 2 vols (London: Williams & Norgate, 1864-7), II, p.253n.

evolutionist, Bölsche makes his position on the foundational role of Darwinian theory clear in his works title: *Die naturwissenschaftlichen Grundlagen der Poesie* (1887; *The Scientific Foundations of Literature*).[18] As with Nordau, there is no opposition in principle between science and literature. The true foundation of the arts, now in crisis, are the sciences, and that means regeneration through Darwinism. But if Nordau is interested in aesthetic reception (or consumption), Bölsche, himself a practising novelist, is interested in production, namely the reform which made out of the then dominant poetic realism that abrasive revitalisation we know today as Naturalism and which ultimately became aesthetic Modernism. Thus Bölsche, himself clearly in this respect in the Romantic tradition, concentrates on the creative role of aesthetic experience in the process of evolution itself. Adaptative modifications in the physiological sense, which produce victory in the struggle for existence are, says Bölsche, nothing less than the 'Genie' (Bölsche, p.55; 'genius') of the animal world. Their analogy in the human world is however something not strictly physical, namely ideas, and the ability to generate ideas as chance mutations of intellectual species is humanity's 'genius'. These ideas, somewhat in the manner of Dawkins's well-known selfish gene – or rather meme – struggle for victory in life only on their own behalf, not that of their progenitor: hence Christ's idea, and Christ's fate. The task of the poet, as of other Darwinian geniuses, is Darwinian truth finding by intellectual struggle, in his case in the world of literature. Literature does, then, have a familiar role as adaptive educator, populariser of the authoritative Darwinian perspective on the individual, society and history. But its chief role is as experimenter, imaginative combiner of Darwinian ingredients – notably inherited traits and notably examining the issue of free will and determinism – under controlled conditions to arrive at truths of equal dignity with those of scientific psychology and to the benefit of both. (And Zola is frequently mentioned.) But that is not all. Literature, for Bölsche, always encompasses cognitively more than merely a limited perspective and strives to encompass 'das Ganze' (p.33; the 'totality'). Literature grasps that even the psychophysical principles of modern natural science disclose no ultimate truths about human nature, so that it can even explore domains beyond earthly life and death.[19] And poetic genius is in any case epistemologically *sui generis*. Thus Bölsche is clearly keen to open out the cognitive horizon of an experimental

---

18 *Die naturwissenschaftlichen Grundlagen der Poesie. Prolegomena einer realistischen Ästhetik*, ed. by Johannes J. Braakenburg (Munich: Niemeyer, 1976) (1887), pp.1-65.

19 Bölsche, p.33. Compare the account of Alfred H. Kelly, *Between Poetry and Science: Wilhelm Bösche as Scientific Popularizer*, unpublished doctoral thesis, University of Wisconsin 1975, pp.23-24.

literature – in the spirit of Darwinism, but *beyond* that limit. This, however, he manages only at a price: that of importing into Darwinist aesthetics something which Nordau, his modern successors, and certainly Darwin himself would have rejected: the spiritualising and teleologising, ultimately Romantic concept of *Genie*. He is unwilling to abandon the foundational status of Darwinian natural science. But he is also evidently unwilling to make autonomous literature into a servant discourse, and he resolves the conflict of Darwinistic science and literature by *aestheticising Darwinism* and hypostasing an absolute creative principle. Of course that soon led Bölsche into the monistic aestheticism characteristic of the high *Fin-de-siècle* and the lifelong close alliance with his friend Ernst Haeckel. Bölsche's idealised Darwinism (following Haeckel, no doubt, but also, ultimately, Eduard von Hartmann's philosophy of the creative unconscious) is typical of one major strand of the German reception of Darwinism.[20] Here again, then, we find key issues posed and key debates in contemporary evolutionary theory prefigured in an early phase of Darwinian reception.

Now Bölsche and to a lesser extent Nordau also wrote fiction in which Darwinian issues were treated. But if they were the revolutionaries and pioneers of evolutionary literary theory and practice *stricto sensu* in Germany, an earlier generation still, of (more or less) poetic realist writers, reflected just as deeply on the consequences of the new Darwinian anthropology, not in their theory but concretely, in fiction.[21] Prominent amongst these were the close friends and bitter rivals Wilhelm Raabe (1831-1911) and Wilhelm Jensen (1837-1911), popular novelists and competitors in the literary marketplace both. Indeed Jensen was one of the most widely published (and

---

[20] On Bölsche and aestheticism see the (frankly unsympathetic) Jutta Kolkenbrock-Netz, *Fabrikation–Experiment–Schöpfung. Strategien ästhetischer Legitimation im Naturalismus* (Heidelberg: Winter, 1981), pp.252-85; and Walter Gebhard, *'Der Zusammenhang der Dinge': Weltgleichnis und Naturverklärung im Totalitätsbewußtsein des 19. Jahrhunderts* (Tübingen: Niemeyer, 1984), especially pp.330-428. On the German idealistic reception of Darwinism see Eibl, 'Darwin, Haeckel, Nietzsche'; on the more sceptical strand of German reception see Ajouri, *Erzählen nach Darwin*, pp.94-142.

[21] See on Darwinism in nineteenth-century German literature Peter Sprengel, *Darwin in der Poesie: Spuren der Evolutionslehre in der deutschsprachigen Literatur des 19. und 20. Jahrhunderts* (Würzburg: Königshausen & Neumann, 1998); Werner Michler, *Darwinismus und Literatur. Naturwissenschaftliche und literarische Intelligenz in Östereich 1859-1914* (Vienna: Böhlau, 1998); Monika Ritzer: 'Darwin und der Darwinismus in der deutschsprachigen Literatur des 19. Jahrhunderts', in *Weltanschauung, Philosophie und Naturwissenschaft im 19. Jahrhundert*, ed. by Kurt Bayertz, Myriam Gerhard and Walter Jaeschke, 3 vols (Hamburg: Meiner, 2007), II: *Der Darwinismus-Streit*, pp.154-85.

read) authors of his epoch, with over 150 volumes to his name.[22] Raabe's creative reception of Darwinism from the 1860s on has in recent years been widely recognised as a central focus of his critical portrait of the age.[23] Following Gillian Beer's lead, I recently sought to show how in one of his first mature works, *Drei Federn* (1865; *Three Pens*), the generative energies implicit in Darwin's texts inspire Raabe to his own formal narrative innovations, so that one can truly speak of an, in this case anti-Social Darwinist, evolutionary aesthetic or literary anthropology sedimented in Raabe's novel.[24] Darwin was also a central focus for Jensen. Like Raabe (and Dickens and Eliot), he aimed to combine the exoteric and the esoteric, popular appeal (he was dependent on his publishers' honoraria) with respectable elevation of matter and (to those with eyes to see) profundity of comment on the great issues of the day. His novels, then, intentionally function in this dual modality of receptive appeal, to light and serious readers. Perhaps even more than Raabe, Jensen was a persuaded Darwinist, and their correspondence is packed with allusions to the new theory and its contemporary recipients (Vischer, D.F. Strauß, Hartmann). A qualified medical doctor, Jensen had studied medical sciences at the leading faculties of Kiel, Breslau and Würzburg 1857-60 before turning after a crisis of mission to literature, and he was personally well acquainted with both Haeckel and Nordau. Unsurprisingly against this background, much of his fiction not only reflects on some dimension of the relation between literature and the life sciences on their new Darwinian basis, but also combines this with frankly autobiographical elements.

So it is that Jensen treats the acute post-Darwinian issues of racial identity, racial hybridity and racial extinction in early novellas such as *Unter heißerer Sonne* (1868; *Under a Hotter Sun*) and *Die braune Erica* (1869; *Dusky Erica*).[25] Indeed Jensen wrote a clutch of works around this time which

---

22 See on Jensen Nicholas Saul, 'Wilhelm Jensen: Literary Value, Evolutionary Aesthetics and Competition in the Marketplace (*Sanhita*, *Fragmente*)', in *The Tradition of the Bestseller: Success, Identification and Complexity in the Nineteenth-Century German Novel*, ed. by Benedict Schofield and Charlotte Woodford (forthcoming, Camden House, 2010).

23 See *inter alia* Eberhard Rohse, '"Transzendentale Menschenkunde" Im Zeichen des Affen: Raabes literarische Antworten auf die Darwinismusdebatte des 19. Jahrhunderts', in *Jahrbuch der Internationalen Raabe-Gesellschaft* (1988), 168-210; Sprengel, *Darwin in der Poesie*, pp.117-34; Katharina Brundiek: *Raabes Antworten auf Darwin. Beobachtungen an der Schnittstelle von Diskursen* (Göttingen: Universitätsverlag, 2005).

24 See Nicholas Saul, 'Raabes Geschichte(n): Realismus, Erzählen, Historie, Prähistorie', in *Jahrbuch der Internationalen Raabe-Gesellschaft* (2009), 22-39.

25 See Nicholas Saul, *Gypsies and Orientalism in German Literature and Anthropology of the Long Nineteenth Century* (Oxford: Modern Humanities Research Association, Maney, 2007), pp.106-20.

meditate on a range of implications of the new Darwinian anthropological paradigm. Another narrative from this epoch, *Das Erbtheil des Blutes* (1869; *The Legacy of Blood*) is of particular interest for our concern, for it combines a focus (as its none-too-subtle title promises) on the classic evolutionary themes of inheritance, personal destiny, free will and determinism, with a model example of an early German evolutionary aesthetic to match that of Raabe. A country pastor, Eduard Baumholz, over 60 years old at the time of writing and climax of the action in 1855-56, has married the much younger Sophie, now in her mid-thirties, relatively late in life and, following the death in childbirth of their only son, also Eduard, adopted two children. The elder, Hans Reinhard, is the son of a colleague, who was involved in a mysterious, apparently counterfeit, cloak-and-dagger marriage ceremony, and died shortly after with symptoms of poisoning. If Hans has been adopted on foot of a deathbed promise, the younger child, Judica, is adopted as the result of his natural son's death. Shortly after this disaster a baby, accompanied by the imploring note of a mother in distress, is abandoned on the pastor's doorstep. Under pressure from Sophie, the pious Baumholz decides that the appearance of the foundling is God's will: compensation for his own, and especially Sophie's, loss. Implicitly accepting that they will have no more children of their own, they take the mysterious child in. Thus Judica is named for the day on which she was found, in the German Lutheran calendar *Sonntag Judica* or Passion Sunday.[26] Of course the unfolding narrative investigates the conflict of nature and nurture in the upbringing of these two children, and sustains suspense for the popular reader through the concomitant investigation of Judica's mysterious descent and the equally mysterious death of Hans's father. As the novel's analytical form unfolds, these twin cognitive interests link the theme of Darwinian inheritance with that of the function of art in an original and interesting manner.

A good deal of the narrative interest is generated by the situation: Jensen's well-calculated disproportion between what is inherited in the children's character and the ability of their unrelated parents – Baumholz in particular – to channel that innate, yet alien disposition into the life paths they prefer. The autobiographical moment is immediately evident. Both children, like Jensen himself, are orphans. They seem moreover to reflect the split halves of Jensen's own career: one follows a path into the medical profession, the other has aesthetic yearnings. Hans is dogged, methodical and courageous, with a liking for exploring the natural world. Easy to guide, motivated doubtless by witnessing his father's death struggle, he follows

---

26 The reference is to the *Introitus*: 'Judica me, Deus, et discerne causam meam de gente non sanctis ...' ('God give me justice and plead my case against an unholy nation ...').

both his early inclinations and the advice of the local Dr Fabri to study medicine at the regional capital (Berlin, perhaps), and does at last become a doctor. Judica, by contrast, is interested less in nature than art, has 'weniger Sinn für das wirkliche Lebendige als Eifer [...] aus todten Gegenständen sich ein lebendes Ding zu machen' (p.71); 'less inclination towards what truly lives than eagerness [...] to make something living for herself out of dead objects'). Now art from Baumholz's theological standpoint is an *adiaphoron*, an ethically indifferent thing. As such, it may be used for good or ill, but intrinsically represents a temptation, and is legitimate only when serving higher ends. Alas a travelling theatre troupe – presenting a production with no edifying moral – finally awakens Judica's embryonic artistic tendencies. Thereafter her 'phantastischer Sinn' ('fanciful disposition') and 'Hang zum Abentheuerlichen' (p.91; 'propensity for adventure'), as Baumholz puts it, unfold irrespective of his counter-pressure. She grows into an emblematic beauty, reads and plays Shakespeare, Goethe and Schiller and falls readily into the Lucretia-Tarquinius script when her virtue is threatened by the dissolute Albert, adoptive son of the local Baron Hochseß. When at Easter 1855 Judica refuses confirmation (she has read too much unChristian *Faust*), Baumholz reads her a public pulpit lesson on religious duty as the sole justification of feminine beauty. As a result she – and her beloved Hans – promptly leave for the distant city.

Baumholz, it is clear, wholly fails to comprehend the development of Judica's nature. Likeable and well-intentioned as he is, he is presented by Jensen as a man immersed in problems of education and development, yet who is in truth incompetent and behind the times. A typical product of Jensen's visceral anti-clericalism, he serves as the foil to clarify the novel's argument *e contrario*. As his family name – an ironic allusion, perhaps, to Darwin's model for the evolution of species – suggests, Baumholz is a lover and cultivator of trees, possessed of his very own, emblematic arboretum. Yet he is limited to an emphatically pre-evolutionary state of understanding. Deploying the obvious organic metaphors offered by the Bible (the sower and his seeds), he likes to theorise within his received theological framework on the relation of innate characteristics and external circumstances as determinants of fate. Having observed how trees regenerate their foliage despite even the severest storm damage, he is convinced that no created being can change its given nature and will always express that. This is equally true of people: 'Was einmal im Menschen liegt, ist nicht heraus zu treiben' (p.1;'Once in human nature, a thing cannot be driven out). It is passed unchangeably from parent to child. He had thus imagined that his stillborn son would turn out just like his father. Otherwise, Baumholz trusts blindly in divine Providence to manipulate the course of earthly lives for the best, whatever about appearances. It is no surprise, then, that he at last capitulates before the riddle of Judica's inherited character, recognizing, as he sees it, the invincibility of nature by nurture. But of course Baumholz is ironised by Jensen. In the age of evolutionists from Goethe to Lamarck and Lyell (and Darwin), Jensen gives him as his source of authority in developmental theory none other than Georges Léopold Cuvier (1769-1832), certainly one of the most distinguished pre-Darwinian natural scientists and founder of palæontology.

Faced with the riddle of Judica's character, Baumholz can only envy Cuvier's brilliant ability by divination to (as it were) reverse extrapolate the complete form of extinct antediluvian animals from an examination of minute fossilised bones, and wishes he could do the same to identify Judica's parents from the evidence of her person. Cuvier, however, was renowned not only for the functionalist methodology and divinatory talents which allowed him to identify, differentiate and reconstruct species preserved in sedimented strata, but also for the benighted denial of descent by modification and defence of the creationist theory of the origin of species which seemed to follow from that.[27] Thus whilst he rightly argued for the extinction of past life forms by natural disasters, he also, as Darwin noted, taught something equally congenial to Baumholz: the 'immutability of species'.[28] It is typical of Baumholz's mentality that he refuses to sanction a post mortem examination of Hans's father, casting doubt on its (obviously true) findings and relying on God's will for resolution rather than a judicial investigation. Baumholz, then, is Jensen's symbol of the passing epoch, in a novel which aptly situates the mid-nineteenth-century German mentality on the cusp of a transition from the old, theocentric age to the new,

---

27 A likely source is Georg Cuvier, *Die Erd-Umwälzungen*, ed. by C. G. Giebel (Leipzig: Abel, 1851). This is an accurate translation with appended updatings of 1812's *Discours sur les révolutions de la surface der globe, et sur les changements qu'elles ont produits dans le règne animal.* Par M. le Baron G. Cuvier (Paris: G. Dufour, Ed. D'Ocagne; Amsterdam: Maison de Commerce, 1826). The first German translation appeared in 1822. See on Cuvier's reconstructive method *Erd-Umwälzungen*, pp.1ff., 51-58 (*Discours*, pp.1ff., 47-53); on his argument therefrom against evolution, for strictly fixed types and hence the likelihood of creationism *Erd-Umwälzungen*, pp.63-71 (pp.58-65). On Cuvier's place in the history of evolutionary theory see Thomas Junker, Uwe Hoßfeld, *Die Entdeckung der Evolution. Eine revolutionäre Theorie und ihre Geschichte*, 2nd edn (Darmstadt: Wissenschaftliche Buchgesellschaft, 2009 [2001]), pp.57f. For the status of Cuvier's theory in the immediately post-Darwinian German debate see the drastic comments by Friedrich Albert Lange and Carl Vogt, in Lange's widely-read *Geschichte des Materialismus und Kritik seiner Bedeutung in der Gegenwart*, 2 vols, 2nd edn (Iserlohn: Baedeker, 1873-5 [1866]), pp.314-15.

28 Charles Darwin, *On the Origin of Species by Means of Natural Selection, or the Preservation of Favoured Races in the Struggle for Life* (London: Murray, 1859), p.310.

anthropocentric 'age of science' (Werner von Siemens) – here represented by the medical characters.

At this point in the text, after a first section comprising Baumholz's journal, the narrator changes to a figured, but otherwise conventional, omniscient voice, for what follows – ultimately a narrated evolutionary theory of the function of art – is clearly even more beyond the pastor's comprehension than the events of part one. One of the strangest events in the first half of the novel had been the encounter with travelling players which awakened Judica's sense of artistic mission, for they put on a play, which the pastor inevitably dismisses as a farrago of knights, robbers and Gypsy fantasy, but which, treating of a mother who has lost her child early in life, searches everywhere for it, and at last finds it in the strangest circumstances, ultimately reveals everything about Judica's life – without recourse to a Cuvier of literature. The strangeness lies in Judica's response. It is of course not unknown for a six-year-old child to be so enchanted by a theatre performance that she spontaneously joins in, and this is just what Judica does. The curious thing is, however, that she takes on the role of the lost child precisely on cue as the stage mother and child are reunited, *as if she knew the role already* – much to the annoyance of the appointed performer in the wings, but much to the edification of the audience, for the realistic quality of the acting, particularly by the stage mother, has been particularly emotive. It is as if the 'script' of this behaviour is already innate in Judica, as it were programmed by inheritance into her, from a tragic past of loss and separation to a utopian and healed future. Judica falls into a fever for days after the event, but on awakening still has the script by heart. Of course this presentation of play's striking interleaving of real and aesthetic experience is intended by Jensen as more than the mere occasion of *Bildung*. By the time Judica leaves home it is clear to the reader that Judica's stage mother is indeed her birth mother, the beautiful and gifted actress Cölestine. Jensen's suggestion, so it seems at this stage of his argument, is that the 'script' of nature in Judica, asserting itself on the encounter with the travelling players, 'overwrites' the biblical texts with which the pastor has habitually but ineffectually attempted to guide her development. The reader can also easily establish from the other indicators skilfully embedded in the text the identity of Judica's natural father (Judica's and the Baron's uniquely shared timbre of voice, a visit by the players to the *Schloß*), and so fully reconstruct the true legacy of blood. Cölestine has been tricked into a fake marriage (the one enacted by Hans Reinhard's father, of course), has thus become the victim of seduction by the Baron, has been abandoned by him and in turn abandoned their child to Baumholz's care, and pursued her acting career as best she can. She has watched her daughter's progress from afar, but she has also intervened. It was she who organised the performance of the play and invited the Baumholz family to attend in the first place. And it is she who sent her the Shakespeare, Goethe and Schiller which reinforce Judica's sense of artistic vocation and so occasion the conflict with Baumholz's Bible-centred world view.

This, however, is only the first iteration of Jensen's implicit evolutionary aesthetic theory. So far, he has in fact offered the reader little more than a suggestion as to how – rather as Joseph Carroll argues – literary texts (the Bible, the play) tend as aesthetic experience to awaken and reinforce inherited dispositions of behaviour. Not only, of course, does Judica respond to the travellers' play, she also, as we recall, tends in conflict situations to act out in reality other literary scripts cognate with her inherited self-understanding (the Lucretia role when sexually molested by the Baron's adoptive son Albert, the Faust role when rejecting Christianity at Easter). So forceful is the legacy of blood, the inherited 'script' of behaviour, on this momentary presentation of arguments, that it seems able to resist counter-adaptive modification by inappropriate alternative scripts. Indeed, irrespective of the extent to which the narrator of *The Legacy of Blood* may share Baumholz's theocentric perspective (and he does not), the frequency with his characters display deterministic, inherited behaviour in this vein is positively the dominant feature of the second part of the narrative. Thus we find that just as Judica has unconsciously (as it were) inherited her mother Cölestine's acting talent, so she also, as if programmed to do so by the innate script, tendentially repeats all her mother's other actions. After she has run away from home, she too begins an acting career in the *Residenz*. And just as her beguilingly beautiful mother was pursued by the Baron, so his son Albert, who first molested her in the village, himself repeating the Baron's 'script,' now, with her acquiescence, pursues her in the city. Albert, indeed, repeats his parent's behavioural paradigm so exactly that he too attempts to enact a sham marriage with Judica, much as the Baron did with Cölestina. Life stories, on this construction, seem like an inherited nightmare of eternal deterministic recurrence, reinforced by literature, be it the Bible or *Faust*. But there is, it seems, an escape from this, and it too involves a now truly emancipatory use of art.

It is Judica herself who breaks the vicious cycle of repetition, literally by stepping out of her role in a moment of meta-aesthetic catharsis. Some readers of this essay may already have deduced that the title of the play in the village which had such an effect on Judica and which poor Baumholz could not recall is *Preciosa* (1820), a little play by the German Biedermeier writer Pius Alexander Wolff, which was based of course on Cervantes's novella *La gitanilla* (1613; *The Little Gypsy Girl*), and, equipped with an overture by Carl Maria von Weber, had enormous success on the nineteenth-century German stage.[29] Its plot fits closely enough for Jensen's ends the bare existential facts of Judica's case. More than that, it transpires that there is yet more repetition: Judica's first acting role in the city is none other than the title role of *Preciosa*, playing herself, as it were, a second time, but now as an adult. This time, however, she fails to complete her self-performance as Preciosa. A professional critic in the audience equipped with spectacles of keen percipience makes a trenchant and fatal attack on her performance. Crushed by the audience's vocal displeasure after what is objectively an untrue performance, she allows the last words of act one – "'Leb' wohl, Madrid ...'" (173; "'Fare well, Madrid ...'") – to serve as the last time she allows her personality to be constructed by the role, and emblematically tears up her Gypsy girl costume before a mirror. Thus is destroyed the illusory self-identification which the Gypsy costume embodies, and the aesthetic script which had emancipated her from biblical determination only again to entrap her is itself effaced. She forthwith abandons her acting career: "'Ich war eine Närrin, und alles Treiben meiner Kindheit war Possenspiel und Wahnsinn'" (p.185; "'I was a fool, and all that work in my childhood was pantomime and madness'"). It would have been better, she concludes with bitter irony, to have played her role to the end, when she would have stepped out of it – as in the original – a Gypsy no longer, of course. Thus just as the literary script can suggest, prescribe and liberate, so it can also, with equal power and thanks to its charisma, prompt *false* self-understanding. Judica's error is aestheticism, to confuse art with life, not to understand that the function of art – and this is Jensen's lesson here – is to serve life, to serve life in the unfolding of whatever script *might still be written* for the years to come. So this is a theory of the evolutionary, developmental function of art which acknowledges art's power and role in the process, but also its limits. Art is a *therapy*, which may however go wrong.

In this case the script to be written is the total triumph of the medical profession, emblem of the new scientific age, symbolised by the novel's closures and final, sentimental and crowd-pleasing (poetic realist) tableau five years after the main action, in 1861, on the threshold of the Darwinian age. The religious line of Baumholz will die out. Baumholz's fate in this mirrors that of his childless Goethean namesake, Baron Eduard of *Die*

---

29 Pius Alexander Wolff, *Preciosa: Schauspiel in vier Aufzügen* (Leipzig: Reclam, n.d.); *The Little Gypsy Girl* (*La gitanilla*), in Miguel Cervantes de Saavedra, *Exemplary Stories*, trans. by C. A. Jones (Harmondsworth: Penguin, 1972), pp.19-84. See on the relation and significance of the texts Saul, *Gypsies and Orientalism*, pp.22-25.

Wahlverwandtschaften (1809-10; *Elective Affinities*). His son's lack of vitality and his continuing childlessness are doubtless authorial comment on his world view and the evolutionary validity of his disproportionate union with Sophie, daughter, need it be said, of the *Stadtschreiber* (p.33; town clerk, copyist).[30] The medical line will flourish and dominate the new age. Cölestine's marriage to the Baron is discovered to be valid, so that his subsequent marriage is void. She is the true Baroness, and Judica's final legacy of blood the aristocratic estate. But she has married the bourgeois Hans, now a practising medical doctor, so that in the spirit of 1848 the age of the aristocracy is past and a *Bürger* the true lord of the manor. Indeed Judica herself is now not only with child, but also herself practises medicine. Even Dr Fabri is included in the tableau, and the only doctor in the volume who does not figure is the courageous unnamed city doctor who performed the post mortem on Reinhard which Baumholz denied. Baumholz, for his part, continues to write his theocentric journal, but this document which fails to capture the truth is merely the contrasting immanent reflection of the document which does: Jensen's novel of evolutionary aesthetics *The Legacy of Blood.*

To sum up: The contemporary debate on evolutionary aesthetics is not over, yet it appears on this account that there are still issues of legitimation to be resolved: in particular the foundational status of Darwinian evolutionary discourse and its conflict with the autonomy and the cognitive licence of literature. It is also noticeable that many of the features of today's discussion recapitulate those of the pioneering German reception in Nordau and Bölsche, and even earlier reception by Raabe and Jensen, both of whom suggest that truly evolutionary aesthetics involve the active participation by aesthetic discourse in the furtherance of life. Thus the contemporary debate on evolutionary asethetics might gain from examining how the dynamic form of narratives may contribute to art's relation with Darwinism; and from an encounter with the history of literary encounters with Darwinism.

30 Baumholz is engagingly blind to the reason for the large number of young men who enjoy socializing with his young wife (pp.90-91).

'Live like a Man and not like a Monkey': Nietzsche's Philosophic Vitalism and Darwin's Theory of Evolution

John A. McCarthy

Abstract: This essay asks: How does the creative act that is literature align with a notion of evolution in biological and geological terms? What role did specifically Darwin's theory of evolution and its adaptations play in the evolution of literature around 1900? My focus on Nietzsche reveals an essential quality of 'evolutionary literature' because his vitalism frequently conflated the organic and inorganic operations of nature. After sketching the relevant scientific environment that shaped Nietzsche's view of the world, I seek to define his understanding of the creative act and demonstrate how Nietzsche's own literary production mirrors prevalent evolutionary ideas.

Satirical caricature (*The Hornet*, 1871)

Häufiger schon geschieht es, dass [...] der wissenschaftliche Kopf auf einen Affenleib [...] gesetzt ist. (*Beyond Good and Evil,* No.26)

The Scientific Scene

In 1855, four years before Charles Darwin (1809-82) published his *Origin of Species*, Friedrich Karl Christian Ludwig Büchner (1824-99) had published his seminal *Kraft und Stoff. Empirisch-naturphilosophische Studien* (*Force and Matter: Empirico-Philosophical Studies*, 1855). In 1859 Nietzsche was a precocious fifteen-year old in his second year at Schulpforta, one of the most distinguished secondary schools in Germany. Darwin followed his *Origin of Species* with *The Descent of Man* (1871), Büchner his *Force and Matter* with *Der Fortschritt in Natur und Geschichte im Lichte der Darwinischen Theorie* (*Progress in Nature and History in the Light of Darwinian Theory*, 1884). The latter points to adaptations of Darwin's nature observations to an ordering of historical-cultural developments as well as biological ones. This is not astonishing in light of inroads that positivism made into the social sciences and literary practice in the second half of the nineteenth century. Here we need only recall the work of Auguste Comte (1798-1857) and Hippolyte Adolphe Taine (1828-93) with their respective works *Cours de philosophie positive* (*Course in Positive Philosophy*, 1830-42) and *Essais de critique et d'histoire* (*Critical and

Historical essays, 1857). Darwin and Büchner were just two of the leading thinkers of the day to have caught Nietzsche's attention. In the late eighteen seventies, Nietzsche remarked: 'A truly burning thirst took hold of me: henceforth I really pursued nothing more than physiology, medicine and natural sciences.'[1] Nietzsche is considered to be a 'naturalist,' a 'profound physiologist,' and the first philosopher to have mirrored Darwinism in his concept of vitalism and in his theory of ethics.[2] Ludwig Büchner thought highly of Nietzsche.

Nietzsche's vitalism amounts to a 'radical reality' and his works a 'school for suspicion' as he puts it in *Human, All Too Human* 1 (Preface 1; 1886).

---

1 Friedrich Nietzsche, *Ecce Homo* III (KSA, VI); *Human, All Too Human* 3 (KSA, II). All references to Nietzsche's works in the original are to *Sämtliche Werke*, Kritische Studienausgabe, ed. by Giorgio Colli and Mazzino Montinari, 15 vols (Munich: de Gruyter, 1988), cited as KSA, by volume and page number. Readily available English translations are cited, usually Hollingdale. Whenever feasible I also include reference to section or aphorism number in the quoted work.

2 C. U. M. Smith, 'Zarathustra's Evolutionary Epistemology', *Journal of Social and Evolutionary Systems*, 15 (1992), 75-85. Smith was the first, as far as I can tell, who shows how Nietzsche's work carries Darwinian ideas into the domain of philosophical anthropology in the development of his evolutionary epistemology. Brian Leiter, *Routledge Philosophy Guidebook to Nietzsche on Morality* (London: Routledge, 2002), speaks of Nietzsche as a 'naturalist' rather than as a postmodernist. What Leiter means, however, is essentially that Nietzsche was more like Hume than Foucault. For Leiter, 'naturalist' designates the thinker as a philosopher of human nature. For Leiter Nietzsche's naturalism is primarily methodological; that is, the methods of philosophy ought to be continuous with the quantifiable/verifiable methods and results of the empirical sciences. Leiter devotes chapter 2 of his study (esp.pp.63-72) to retracing the naturalist arguments leading up to Nietzsche. He rejects any explanation in terms of non-natural causes but simultaneously opposes a materialistic reduction of all phenomena to physical phenomena, reserving space for free will, for example. That is not quite my meaning in the following. I adhere much more closely to the materialistic views of nature and of human nature prevalent in the second half of the nineteenth century than Leiter does, even though Leiter's 'characterization of Nietzschean naturalism in connection with morality' – his focus is on the *On the Genealogy of Morals* – is considered by some to be 'the most systematic and compelling to date'. See Bernhard Reginster's review of Leiter's book in *Notre Dame Philosophical Reviews* 2 January 2003<http://ndpr.nd.edu/review.cfm?id=1157)> [accessed 13 September 2010]. Also compelling is Gregory Moore's delineation of Nietzsche's theory of evolution in the wake of Darwin in *Nietzsche, Biology, and Metaphor* (Cambridge: Cambridge University Press, 2002), although I do not limit myself here to biological or organic phenomena. Richard S. G. Brown associates Nietzsche's dictum that Buddha was a profound physiologist with Nietzsche himself. The body and the environment determines everything. That is my meaning in the following. See R. S. G. Brown, 'Nietzsche: "That Profound Physiologist"', in *Nietzsche and Science*, ed. by Gregory Moore and Thomas H. Brobjer (Aldershot: Ashgate, 2004), pp.51-70 (63-68).

Radical reality raises suspicions about the essence of existence by essentializing naturalism and making use of what Spanish philosopher José Ortega y Gasset (1883-1955) designated radical reason or *razón vital.* Ortega y Gasset drew inspiration from Nietzsche for his conception of radical reason, even as Nietzsche had drawn inspiration from others. The concept stood in opposition to traditional rationalism that had led to the construction of metaphysical systems (e.g., Kant, Hegel). 'Radical reason' and 'radical reality' in an expanded, holistic sense place the body at the center of deliberations without seeking to privilege the inquiring human mind as an anomaly in the physical world. In his study, *Human Life is Radical Reality* (2005), Howard Tuttle sums up the concept when he states:

> Through the absorbing of circumstances the human being constructs the world. By the use of a life form which ancient Greek philosophy designated as pragmata, human beings perform the actions which 'must be done' for the uses and needs of human life. Such humanization of circumstances allows us to develop our lives in the 'system of importances' which surrounds us. Our life is a continual reabsorption of circumstances, a continual working out of our lives through them. In this respect human beings are doomed to achieve themselves through something else, to become human through the absorbing of the non-human.[3]

Noteworthy in my current context is Tuttle's emphasis not only on the interconnections between the organic and inorganic, between the human and non-human, but also his highlighting of the qualitative aspects of human existence that give rise to meaning.[4] The proximity of this conception to Nietzsche's vitalism will emerge by and by. Although Tuttle does not include Nietzsche in his study, the latter's famous will to power lies at the core of radical reality.

Nietzsche's will to power is also at the core of John Richardson's *Nietzsche's New Darwinism*, published the same year as Tuttle's investigation. Richardson interprets Nietzsche's will to power as an internal revision of Darwinism, placing Nietzsche's concept in opposition to the neo-Darwinist emphasis on survival of the fittest as a teleological process.[5] This

---

3 Howard N. Tuttle, *Human Life is Radical Reality: An Idea Developed from the Conceptions of Dilthey, Heidegger, and Ortega y Gasset* (New York: Lang, 2005), pp.97-100, 153-59 (p.158).

4 The latter notion appears as qualia in contemporary investigations into the nature of consciousness. Cf. Christof Koch, *The Quest for Consciousness: A Neurobiological Approach* (Engelwood, CO: Roberts, 2004) and Eric R. Kandel, *In Search of Memory: The Emergence of a New Science of Mind* (New York: Norton, 2006).

5 Richardson, *Nietzsche's New Darwinism* (2005), pp.24, 35, 46, 78 et passim. Hans Schwarz, 'Darwinism between Kant and Haeckel', *Journal of American Academy of Religion*, 48 (1980), 581-602, argues that Haeckel sought to bridge the gap between

'grounding' of creativity, as Babette Babich has called it, is also central to Gregory Moore in his 'Art and Evolution' where he highlights Nietzsche's response to the notion of *Kunsttrieb* derived from Blumenbach's *Bildungstrieb* and popularised by Ernst Haeckel in the nineteenth century.[6] Richardson and Moore interpret Nietzsche to be much more indebted to Darwin than the philosopher cared to admit. Most critics, whom Richard Brown labels 'the Unmentionables' because they have found it of little use to reflect upon physiology in connection with Nietzsche's thought or to consider the interconnections between scientific advances and developments in the world of arts and letters. The widely used *Cambridge Companion to Nietzsche* (also 2005) surprisingly contains no essay on Nietzsche and science, nor does the index list Darwin, Büchner, Haeckel, or Lange.[7] When critics do pay attention, they are not always as attentive to Nietzsche's reaction to Darwin and Nietzsche's reaction to someone else's use of Darwin.

Darwin's *Origin* occasioned wide debate on such issues and concepts as adaptability, gene dominance, geological patterns, migratory habits, competition for resources, gradations in perfection, variability of organs and instincts, natural selection, and survival of the fittest. Common to these several fields of inquiry is the insight that a complex network of interconnected links is central to the evolutionary process. Darwinian evolutionary theory is still a hot topic today, as is evident in David Sloan

---

monism and dualism through a comprehensive monistic view of reality. This is also the approach adopted by Ludwig Büchner and Nietzsche. It leads to the concept of consilience invoked by Edward O. Wilson in *Consilience: The Unity of Knowledge* (New York: Knopf, 1998).

6 Babette E. Babich argued for an 'ecophysiological' grounding of Nietzschean thought in her *Nietzsche's Philosophy of Science: Reflecting Science on the Ground of Art and Life* (Albany: SUNY Press, 1994). It points toward the concept of radical reality (see pp.77-134). Gregory Moore, 'Art and Evolution: Nietzsche's Physiological Aesthetics', *British Journal for the History of Philosophy* 10 (2002), 109-126 (pp.112-15). My own *Remapping Reality: Chaos and Creativity in Science and Literature (Goethe—Nietzsche—Grass)* (Amsterdam: Rodopi, 2006) devotes chapter 3 to 'Grounding Creativity: Nietzsche and the New Universe' (pp.110-34). My study was written without the benefit of Richardson, Babich, or Moore. While I approach the question from complexity theory, I arrive at similar conclusions.

7 *The Cambridge Companion to Nietzsche*, ed. by Bernd Magnus and Kathleen M. Higgins (Cambridge: Cambridge University Press, 2005). One could, of course, explain the absence of 'radical reality' here by citing the simultaneity of the publications; yet Richardson had published *Nietzsche's System* (Oxford University Press, 1996) much earlier and Babich's *Nietzsche's Philosophy of Science* appeared in 1994. In addition to Babich, Richardson, Moore, Leiter, and the various contributions to *Nietzsche and Science*, other notable studies devoted to Nietzsche and materialism include Jörg Salquarda, 'Nietzsche und Lange', in *Nietzsche-Studien* 7 (1978), 236-60; and George J. Stack, *Lange and Nietzsche* (Berlin: de Gruyter, 1983).

Wilson's acclaimed *Evolution for Everyone: How Darwin's Theory can Change the Way We Think About Our Lives* (2007) and in leading articles in news weeklies on the 'evolution revolution'.[8] Foregrounded in the following is Darwin's description of the natural system as 'a genealogical arrangement' and his view of the interconnectedness of all organic matter as constituting one grand natural system.'[9] Like Darwin, Nietzsche saw no need for assuming an intelligent design or a grand plan in the evolution of nature and the universe. It is all a matter of endless combination, recombination, and chance in a blind search for the right arrangement to ensure survivability and 'thrivability'. Darwin identified specific laws of growth with reproduction that include inheritance and variability in the struggle for life.

Darwin's is an optimistic view that holds great promise for the whole new discipline of psychology. 'In the distant future,' Darwin states in the summation of his *Origin of Species*, 'I see open fields for far more important researches. Psychology will be based on a new foundation, that of the necessary acquirement of each mental power and capacity by gradation. Light will be thrown on the origin of man and his history.'[10] This latter comment points directly to Nietzsche's project in *Jenseits von Gut und Böse* that proposes a reconceptualisation of psychology as 'a genuine physio-psychology'. Psychology is the ultimate path to solving the essential problems of understanding ('Grundprobleme').[11] Yet, we should not assume that Nietzsche uncritically embraced Darwin's conception of the importance of classification or emphasis on common descent. Nietzsche's overman is not classifiable, is not subsumable in the herd mentality, and is not a member of an extinct species. And yet the *Übermensch* is critical to the vitality

---

8 Malcolm Jones, 'Who was more Important: Lincoln or Darwin?', *Newsweek* July 7-14 2008; Sharon Begley, 'Beyond Stones and Bones,' *Newsweek*, March 19, 2007, 53-58; Michael D. Lemonick, 'The DNA Revolution: A Twist of Fate', in: *Time*, 3 March 2003, 41-50.

9 'The fact, as we have seen, that all past and present organic beings constitute one grand natural system [...]. The real affinities of all organic beings are due to inheritance or community of descent. The natural system is a genealogical arrangement, in which we have to discover the lines of descent by the most permanent characters, however slight their vital importance may be.' Charles Darwin, 'Recapitulation and Conclusion', *The Origin of Species*, ed. by Gillian Beer (Oxford: Oxford University Press, 1996), p.394.

10 Darwin, Origin, p.394.

11 Friedrich Nietzsche, *Beyond Good and Evil: Prelude to a Philosophy of the Future*, trans. by R. J. Hollingdale, introduction by Michael Tanner (New York: Penguin, 1990), No.23, p.53. German original in: *Jenseits von Gut und Böse: Vorspiel einer Philosophie der Zukunft* (1886), *Sämtliche Werke*, ed. by Giorgio Colli and Mazzino Montinari, 15 vols (Munich: de Gruyter, 1988), V, p.38. Hereafter cited as KSA, with volume and page number.

of the community.[12] In *Literary Darwinism* (2004), Joseph Carroll seeks to distinguish 'adaptionist' literary study from other forms of 'evolutionary' literary study by emphasizing what he sees in them as a 'simple causal' sequence: (1) the mind has evolved through an adaptive process; (2) the adapted mind produces literature. Carroll contrasts this 'adaptionist' model with three other paradigms: (1) cosmic evolution; (2) evolution taken as an analogical model; and (3) evolution taken as a normative value. As we shall see, Carroll's oversimplified, teleologically informed model does not apply to Nietzschean radical reality. Christian Emden argues more persuasively in regard to the neurophysiological evolution of consciousness.[13]

In any event, both Darwin and Nietzsche point to psychology as a gateway to future understanding of evolutionary processes; their insistence resonates with contemporary inquiries in neuroscience. The human brain bears the marks of evolutionary processes and appears to be the ultimate frontier to many researchers, for it seems as vast, as complex, and as strikingly interconnected as the universe itself.[14]

Like Darwin's *Origin*, Büchner's *Kraft und Stoff* also caused quite a stir. Ultimately, it cost the author his professorship at the University of Tübingen because of its radical grounding of everything in natural processes. Philosopher, physiologist, physician (and brother of Georg Büchner), Ludwig Büchner became one of the chief exponents of nineteenth-century scientific materialism. The purpose of his study was to demonstrate the ultimate reality of physical force, the interchangeability of matter and energy, and the indestructibility of energy. Nature for him was purely physical with no purpose, no will, and no grounding in an extraneous authority. Matter is

---

12 Moore suggests a strict binary that is little convincing ('Art and Evolution', 121). He also focuses too one-sidedly on evolution in the organic realm as if the universe were not evolving too.

13 Joseph Carroll, *Literary Darwinism: Evolution, Human Nature, and Literature* (New York: Routledge, 2004), p.xii. No.2 and No.3 do not apply to Nietzsche's use of Darwin because Nietzsche also had inorganic matter in mind and because everything in human valuation was metaphor for Nietzsche. Nor was Nietzsche normative in the usual sense. Carroll mischaracterises Nietzsche's notion of dominance through violence. Christian J. Emden points out that Nietzsche was a keen reader of neurological studies (E. Hitzig, Lange, A. Bain) in his 'Metaphor, Perception and Consciousness: Nietzsche on Rhetoric and Neurophysiology', in *Nietzsche and Science*, pp.91-110.

14 See, e.g., Karl R. Gegenfurtner, *Gehirn und Wahrnehmung*, 2nd edn (Frankfurt: Fischer, 2004); *Neuroscience and Religion: Brain, Mind, Self, Soul*, ed. by Volney P.Gay (Lanham: Lexington, 2009).

immortal, indestructible and is subject to constant and continuous metamorphosis.[15]

While Büchner was not always clear on the relation between matter and force, he nonetheless posited natural and spiritual forces as indwelling in matter. Drawing an analogy to the steam engine that produces motion via the energy of fire, water, and air pressure, he argued that motion in organic and inorganic nature results from the unseen dynamics of energy exchange that produce via the process of *Stoffwechsel* ('phase of matter') a total sum of effects.[16] He called these internal workings in animate beings mind, soul, or thought. One of his central postulates was that force/energy and mind emanate from original matter. His view of nature, therefore, was a materialistic monism. At other times, however, he was not so sure. Mind and matter sometimes seemed to him to be two different aspects; the mind in that case was not necessarily rooted in matter (*Kraft und Stoff*, pp. 97-138). Those views resonate, at least in part, with Nietzsche's conceptions of the eternal return and will to power. They live on in contemporary fields of cosmology and particle physics, even if Büchner has not entered the contemporary debate by name as Darwin has.[17]

Between 1855 and 1898 *Kraft and Stoff* went through nineteen editions. Nietzsche owned the fifteenth edition of 1883, but it is quite possible that he read the book earlier. Büchner drew upon the work of others who were well known to Nietzsche: the pathologist Rudolf Virchow (1821-1902), the

---

15 'Matter is immortal, indestructible. There is not an atom in the universe, which can be lost. We cannot, even in thought, remove or add an atom without admitting that the world would thereby be disturbed and the laws of gravitation and the equilibrium of matter interfered with. It is the great merit of modern chemistry to have proved in the most convincing manner that the uninterrupted change of matter which we daily witness, the origin and decay of organic and inorganic forms and tissues, do not arise, as was hitherto believed, from new materials, but that this change consists in nothing else but the constant and continuous metamorphosis of the same elementary principles, the quantity and quality of which ever is, and ever remains, the same. Matter has, by means of the scales, been followed in all its various and complicated transitions, and everywhere has it been found to emerge from any combination in the same quantity as it has entered. The calculations founded upon this law have everywhere proved to be perfectly correct.'Ludwig Büchner, *Force and Matter: Empirico-Philosophical Studies Intelligibly Rendered*, trans. by J. F. Collingwood. ed. by Jerome S. Arkenberg (London, 1870) <http://www.fordham.edu/halsall/mod/-1855buchner.html> [accessed 13 September 2010]. For the German see L. Büchner, *Kraft und Stoff. Empirisch-Naturphilosophische Studien* (New York: Steiger, 1872), pp.9-10.

16 *Kraft und Stoff*, pp.11, 122; *Force and Matter*, pp.11, 132.

17 See, e.g., Lisa Randall, *Warped Passages: Unraveling the Mysteries of the Universe's Hidden Dimensions* (New York: HarperCollins, 2005); George Smoot and Keay Davidson, *Wrinkles in Time* (New York: Morrow, 1993).

mathematician and physicist Sir William Thomson, first Baron Kelvin (1824-1907), and the biologist Ernst Haeckel (1834-1919).[18]

Central to my present purposes is the similarity between Darwin's idea of nature as a genealogical arrangement and Büchner's conception of nature as matter and energy that does not approach entropy and that requires no authentication by an authority external to the process itself. In 1884 – at the very time that such issues occupied Nietzsche – Büchner published his treatise on the concept of progress in the wake of Darwin: *Progress in Nature and History in the Light of the Darwinian Theory*. These concerns with natural selection, the nexus of matter and energy, and the notion of progress are dominant in Nietzsche's philosophical musings. If the popular science writer Wilhelm Bölsche could claim in 1894 that Nietzsche was defenseless against the natural sciences and that not Zarathustra – whose promised redemption had not been served up – but Darwin dominated intellectual life at the end of the nineteenth century, we are more than 100 years later in a better position to judge whether Nietzsche was really 'ohne Wappnung'.[19]

Nietzsche and Science Modelling

Nietzsche's specific project in the 1880s was the writing of a natural history of the higher man.[20] When he was called to the University of Basel in 1869 at the age of 25 to assume a position in classical philology, he had in fact considered a move to Paris to study chemistry. As a philologist, Nietzsche undoubtedly knew that chemistry is etymologically related to alchemy, which in turn is derived from the Arabic for the art of changing one thing into another. By the time he began composing *Zarathustra*, he was conversant with the essential issues involved in evolutionary biology and used them to

---

18 See Angèle Kremer Marietti, 'Nietzsche's Critique of Modern Reason', in *Theories of Knowledge, and Critical Theory, Nietzsche and the Sciences I*, ed. by Babette Babich and Robert S. Cohen (Dordrecht: Kluwer, 1999), pp.87-102, (p.101 n.50). A revised version of this essay is available on-line as 'Nietzsche and the Ontogeny of Truth': <http://dogma.free.fr/txt/MariettiNietzscheOntogenyTruth.htm> [accessed 13 September 2010]. Although Marietti also names the physicist William Robert Grove (1811-96) and the physiologist Jakob Moleschott (1822-93), I find no reference in KSA, to either. Thomas H. Brobjer remarks in his useful overview of 'Nietzsche's Reading and Knowledge of Natural Science' in *Nietzsche and Science* (2004), pp.21-50, that Nietzsche's views on science were probably more indebted to philosophers of science than to actual reading of scientific works (p.22).

19 Wilhelm Bölsche, 'Das Geheimnis Friedrich Nietzsches', *Neue deutsche Rundschau (Freie Bühne)*, 5 (1894): 1026-33, here 1033. Cited in Theo Meyer, *Nietzsche und die Kunst* (Tübingen: Francke, 1993), pp.224-25.

20 Leiter makes this his major purpose in *Routledge Philosophy Guidebook* (2002) with its focus on *Genealogy* (chapters 5-8).

shape his own thinking about being and the self.[21] We might even view Nietzsche as an alchemist of the spirit, for he did provoke much change in the way we think about things.

His reflections on natural science are part and parcel of his philological studies. His unpublished notes and fragments from the 1870s contain numerous references to Thales, Anaximander, Heraclites, Anaxagoras, Empedocles, Democritus, Pythagoras, and Parmenides. The topics covered range from chemical phenomena to process as a symptom of transitoriness, atomism, a theory of dynamics, the interaction of attraction and repulsion, the dialectic of somethingness and nothingness, *actio in distans*, the concept of substance, the notion of time, the centrality of motion, and much more.[22] These philological studies were in turn further developed by Nietzsche's attention to contemporary nineteenth-century science.

In 1866 young Nietzsche hit upon Friedrich Albert Lange's *Geschichte des Materialismus und Kritik seiner Bedeutung in der Gegenwart* (*History of Materialism and Critique of its Present Importance*, 1865). The book proved decisive in awakening his interest in atomism, chemistry, and physiology. He viewed Lange's work as 'undoubtedly the most significant philosophical work to have appeared in recent decades,' calling it in a letter of 1868 'a real treasure-house' of the 'materialist movement of our times'.[23] Lange, in turn, had assiduously read Büchner's *Kraft und Stoff*, reacting positively to its interpretation of nature as being purely physical.

Nietzsche also read and reacted to the French naturalist Jean Baptiste Pierre Antoine de Monet, Chevalier de Lamarck (1744-1829), who proposed the notion of inherited characteristics in his *Philosophie zoologique* (*Zoological Philosophy*, 1809). The essence of Lamackism is the insight that all changes in the inorganic and organic worlds are the result of laws, not of miraculous intervention, that the physiological needs of organisms resulted from their interactions with the environment. Darwin, of course, while in general rejecting Lamarck's notion of adaptive modification, nonetheless felt inspired by Lamarck. For his part, Nietzsche was skeptical of Darwinian natural selection, preferring the Lamarkian allowance for cultural and

---

21 Robert C. Holub, *Friedrich Nietzsche* (New York: Twayne, 1995), p.58. Duncan Large specifically addresses the metaphorical role of chemistry as 'the science of transformations' in 'Nietzsche's Conceptual Chemistry,' in *Nietzsche and Science*, pp.189-96.

22 Cf. KSA, VII, pp.571 passim.

23 Quoted in Leiter, *Philosophy Guidebook to Nietzsche*, p.65, and George J. Stack, *Lange und Nietzsche* (Berlin: de Gryuter, 1983), p.10. Original source: *Nietzsche Briefwechsel: Kritische Gesamtausgabe*, ed. by G. Colli and M. Montinari (Berlin: de Gruyter, 1975), I, p.2. *Briefe an Hermann Mushacke (Nachschrift)*, November 1866, 184 (*Archidatenbank des Goethe- und Schiller-Archivs*, GSA 71/BW 290, 7-8).

intellectual inheritance, while adding his own distinctive notion of an internal creative force, the will to power.

The German natural philosopher Karl Eugen Düring (1833-1921) caught Nietzsche's eye with his *Cursus der Philosophie als streng wissenschaftlicher Weltanschauung und Lebensgestaltung* (*Introduction to Philosophy as Rigorously Scientific World View and Form of Life*, 1875), an attempt to ground philosophy in natural processes. It contains a protracted critique of Darwinism and what he labels 'Darwinismusspielerei'. Düring rejected Darwin's presumed one-sided insistence on the idea of sexual binarism that reduced both female and male to a brutal struggle for dominance. Instead of focusing on those qualities that came to dominate, Düring, like Lamarck, suggested that one should reflect more on the negative qualities that lost out in the struggle. Differences make a difference in the evolution of a species. Associating beauty and symmetry with women and brutal force with men, Düring expressed skepticism about Darwin's notion of perfectibility of the species solely through the 'Kampf um das Recht des Lebens'.[24] On Nietzsche's reading list we encounter English national economist Thomas Robert Malthus' (1766-1834) *An Essay on the Principle of Population* (1798). Malthus offered an inquiry into the relationship between human instinct for survival, population density, and the availability of food sources. Because of his thesis that vice, disease, and death maintain a balance between population density and the limits of food production, he seemed to Nietzsche to be 'der Pessimist der Socialphilosophie'. In the expanded edition of his *Essay* (1803), Malthus added a moral regulatory instance to those of vice and disease: sexual abstinence. Nietzsche read Düring and Malthus simultaneously, critiquing their privileging of the landed and wealthy classes over the proletariat. Aggression was more in tune with the struggle for existence for Nietzsche.[25]

---

24 Karl Eugen Düring, *Cursus der Philosophie als streng wissenschaftlicher Weltanschauung und Lebensgestaltung* (Leipzig: Koschny, 1875), p.2. Abschnitt, Kapitel 3, pp.100-127, 247 passim. In *Nachgelassene Fragmente* (1875) Nietzsche contended: 'Beim Kampf um das Recht des Lebens ist es ein Mittel ohne Bedenklichkeit. Es giebt keine andre letzte Garantie des Rechts als die Einsetzung der physischen Gewalt' (KSA, VIII, p.169).

25 For Nietzsche's reaction to Malthus see, e.g., KSA, VIII, p.167-70. Holub suggests that Nietzsche reacted more favorably to the biological theory of Lamarck because Lamarckian evolution was less externally deterministic than Darwinian evolution. Whereas Darwin believed that acquired characteristics could be inherited, he nonetheless paid great attention to the mechanisms by means of which organisms survive and change. By contrast, Lamarck did not emphasise 'survival of the fittest' as the chief attribute of evolutionary change; he allowed greater room for the exercise of individual will (Holub 1995, 168-69). Holub argues that Nietzsche (like Düring) found Darwin's explanation problematic of how species evolve and thrive and how forms progress to perfection (Holub 58). Darwin had found inspiration for

The Scottish physicist Balfour Stewart's (1828-87) *The Conservation of Energy* appeared in 1875, and by 1900 had gone through nine editions. Nietzsche acquired it in translation as *Die Erhaltung der Energie* as early as January1875, and even transcribed the first twenty-two pages of it in which Stewart develops a modern version of Democritus's atomism (KSA, VIII, pp.181-85). Adopting Stewart's idea of the universe as a machine comprised of atoms in constant motion, Nietzsche averred that the laws of energy determine its operation, seeing in its interactions an immense struggle of atoms contending for position in a constantly shifting configuration of being (KSA, VIII, p.181). The law that governs all individual phenomena remains invisible to us. An apt metaphor is that of a huge battlefield governed by a complex dynamic of the human and non-human. What we see in the end is just the result of the battle. Although our knowledge of inorganic molecules is quite limited, Nietzsche notes, we nevertheless recognise that we are intimately linked to an entire universe of creatures and events.[26] All of the above help us understand Nietzsche's claim in *Die fröhliche Wissenschaft* that we should want to become what we already are: new, unique, incomparable, self-determining, self-creative. In order to do we have to be excellent learners and discoverers of everything in our environment. In short, 'we must be physicists in order to be creators in that sense'.[27]

While Nietzsche refers to Darwin far more frequently than to such thinkers as Düring, Malthus, or even Ludwig Büchner, their impact is nonetheless felt. In responding to them, however, he is reacting to Darwinism. If Düring seemed to him to be the 'pessimist of social theory,' Büchner and David Friedrich Strauss – whom he at times considered jointly – appeared disparagingly as 'classic mass thinkers' ('Klassiker des Pöbels'). Büchner was singled out as 'the evil power-man' ('der böse Kraftmensch') and 'fanatic advocate of matter' ('jener fanatischer Freund des Stoffes'; KSA, VII, pp.596, 740). Most famous is Nietzsche's reaction to Schopenhauer and

---

his theory of selection and survival in Malthus's explanation of *competition* among populations for food and natural resources. Leiter (2002, p.64) cites the work of physiologist Jacob Moleschott on nutrition that surely influenced Nietzsche as well: *The Physiology of Food* (1850), *The Theory of Food: For the People* (1850), and *The Cyclical Course of Life* (1852).

26 Cf. KSA, VIII, pp.182-83: 'So sind wir mit einer ganzen Welt von Geschöpfen auf's Innigste verknüpft und kennen sie nicht besser als die Bewohner von Mars' (8, p.182).

27 KSA, III, p.563: 'Wir aber *wollen die werden, die wir sind*, – die Neuen, die Einmaligen, die Unvergleichbaren, die Sich-selber-Gesetzgebenden, die Sich-selber-Schaffenden! Und dazu müssen wir die besten Lerner und Entdecker alles Gesetzlichen und Nothwendigen in der Welt werden: wir müssen *Physiker* sein, um, in jenem Sinne, *Schöpfer* sein zu können.'

his disciples. Nietzsche rejected Schopenhauer's notion that there is but one Will, lamented the denial of the importance of the individual, scoffed at his enthusiasm for the so-called genius, and belittled his grounding of morality in compassion (*Fröhliche Wissenschaft*, KSA, III, p.454).

The most direct references to Darwin are found in *Twilight of the Idols* (1888) in a section titled 'Anti-Darwin' (*TI*, p.522-3) and in his notebooks from the years 1887-8. The oft-cited reference to *Zarathustra* (that states there is much of the worm and the monkey in man) does not elucidate the real issues in evolutionary theory.[28] In *Götzendämmerung* Nietzsche complains that Darwin's concept of struggle for life is focused too narrowly on survival of the fittest rather than on competition for power (Macht'; KSA, VI, p.120), which opens up a space for the female sex. In his evolutionary theory Darwin overlooks, moreover, the role of 'Geist'. This quality marks the less vital members of the human race as it has evolved by the end of the nineteenth-century. For him spirit connoted 'caution, patience, cunning, dissembling, self-control and everything that belongs to mimicry'. In other words, he names almost everything that goes under the name of 'virtue'.[29] Nietzsche's iconoclasm is evident here. As if we hadn't noticed, Nietzsche explains in *Ecce homo* that he intended the term 'Götze' to designate truth as it was hitherto understood; that is, as representation (KSA, VI, p.354). It marked the 'weaker' humans who dominated Nietzsche's own day. And of course he railed against these false free spirits incessantly. This is the quality that he never tired of unmasking as an 'aesthetic' operation.

In his notes from 1873 we find a clear reference to geological and Darwinian processes ('geologischen und Darwin[istischen] Vorgängen'; KSA, VII, p.597) that, as Nietzsche laments, privilege the subject. Rather, he thinks one should consider philosophy as analogous to astrology. In other words, while one should see the fate of man as tied to the fate of the physical universe, one should not view human evolution as the raison d'être of evolution per se. One should not privilege man as the presumed crown of creation. This anthropomorphizing trait is the dominant tendency in philosophical treatises. For this reason, Nietzsche maintained that mankind first destroyed religious beliefs and was now destroying the sciences.[30]

---

28 Nietzsche, *Zarathustra*, 'Vorrede' 3, KSA, IV, p.14.

29 Nietzsche, *Götzendämmerung*: 'Ich verstehe unter Geist, wie man sieht, die Vorsicht, die Geduld, die List, die Verstellung, die grosse Selbstbeherrschung und Alles, was mimicry ist (zu letzterem gehört ein grosser Teil der sogenannten Tugend)' (KSA, VI, p.121).

30 Nietzsche, *Nachgelassene Fragmente*, 1873: 'Die Philosophie anzuschauen wie die Astrologie: nämlich das Schicksal der Welt mit dem des Menschen zu verknüpfen: d.h. die höchste Evolution des *Menschen* als die höchste Evolution der *Welt* zu betrachten. Von diesem philosophischen Triebe aus empfangen alle Wissenschaften

A case in point is David Friedrich Strauss's misuse of Darwinism that transformed it into a simplified 'cosmodicy' with a popular appeal ('Kosmodicee'; KSA, VII, p.597). In his notes we find numerous derogatory comments on Strauss as thinker, would-be iconoclast, and prose stylist. A full characterisation is found in *Unzeitgemässe Betrachtungen I*, where, as far as his moral theory goes, Strauss is read as an amateurish disciple of Darwin.[31] Essentially, Strauss merely replaced the concept of God with a concept of universal, undifferentiated Nature. It seems Strauss read only bad books, did not think very deeply about Darwinism, and reacted Dervish-like to Hegel, (7, p.587-96; 'die Hegelei [...] machte ihn wirblicht,' 595). His whole way of thinking has nothing lasting about it; even the protests against his work he mistook as a sign of popularity. Nietzsche remarks that the masses are notoriously unphilosophical, and Strauß is one of them (7, p.593). Nietzsche's critique echoes that of Düring aimed at practitioners of 'Darwinismusspielerei' (toying with Darwin's ideas).[32]

There is, nonetheless, a passing reference in Nietzsche's devastating critique that proves to be a golden nugget: the word 'wirblicht' (whirling) to describe Strauß's state after having encountered Hegelian thought is a direct reference to Democritus's theory of atomism that operates on the new principle of a dialectic between void (non-being) and being. The two concepts translate into non-being and material particles, intrinsically endowed attributes that one associates with the Parmenidean monistic concept of One Being (the latter did not allow for non-existence as the opposite to being). While Strauss's universe cannot account for decline, irrationality, and evil of all kinds, it does allow for vortices of attraction and repulsion, and the fusion of atoms in an apparently arbitrary manner. Hence, 'phenomena [...] are nothing more or less than the epiphenomenal effect of the distribution of atoms in eternally substitutive and seemingly random motion within a cosmic whirl.'[33] Moreover, Democritus' atomism had an intellectually subversive quality that appealed to Nietzsche's sense of daring. His affirming reaction to it was quite the opposite of his dismissive response of Strauss, Schopenhauer, and even Darwin. To his mind, all of them tended to posit universal laws of nature that provided a metaphysical layer of explanation of phenomena that ran contrary to Nietzsche's own radical

---

ihre Nahrung. Die Menschheit vernichtet erst die Religionen, dann die Wissenschaft' (KSA, VII, p.466).

31 KSA, I, pp.159-242, (p.196).

32 Düring, *Cursus* (1875), p.127.

33 James I. Porter, *Nietzsche and the Philology of the Future* (Stanford: Stanford University Press, 2000), p.87.

realism. Indeed, atomism is thought to have engendered normative cultural histories and moral genealogies.

Atomism resonated clearly with Nietzsche's view of the world. It was augmented by insights into the inner working of nature, in particular the importance of perspective, of relationality, and the principles of individuation. By the latter, Nietzsche means the complementary Apollonian and Dionysian principles that structure reality.[34] Taken literally and applied to its fullest consequence, atomism's agonal character can also turn destructively on itself so that one puts oneself at risk in accepting an atomistic point of view (Porter, p.91). How familiar Nietzsche was with atomistic thinking in Antiquity is clear from his lecture notes on the Pre-Socratic philosophers, lectures that he gave at Basel in four different semesters between 1869 and summer 1879 and from the major works of his middle period, the 1880s. But it is already evident in his early essay, 'Die Philosophie im tragischen Zeitalter der Griechen' (1873) and in the passages devoted to Socrates and his famous dialectic method found in *Twilight of the Idols*. Nietzsche labels the Socratic style a new kind of agon ('eine neue Art Agon'; KSA, VI, p.71). His reading of Balfour Stewart's *The Conservation of Energy* added to his deepening understanding of atomism as a dialectical process involving opposing forces.

Nietzsche's Aesthetic Turn: Life as Literature

Of particular interest because of its move toward the aesthetic is Nietzsche's succinct essay, 'Über Wahrheit und Lüge im aussermoralischen Sinne' ('On Truth and Lies in an Amoral Sense,' 1873). Although not published at the time, it presents ideas readily found in *Gay Science*, *Thus Spoke Zarathustra*, and *Beyond Good and Evil*. In this little essay Nietzsche formulated his views on the nature of truth being fundamentally aesthetic; that is, as the product of metaphor, anthropocentrism, and the imposition of numbers on nature to quantify and coordinate the universe. He characterises man as falsely viewing himself to be the center of everything as if the 'eyes of the universe [were] fixed telescopically on his actions and thoughts' (p.876). Man emerges as an engineering genius ('Baugenie,' p.882) who has succeeded in erecting fantastically complicated conceptual dome-structures; unfortunately on a highly unstable foundation, on flowing water. By imposing himself on nature, man morphs nature into himself. In doing so he demonstrates his competence as an 'artistically creative subject' ('künstlerisch schaffendes Subjekt,' p.883). Yet none of what man has created has anything to do with the thing-in-itself. 'Truth' is revealed as nothing more than a mobile army of

---

34 See Nietzsche, *Die Geburt der Tragödie aus dem Geist der Musik* (KSA, I, p.28).

metaphors, metonymies, anthropomorphisms, and other artistically and rhetorically embellished relations.[35] Essentially, the relationship between subject and object is an aesthetic one. The long-suspected *Ding an sich* does not exist. Hence, art for Nietzsche is synonymous with everything that has to do with the production of art, not just with the product itself. What he means, then, is the creative act itself.[36] There is no need to flee to metaphysics for an explanation, he avers; one need only be attentive to the phenomenon of emergence (to be sure, Nietzsche speaks of 'werdende Kultur').[37]

In order to get at the presumed essence, one would have to adopt the right perspective since perspective is the crucial factor in gaining knowledge. Birds, insects, worms, animals all have their particular perspectives linked to their physical beings. Scientific advances have taught us to look telescopically further into the heights and microscopically into the depths of the phenomenal world in search of laws on all scales of nature. Of course, we find confirmation of what we seek because we start out on the false premise that numbers clarify and explain reality rather than fabricate it. The only adequate perspective for humankind would be one that mirrored the *object in* the subject, not one that simply projects the *subject onto* the observed object. The latter Nietzsche likens metaphorically to the spider that spins its intricate web from within itself (1, pp.884-85). What bothers him about this approach is the equation of lawfulness and regularity ('Gesetzmässigkeit') to the concept of universality that affords too little or no room for individual exception, for its antinomy, for its opposite. Dionysus as well as Apollo must be accounted for those complementary principles of growth and destruction evident in the operations of nature (cf. *Geburt der Tragödie*, KSA, I, pp.25-31; 'jene Kunsttriebe der Natur'). He emphasises that all regularity, which we find so impressive in the movement of stars and in chemical reactions, essentially stems from our penchant to read qualities into the operations of nature that are not present in them. We simply mirror ourselves.[38]

---

35 Nietzsche: 'ein bewegliches Heer von Metaphern, Metonymien, Anthropomorphismen, kurz eine Summe von menschlichen Relationen, die, poetisch und rhetorisch gesteigert, übertragen, geschmückt wurden' (KSA, I, p.880).

36 Theo Meyer, *Nietzsche und die Kunst* (Tübingen: Francke, 1993), speaks specifically of art as 'kunstschöpferischer Akt' (70). Cf. Arthur Koestler, *The Creative Act* (1964) for a thoroughly interdisciplinary examination of the concept.

37 Nietzsche, *Nachgelassene Fragmente* 1873: 'Ihr sollt nicht in eine Metaphysik flüchten, sondern sollt euch der *werdenden Kultur* thätig opfern! Deshalb bin ich streng gegen den Traumidealismus' (KSA, VII, p.467).

38 Nietzsche, 'Über Wahrheit und Lüge': 'Alle Gesetzmässigkeit, die uns im Sternenlauf und im chemischen Process so imponiert, fällt im Grund mit jenen Eigenschaften

In this injection of self into the object, we find a hint of Democritus' call for non-being as a necessary counterpoint to being, chaos, and order. Like the cosmic constant, the residue of metaphor (as projection rather than being) is everywhere evident as background noise. (It is present even in Nietzsche's own critique of metaphorical style.) As long as man can forget that he has created the metaphysical structures to explain *physis*, he can be happy. But then Nietzsche called for wakefulness. And wakefulness leads to an awareness of the middle ground between subject and object and of 'reality' as a human construct marked by icons and metaphors ('Götzendämmerung,' KSA, VI, p.57). Wakefulness calls for the unmasking of the metaphors we use to embellish our understanding of the processes of nature and hence of ourselves.

Bearing the foregoing discussion of the creative act in mind, we turn now to Nietzsche's formulation of philosophy as life that equates life to literature. We will view this notion of philosophy as Nietzsche's response to evolutionary and atomistic theory. 'Literature' is a stand-in for art in the broadest aesthetic sense. Nietzsche's asceticism is the natural complement to his perspectivism.[39] I might have labeled my topic 'Poetry in Atomism' after Democritus (cf. KSA, III, p.336), for there is 'a magnificent poetry to the atomistic conception, a perpetual raining of diverse, minute bodies, which fall in manifold ways and in falling interlock, so that a cosmic whirl or 'vortex'] comes into being'.[40]

Will and Energy: The Total Economy of Life

All movement, all phenomena, and all so-called 'laws,' Nietzsche averred, harking back to Heraclitus, are in reality just 'symptoms of an inner event' (*Will to Power*, aphorism No.619). Everything, from start to finish, is a matter of dynamics. To ground creativity, we must look at the whole picture, not just at linear causality. With his views Nietzsche fits in well into the scheme of increasing complexity and innovation that is evident in Lamarck, L. Büchner, Düring, and Darwin. In fact, his biological conception of evolution seems at times to be consciously based on the principles of energy conservation in the universe. He saw not only the positive side of the

zusammen, die wir selbst an die Dinge heranbringen, so dass wir damit uns selber imponiren' (KSA, I, p.886).

39 Alexander Nehamas, *Life as Literature* (Princeton: Princeton University Press, 1985), pp.1-10 et passim. Also adopting a literary approach to Nietzsche are Sarah Kofman, *Nietzsche and Metaphor*, trans. by Duncan Large (Stanford: Stanford University Press, 1993); Gary Shapiro, *Nietzschean Narratives* (Bloomington: Indiana University Press, 1989), and Kathleen M. Higgins, *Nietzsche's Zarathustra* (Philadelphia: Temple University Press, 1987).

40 Porter, p.84 = BAW 3, p.332, 4, p.44.

principle of conservation as formulated by Ludwig Büchner, Julius Robert Mayer and Hermann von Helmholtz, but also the negative side (as Düring had urged with regard to evolution). Energy conversion is not simply the dissipation of difference in temperature and energy levels (= being); it is also the creation of new differences (= becoming). Hence, *différance* proved to be a seminal characteristic of physical existence. And *différance* became a cornerstone of Nietzsche's philosophical conception of the struggle for life. This is a connection that Alfred Döblin drew in 1902 in his study of Nietzsche's will to power and self-preservation.[41] It is what links Nietzsche to the Darwin debate in Germany in his own era.

The overarching theme of Nietzsche's opus in the 1880s is the 'total economy of life' that embraces the full range of human emotion and instinct (*BGE,* No.23). After spring 1885 Nietzsche incorporated ever more the physical concept of force/energy in the initial formulation of his will to power. In a fragment from June/July 1885, he specified that will to power is manifest in 'every combination of force' ('in jeder Kraft-Kombination'; KSA, XI, p.560). Three years later he spoke of the 'will to accumulation of force' and wondered whether this will to increased energy should be hypothesised for chemical reactions and cosmic order as well ('der Wille zur Accumulation von Kraft,' KSA, XIII, p.261). Thus, he was thinking in terms not only of a constant level of energy, but also of energy fluctuation with periodic knots of energy vying with one another for superiority. He labeled the concept 'maximizing an economy of energy use' ('Maximal-Ökonomie des Verbrauchs'; KSA, XIII, p.261).[42]

In the preface to the second edition of *Gay Science*, Nietzsche conjectured that 'behind the loftiest estimates of value by which the history of thought has hitherto been governed, misunderstandings of the bodily constitution, either of individuals, classes, or entire races are concealed' (*GS*, No.2). The reference to evolutionary theory could not be more evident. His purpose in *Beyond Good and Evil* is to confront the 'terrible basic text *homo natura*' in a rigorously scientific manner, the way that researchers examined the rest of nature in the age of positivism (*BGE,* No.230). His purpose is to recoup the text that has 'disappeared beneath the interpretation' (*BGE,* No.38), to ground existence in 'the reality of our instincts' (*BGE,* No.36), and to

---

41 Alfred Döblin, 'Der Wille zur Macht als Erkenntnis bei Friedrich Nietzsche' (1902). Cf. my *Remapping Reality* (2006), pp.115-16.

42 The passage reads in its entirety: 'nicht bloß Constanz der Energie: sondern Maximal-Ökonomie des Verbrauchs: so daß das *Stärker-werden-wollen von jedem Kraftzentrum* aus die einzige Realität ist – nicht Selbstbewahrung, sondern Aneignung, Herr-werden- Mehr-werden, Stärker-werden-wollen' (KSA, XIII, 261; emphasis in original). G. J. Stack speaks of a 'force point world'. *Lange and Nietzsche* (1983), p.124.

disclose the essence of every living thing as 'a fundamental organic function' (*BGE,* No.259). Life appeared to him to be at its very core nothing but 'appropriation, injury, overpowering of the strange and weaker, suppression, severity, imposition of one's forms, incorporation, and at the least and mildest, exploitation'. Translated to the body politic, he continues, this life-drive 'will want to grow, expand, draw to itself, gain ascendancy – not out of any morality or immorality, but because it *lives*, and because it *is* will to power' (*BGE,* No.259). Genius and the creative act are part and parcel of this compulsion to emergence. The individual human is 'without content' and without a prescribed 'self' (*BGE,* No.207). S/he is a mere vehicle for giving shape and form to experience and sensation. It would be a mistake to see the human being as an end or the teleological purpose of 'the *rest* of existence' or even as 'a beginning, a begetting, and a first cause' for the rest of existence (*BGE,* No.207).

On the other hand, the human being appears to be something much more than a medium. The most alive individual is a creator of values, a commander and lawgiver, whose knowing is creating and whose will to truth is will to power (*BGE,* No.211). Greatness is defined as the ability to be as complex and encompassing as life itself (*BGE,* No.212). Such an individual lives life imprudently, unwisely, dangerously (*BGE,* No.205). In external nature, greatness is expressed as the venting of energy (e.g., tornado, earthquake, volcanic eruption), whereas in human nature greatness (or genius) appears as ordering, dominance, and appropriation. In the preface to his *On the Genealogy of Morals* Nietzsche laments the dominant exaggerated sense of pity that has repressed human flourishing and prevented the rise of 'the highest power and splendor of the type 'man' (*GM*, KSA, V, p.253). To understand his purpose, Nietzsche requires a renewed art of reading and interpreting (*GM*, KSA, V, p.256). The interpretive act amounts for Nietzsche to a lie by means of which humanity derives meaning from the surface play of forces. Calling this ability 'the creative talent par excellence of humankind,' he labels the creative individual a 'genius of the lie' (KSA, XIII, p.193), discerns his/her source in the very operations of nature. The true artist imitates nature by *acting out* the dynamic forces of nature rather than merely copying its artifacts. These are also traits of Nietzsche's philosopher of the future. The mere copyists, on the other hand, are 'false free spirits'; they are tireless scribblers in the service of the mass reading public (*BGE,* No.44) and hence advocates of the herd mentality. While a matter of the will, genius is not wholly dependent upon *human* volition. Something – an 'it' – wills in him, even as 'it' thinks in him. Nietzsche succinctly characterised the genius in *Human, All Too Human* when he metaphorically suggested:

> The way in which a prisoner uses his wits in the search for a means of escape, the most cold-blooded and tedious employment of every little advantage, can teach us what instrument nature sometimes makes use of to bring genius into existence – a word I ask to be understood without any flavor of the mythological or religious: it takes it and shuts it in a prison and excites in it the greatest possible desire to free itself. – Or, to employ a different image: someone who has completely lost his way in a forest but strives with uncommon energy to get out of it again sometimes discovers a new path which no one knows: that is how those geniuses come about who are famed for originality. – It has already been remarked that a mutilation, crippling, a serious deficiency in an organ offers the occasion for an uncommonly successful development of another organ, the reason being that it has to discharge not only its own function but another as well. In this way one can suppose many a glittering talent to have originated.[43]

The three images invoked – the metaphor of the imprisoned freedom-seeker, the soul hopelessly lost in the forest, the crippled organ – draw attention to both the evolutionary struggle for existence and the agonistic orchestration of matters to bring about a creative result. One organ, one talent – or a concert of talents – is stimulated by external circumstances and inner forces to compensate in agonistic fashion for the absence of something else. The genius of being is born of necessity, of absence, of non-being. The traditional, philosophical distinction between being and becoming dissolved for Nietzsche. Being *is* Becoming. The *venting* of energy is the essential thing, not the having vented energy. The human being is an energy knot ('Kraftcentrum').

This energy knot is encapsulated by the brain. Nietzsche viewed the human brain as the product of evolution and thus as an extension *of* nature. He did not conceive of it as a master organ that reigns over the rest of nature, but one that is fully integrated in natural processes themselves. What distinguish human consciousness are sensations that seem to signify for him what 'qualia' are in contemporary neuroscience. In his unpublished notes from 1873 we find such telling comments that rational constructs gradually arise from sensations ('Die Formen des Intellekts sind aus der Materie entstanden, sehr allmählich' (KSA, VII, p.467). The eye and brain are a kind of 'extra-I-before-itself'. The world is not a projection of the mind. Rather, mind and matter are interrelated. Thinking is not disassociated from physical sensations and feelings, and qualia are decisive.[44] This view of the brain further explains his notion of the aesthetic, that is, of life as literature.

---

43 Nietzsche, *HAH*, trans. by R. J. Hollingdale, I, No.231, pp.110-11.

44 'Unser Gehirn, unser Auge ist bereits ein extra nos oder praeter nos: es ist nicht die Welt eine Gehirnqualität, sondern das Gehirn selbst ist ein Theil dieser Empfindungen und Vorstellungen. Nicht das Gehirn denkt, sondern wir denken das Gehirn: das selbst an sich durch keine Realität hat. Empfindung ist die einzige

In an unpublished note from spring 1888 on the material world that is the object of physics, Nietzsche argued that the interpretation of the world offered by physicists is essentially the same as that offered by philologists, who study written texts. They just do it in a more refined manner. However, he misses in the physicist's explanation of the world something of great importance: the concept of perspective. By means of perspectivism, Nietzsche noted, '*every energy knot – and not just the human being* – constructs the rest of the world from within itself (i.e., measures, handles, and forms it according to its own energy.'[45] Seen in this light, the creative act (or the genius) is not essentially different from other energy impulses in the universe. This interpretation of genius is as much a general characterisation of how things become manifest or emerge as are the interactions of cold and warm fronts, of planets, molecules, and chemical reactions: 'Planets and atoms are different only in magnitude; they obey the same laws' (KSA, XI, p.702). Chemical changes in inorganic nature, Nietzsche conjectured, 'are perhaps artistic processes too' (KSA, VII, p.437). Striking in this regard is Nietzsche's insistence that the activity of interpretation is similar to the 'negotiations' of matter and energy in chemistry, physics, and biology. The classic expression of this redefined creative artist is the Dionysian principle. Dionysus – the god of the dance, the god of wine and exuberance, the genius of the heart – who deconstructs ordered worlds, enters the scene at the conclusion of *Beyond Good and Evil.* He serves as the culmination of Nietzsche's grounding of human life in the total economy of the earth with its organic and inorganic interconnections.

The Enactment of Evolutionary Literature

The entire opus of *Beyond Good and Evil* is a highly nuanced narrative that is essentially designed to introduce the tempter-god, Dionysus. A 'story line' unfolds incrementally in the nine parts of *Beyond Good and Evil* and reaches a crescendo in the rendition of Dionysus as the genius of the heart. Having just sketched out the major theme in the work as the evolution of the creative genius, I wish now to portray *Beyond Good and Evil* as the enactment of evolutionary literature. The work is based on a negative dialectic of unmasking and is inherently related to Nietzsche's definition in *On the*

---

kardinale Thatsache, die wir kennen, die einzige wahre Qualität' (*Nachgelassene Fragmente*, 1873, KSA, VII, p.598).

45 Nietzsche, *Nachgelassene Fragmente*, 1888: 'Und zulezt habe sie in der Constellation etwas ausgelassen, ohne es zu wissen: eben den nothwendigen Perspektivismus, vermöge dessen *jedes Kraftcentrum – und nicht nur der Mensch* – von sich aus die ganze übrige Welt construiert d.h. an seiner Kraft mißt, betastet, gestaltet' (KSA, XIII, p.373, my emphasis.)

Geneaology of Morals of the aphorism as demanding a highly developed art of interpretation (*GM*, Preface No.8; KSA, V, p.255).

Beyond Good and Evil is the sixth and last of Nietzsche's books of aphorisms and is formally the most successful. Widely recognised as a seminal work in Nietzsche's literary production for bringing into sharp focus the ideas and challenges explored in previous works, especially in *Zarathustra*, and to be revisited later in *On the Genealogy of Morals*, *Ecce Homo* and *Will to Power*, *Beyond Good and Evil* suffered the same fate as his earlier publications: it sold badly and went unread. A year after publication in a run of 300 copies, only 114 had been sold. However, in the early1890s three editions did follow. It has not been out of print since then. The book covers almost the entire range of the author's philosophical interests. It is comprised of a preface, nine parts, and an epilogue. Each of the main sections has its own subheading:

Part One: On the Prejudices of Philosophers
Part Two: The Free Spirit
Part Three: The Religious Nature
Part Four: Maxims and Interludes
Part Five: On the Natural History of Morals
Part Six: We Scholars
Part Seven: Our Virtues
Part Eight: Peoples and Fatherlands
Part Nine: What is Noble?

The epilogue is labeled 'From High Mountains: Epode'. Each section builds on the preceding one. *Beyond Good and Evil* requires an 'adaptive' reader.

R. J. Hollingdale is right to stress in the brief introduction to his edition of *Beyond Good and Evil* that the sections should be read in order, not willy nilly.[46] To assume that there is no constraint on the reader because the book is merely a collection of aphorisms is to err both in understanding how Nietzsche used the term and in estimating just how self-sufficient an aphorism is. Part IV is the only section that is consistently aphoristic, containing as it does short, pithy formulations that have profound consequences. The vast majority of the so-called aphorisms in the rest of the book read more like essays or fragments. The proper way to read *Beyond Good and Evil* is to view each part – including the preface and epode – as

---

46 Friedrich Nietzsche, *Beyond Good and Evil*, trans. by R. J. Hollingdale (New York: Penguin, 1981), pp.7-8.

inherent structural elements in the totality of the whole construction. We do best to read them 'atomistically'; that is, to see each of the numbered 'aphorisms' (whether genuine aphorisms or not) as so many interconnected and layered building blocks of constructed reality. Even if I have succumbed to the widespread habit in Nietzsche studies to pull out aphorisms from their contexts and make them stand on their own, I wish here to emphasise their interconnectedness, their being part of a total economy of the whole that is always in motion. To pull out an aphorism – even a genuine one – and make it stand on its own is to play out the scenario Nietzsche invokes in his conclusion when he sees individual thoughts as frozen in time and place lacking life:

> Oh what are you, my written and painted thoughts! It was not long ago that you were still many-colored, young and malicious, so full of thorns and hidden spices, you made me sneeze and laugh – and now? You have already taken off your novelty and some of you, I fear, are on the point of becoming truths.

Only that which has grown weary and is about to wither and lose its fragrance can be captured in static form, he goes on to explain: 'We immortalize that which cannot live and fly much longer' (*BGE,* No.296).

While the first three parts examine in scientifically rigorous fashion the underlying principles of traditional philosophy as seen through the lens of scientific positivism, parts 5-9 explore in greater depth the repercussions of his re-envisioning of reason as *razón vital* for a clearer understanding of the genealogy of morals, the ethos of intellectuals and researchers, the evolution of races and nations, and a reevaluation of the concept of nobility in the new evolutionary scheme of things.

In pursuing the overarching theme of the 'total economy of life' (*BGE,* No.23), Nietzsche moves in Part One: 'On the Prejudices of Philosophers' from turning dogmatic philosophy on its head through a reevaluation of every essential concept in the pursuit of truth in philosophy, science, and religion to the assertion that what we need is a new psychology. The outline of the individual steps is as follows:

No.1: the stressing of perspective as epistemologically grounded (why not untruth?);
No.2: the rejection of antitheses in favor of gradations in difference;
No.3: the grounding of conscious thinking in instincts;
No.4: the exhortation 'to recognize untruth as the condition of life';
No.5: the rejection of the 'hocus-pocus of mathematical form' and abstract reasoning that underlies speculative philosophy and Kant's moral imperative;
No.6: the importance of the 'order of rank' in nature;

No.7: the assertion that all previous philosophy is nothing but prejudice;
No.8: the philosopher's exaggerated self-confidence;
No.9: the acceptance of a life properly lived must run contrary to nature because nature is indifferent to us but we are not indifferent to it;
Nos.10-11: the claim that philosophy and positivism in science are essentially tyrannical
drives to control what is beyond our control (No.10 = an attack on the materialists;
No.11 = a rejection of the idealists);
No.12: a rejection of the soul as indestructible, eternal, indivisible like monad or an atom;
Nos.16-17: a deconstruction of thinking as mere interpretation;
No.19: the reduction of willing to an emotion and a plurality of sensations and ultimately the unmasking of the self as a synthetic concept ('our body is only a social structure composed of many souls');
No.23: concluding assertion that psychology is the science with the greatest promise for the future because of its revelatory potential.

The progression is incremental and logical, although at first sight it might appear to be associative, or even disjointed. In essence, Nietzsche addresses, here, however obliquely at times, the phenomenon of what he labels the will to power. Thus, his brand of psychology is characterised as 'morphology' and 'the developmental theory of the will to power' (No.23). Psychology stands at the end of this opening section because it alone is capable of (re)-presenting 'the total economy of life' (No.23). In between he points out the restrictive nature of language (No.20), the 'innocence of becoming' because 'in real life it is only a question of strong and weak wills' (No.21), and the rejection of equality in nature (No.22). All this means that emergence is beyond the traditional categories of good and evil. The concept of the will to power – first brought up in aphorism No.23, re-emerges in later aphorisms at regular intervals as the core concept (Nos.36, 230, 259). The same is true of the conditioned nature of thinking (Nos.54, 188, 192, 239) and of psychology (Nos.230, 269).

Part Two of *Beyond Good and Evil* reveals a similar dialectical movement. Picking up on the idea of life based on a falsification and simplification of life (No.24) first sounded in No.4, he moves quickly to the notion of experiencing a full life in contrast to what the average man is capable of (No.26), the dangers of being understood only superficially (No.27), to the introduction of the overman (No.32), a need for transvaluation in assessing the genealogy of morals as subjective valuations (No.32), to the fundamental idea of the will to power (No.36), the analogy of shallow interpretation of life as equating so many masks (No.40), a call for complete independence in living life as 'the most dangerous game' rather than seeking refuge in safe havens (No.41), and concludes with an image of the true free spirit (No.44). While these genuine free spirits are not the new philosophers themselves, they are the heralds, precursors and preparers for the philosophers of the future. Of course, these philosophers of the future might simply refer to those who philosophise about the future, who are attentive to emergence.

Nietzsche opens Part Three of *Beyond Good and Evil* with a specific reference to the genuine free spirits as midwives. To be sure, he uses the analogy of the big-game hunter set loose in the wild jungles of life. Part Three is dedicated to an examination of man's religious nature. The 'human soul with its frontiers' is placed in a jungle landscape that tests its ability to survive, Moreover, this 'compass of human inner experience' becomes itself 'the predestined hunting ground for a born psychologist and lover of the "big-game hunt"' (No.45). Hence, both the outer and inner worlds share qualities of danger, chaos, the will to life in competition with others and other species. The way Nietzsche ties the parts of his argument together at the macro and micro levels is impressive. And he continues to weave the threads of his fabric together throughout the entire text.

A detailed analysis of such seminal 'aphorisms' as No.188 in Part Five: the 'Natural History of Morals' would reveal a lucid explication of the fundamental tenets of Nietzsche's philosophy of radical realism/vitalism that emphasises the dual ability both to command and obey as characteristics of the overman. Similarly, scrutiny of aphorism No.262 in Part Nine: 'What is Noble?' reveals the jungle analogy again. Here Nietzsche speaks in Darwinian terms of the rise of a species to being fixed and strong and of its decline due to the ascendancy of other species and types. One prevails or runs the risk of extermination (*BGE*, KSA, V, p.216). Danger proves to be the mother of invention and ultimately of morality. Centers of civilisation such as the Greek polis or the city-state like Venice function as breeding grounds for their inhabitants to discipline themselves in common action against the perceived threat. Successful traits attain the status of virtue, eventually taking on the force of law in customs, the treatment of women, marriage contracts, generational conflict, the penal code, etc. The constant struggle against unfavorable conditions fixes and hardens these codes. But when less dangerous times come around, when life must not be earned and sustained but can simply be enjoyed, the tension bonding the community together lessens. Although the original external conditions have changed, the established code of behavior continues unabated. Yet, it too will one day decay. This view of a dialectic of challenging and less challenging times serves to undermine the concept of the survival of the fittest based on strength and force alone because the inherited code of behavior (master versus servant mentality) continues into the next epoch as a force unto itself.

Drawing overtly on evolutionary theory, and developing the cognitive metaphor of the bow in his preface, Nietzsche characterises the waxing and waning of species – now extended to moral systems – as so many turning points in the genealogy of morals (p.182). 'At these turning points of history,' Nietzsche avers:

> there appear side by side and often entangled and entwined together a glorious, manifold, jungle-like growth and up-stirring a kind of tropical tempo in competition in growing, and a tremendous perishing and self-destruction, thanks to the savage egoisms which, turning on one another and as it were exploding, wrestle together 'for sun and light' and no longer know how to draw any limitation, any restraint, any forbearance from the morality reigning hitherto. It was this morality, which stored up such enormous energy, which bends the bow in such a threatening manner – now it is 'spent,' now it is becoming 'outlived'. The dangerous and uncanny point is reached where the grander, more manifold, more comprehensive life lives beyond the old morality; the 'individual' stands there, reduced to his own law-giving, to his own arts and stratagems for self-preservation, self-enhancement, self-redemption. (Hollingdale, *BGE*, p.182, No.262; = KSA, V, p.216)

A *Herdenmoral* (herd morality) represented by the 'levelers' seeks to restrain the strong-willed individuals and to counteract the *bellum omniorum contra omnes* that gives rise to a *Herrenmoral* (master morality) represented by the free spirits. As we know, Nietzsche protested the herd mentality and bourgeois morality of his own day, announcing the free spirits as promoters of vitality. To be sure, he does not deny the levelers a legitimate function in the total economy of the whole. Yet these levelers do not represent evolutionary progress in Nietzsche's mind. While one can read in some commentaries that the philosopher envisioned a development from ape to man to superman, such a take does little justice to Nietzsche's radical realism. Evolution to the common man of the democratic spirit represented for him devolution from the higher individualism of the self-directed, Dionysian-inspired *Übermensch*. The dialectic ensures an eternal return of the same.

Whereas levelers value mimicry, the overman emphasises value creation. In *Untimely Meditations I*, Nietzsche exhorts his reader: 'Live like a man and not like a monkey or seal' (KSA, I, p.195). Immediately, he adds with a sigh that this imperative is unfortunately unusable and ineffective because the contemporary concept of man is that of a servile slave burdened by all kinds of restrictive demands. Modern man has devolved into monkeys incapable of creative acts. The correct imperative, he avers, is 'live creatively' ('lebe als

Genie'); it is an ideal expression of the type 'man'.[47] Later in *Also sprach Zarathustra* he would speak in similar terms when he remarked that man has progressed from the worm and from the monkey. Yet, there is still much in man that is worm-like, there is more of the monkey in modern man than in monkeys themselves.[48]

In Nietzsche's view Darwin, Mill, and Spencer provided ample evidence for the rise of mediocrity in European taste (*BGE*, No.253; KSA, V, p.196f.). Despite his disagreement with Darwin, Nietzsche concedes that narrowness of focus and conscientious industriousness – such as Darwin demonstrated in his careful and detailed research – may not be an unfavorable disposition (*BGE*, No.253; KSA, V, p.197). The reason for this approbation of tediousness is found no doubt in Nietzsche's earlier insistence that a narrowing of perspective is life advancing. Indeed, it is the basic condition of all life (Aphorisms, No.2, No.188), even if the resultant insights are all 'premature hypotheses' (No.192).

All the same, the genuine philosopher steps outside the known, tried, and true. S/he lives instinctually and runs the risk of a thousand dangers in the venting of his/her energy, in the testing of his/her limits. S/he lives dangerously and risks her very existence. If the first analogy refers to the conservation of energy and the second to its venting, both operations of organic and inorganic nature appear as necessary to life on earth as we know it. Hence, it would seem that Nietzsche did after all retain a Darwinian notion of the survival of the fittest to explain how the master breed emerges.

Clearly both the content and the form of *Beyond Good and Evil* mirror evolutionary processes as analogous to the processes prompted by literature. To live life meant for Nietzsche to live creatively as a man, not in unreflective ape-like mimicry evident in bourgeois behavior. While the first mode is proactive, the second is passive. The *Übermensch* is the ideal goal of the earth where 'earth' is understood as the will to power ('Der Übermensch ist der Sinn der Erde'; *Zarathustra*, KSA, IV, p.14). To be sure, the overman is not a steady state, is not free of regression to a mere man or even to the lesser state of monkey or worm. As he stated later in *Wille zur Macht*:

---

47 Nietzsche, *Unzeitgemässe Betrachtungen I*: 'Lebe als Mensch und nicht als Affe oder Seehund. Dieser Imperativ ist leider nur durchaus unbrauchbar und kraftlos, weil under dem Begriff Mensch das Mannigfaltigste zusammen im Joche geht [...] lebe als Genie, das heisst eben als idealer Ausdruck der Gattung Mensch' (Nietzsche, KSA, I, p.195).

48 Nietzsche, *Zarathustra*, 'Vorrede' 3: 'Ihr habt den Weg vom Wurme zum Menschen gemacht, und Vieles ist in euch noch Wurm. Einst wart ihr Affen, und auch jetzt noch ist der Mensch mehr Affe, als irgend ein Affe (KSA, IV, p.14).

> My idea is that every specific body strives to become master of all space, and to extend its force (– its will to power), and to thrust back all that resists its extension. But it continually encounters similar efforts on the part of other bodies and ends by coming to an arrangement ('union') with those of them that are sufficiently related to it; thus they then conspire together for power. And the process goes on. (*Will to Power*, 340, No.636; KSA, XIII, pp.373-74)

The paradoxical description offered in *Beyond Good and Evil* is, then, the terrible basic text *homo natura*: the will to power in its agonistic wholeness of expansion and contraction. The human being is both reactive and proactive, a mere mold and vessel but also a creator and builder of molds, a conserver and a venter of energy. The alternation between them amounts to an evolutionary cycle. Human life is a continual reabsorbing of circumstances, a continual working out of our lives through them, as Howard Tuttle put it. And through an explosive reaction to them, as Nietzsche concluded. While this cycle was for Nietzsche tantamount to an eternal evolutionary dialectic of progression *and* regression, for Darwin progress seemed more linear and teleological through the privileging of dominant biological traits acquired over generations of reactive experience. Hence Nietzsche's focus on the inorganic environment of organic substances together with his notion of cyclical alternation represent major differences between the two thinkers. Such is the nature of Nietzsche's 'new Darwinism'. By comparison his vitalism is a radical realism because humans are condemned to become human through the absorbing of the non-human. To live like a man demanded confrontation with the monkey.

'Creative Evolution:' Bergson's Critique of Science and its Reception in the German-Speaking World

David Midgley

Abstract: In *L'évolution créatrice* (1907), Bergson reviewed the thinking of the biologists of his day about evolutionary and organic processes, and drew some distinctive conclusions about the nature of human understanding, notably privileging intuition over intellect when it comes to grasping the nature of life and evolution. This chapter examines the character of Bergson's arguments, the circumstances in which his text came to be translated into German in 1912, and the range of responses that the work elicited from German intellectuals of the time. It identifies some senses in which Bergson's arguments were assimilated to expectations moulded by German philosophical traditions, and considers what the reception of Bergson's text reveals about the conflicts at work in German literary and intellectual culture in the decade before the First World War.

Henri Bergson's philosophical reputation rests on the fundamental critique of scientific thought that he conducted in a series of publications, starting with the *Essai sur les données immédiates de la conscience* (1889). But the work that made him the object of international public interest in the years immediately preceding the First World War was *L'évolution créatrice* (1907), in which he turned his attention to biological theories and their implications for our understanding of ourselves as human beings.[1] In the broad perspective of intellectual history the 'Bergson boom,' as it was known at the time, was a transient phenomenon and perhaps no more than a stepping stone towards more significant developments in European thought. But the value of approaching the German literary and intellectual culture of the early twentieth century via the reception of Bergson is that this serves to bring into sharp focus the nature of the issues that presented themselves to German writers in the wake of the impact of Darwinism.

As in the English-speaking world, and indeed in France, the German reception of Bergson was a pervasive phenomenon with no distinct boundaries, and to date there is no single study that encompasses its various dimensions. The Viennese essayist Franz Blei recalled in 1930 that Bergson had been an inescapable presence in the pre-war intellectual atmosphere, but the effect this may have had on particular literary writers is difficult to

[1] The most comprehensive Bergson bibliography – P.A.Y. Gunter, *Henri Bergson: A Bibliography*, 2nd rev edn (Bowling Green State University: Philosophy Documentation Centre, 1986) – indicates that there was a sharp rise in the number of publications about Bergson in 1907-08, followed by a surge of interest from 1910 to 1914, which peaked in 1912 with 235 such publications worldwide. Cf. also R.C. Grogin, *The Bergsonian Controversy in France, 1900-1914* (Calgary: University of Calgary Press, 1988), pp.122-32.

gauge.[2] Readers of Thomas Mann's *Der Zauberberg* (*The Magic Mountain*, 1924), for example, may have good grounds for suspecting a close affinity between the treatment of the subjective experience of time in that novel and Bergson's views on the subject, but they are firmly told in the recent critical edition of the work that Mann did *not* read Bergson (although he is very likely to have read about him).[3] The poet Rilke records having *tried* to read two volumes of Bergson in June 1914, but admits in a letter to Lou Andreas-Salomé, from whom he had borrowed them, that he only partially understood them.[4] Andreas-Salomé for her part undoubtedly had the intellectual capacity to understand Bergson, but she appears not to have committed herself to any views about him in print. Robert Musil (to whose writings Franz Blei was referring when he made his retrospective comment about the pre-war intellectual atmosphere in Vienna) did show an active interest in Bergson for two related reasons: he shared Bergson's interest in intuition and the limits of rationalism, but he saw the cult of Bergson as one phenomenon of the times amongst others that warranted critical analysis. Musil mentions Bergson in a note connected with his planning of a critical essay on the cultural pessimism of Oswald Spengler, which he published in 1921, and it is evident from that note that Musil wanted to give himself an opportunity to explore the importance of intuition as a mode of perception that is not reducible to rational categories, while also attacking the current fashion for anti-rationalism.[5] Looking back in 1941, Musil accepted that some of his own themes invited comparison with Bergson's, but was concerned to distance himself from Bergson's rather dogmatic association of scientific thought with the reduction of experience to spatial categories.[6] All in all, it is likely that in his great epic project, *Der Mann ohne Eigenschaften* (*The Man Without Qualities*, 1930-42), Musil took up certain thoughts of Bergson's, along with those of many other contemporary figures, modulated

2 Franz Blei, *Erzählung eines Lebens* (Leipzig: List, 1930), pp.450-51.

3 Thomas Mann, *Große kommentierte Frankfurter Ausgabe*, 5.2 (Frankfurt: Fischer, 2002), p.62.

4 Rainer Maria Rilke, Lou Andreas-Salomé, *Briefwechsel*, ed. by Ernst Pfeiffer (Frankfurt: Insel, 1989), p.324; cf. p.306. The volumes have been identified as *Matière et mémoire* and *L'évolution créatrice* (pp.578-79.).

5 Robert Musil, *Tagebücher*, 2 vols (Reinbek bei Hamburg: Rowohlt, 1976), I, p.661; II, pp.1163-69. For Musil's essay on Spengler, 'Geist und Erfahrung. Anmerkungen für Leser, welche dem Untergang des Abendlandes entronnen sind ,' see *Gesammelte Werke*, 9 vols (Reinbek bei Hamburg: Rowohlt, 1978), II, pp.1042-59.

6 *Tagebücher*, I, p.960; II, pp.730-31. See also Renate von Heydebrand, *Die Reflexionen Ulrichs in Robert Musils Roman 'Der Mann ohne Eigenschaften'* (Münster: Aschendorff, 1966), pp.213-16.

them, and construed them anew in the fictional setting he works with: Viennese high society on the eve of the First World War.[7]

If we consider the case of Alfred Döblin, however, who is most commonly remembered for his novel *Berlin Alexanderplatz* (1929), we come closer to identifying those features of the German intellectual culture around 1900 with which Bergson's thought most readily resounded. Döblin was a trained psychiatrist who also showed an intense interest in philosophical ideas during his student days in Berlin, between 1900 and 1904. There is clear evidence that, at that time, he was particularly excited by Spinoza's pantheistic view of nature, by Schopenhauer's notion of an underlying dynamic principle at work in the universe (the will), and by Hegel's way of relating the world as we experience it to an underlying spiritual principle (particularly in his *Phenomenology of Spirit*, 1807). In these respects the young Döblin was participating in a dominant trend of the time, which consisted in the quest for a unified world-view (*Weltanschauung*) that was based on perceptions of natural processes and did not resort to theism.[8] That pursuit of a unified view of natural processes is also apparent in the imagery Döblin deploys for the representation of the natural and the human world in his early fiction.[9] In his responses to Nietzsche, on the other hand, Döblin was overtly critical, for reasons which are also directly relevant to our present theme: he recognised how central the idea of evolution was to Nietzsche's discussion of human nature, but strongly resisted what he saw as the biological basis for Nietzsche's 'morality of strength'.[10] Moreover, he firmly

---

7 For a brief account in English of Musil's achievement in *Der Mann ohne Eigenschaften*, see David Midgley, 'Robert Musil: *Der Mann ohne Eigenschaften*', in *Landmarks in the German Novel*, ed. by Peter Hutchinson (Oxford: Lang, 2007), pp.183-98.

8 For this background trend, see especially Wolfgang Riedel, *'Homo Natura'. Literarische Anthropologie um 1900* (Berlin: de Gruyter, 1996), pp.41-77; Walter Gebhard, *"Der Zusammenhang der Dinge". Weltgleichnis und Naturverklärung im Totalitätsbewußtsein des 19. Jahrhunderts* (Tübingen: Niemeyer, 1984).

9 See Helga Stegemann, *Studien zu Alfred Döblins Bildlichkeit: 'Die Ermordung einer Butterblume' und andere Erzählungen* (Bern: Lang, 1978); David Midgley, 'The Early Fiction of Alfred Döblin: The Short Story between Case Study and Parable', in *Kafka and Short Modernist Prose*, ed. by Manfred Engel and Ritchie Robertson (Würzburg: Königshausen und Neumann, forthcoming).

10 See David Midgley, 'Metaphysical Speculation and the Fascination of the Real: On the Connections between Döblin's Philosophical Writings and his Major Fiction before *Berlin Alexanderplatz*', in *Alfred Döblin: Paradigms of Modernism*, ed. by Steffan Davies and Ernest Schonfield (Berlin: de Gruyter, 2009), pp.7-27 (p.12). Döblin's essays on Nietzsche, which date from 1902-03, have been published in Alfred Döblin, *Kleine Schriften* I (Olten and Freiburg im Breisgau: Walter, 1985), pp.13-55. See also Beat Knuttig, *Die Nietzsche-Aufsätze des jungen Alfred Döblin. Eine Auseinandersetzung über die Grundlagen von Erkenntnis und Ethik* (Bern: Lang, 1995).

associated Nietzsche's arguments with the dominance of scientific empiricism towards the end of the nineteenth century, which he was seeking to transcend with his metaphysical vision (pp. 12ff.); and in later essays on the natural world, in the 1920s, Döblin professes a faith in a subject (an *Ich*) that resides above nature, but is inherently organic in character – a notion that invites comparison with the Romantic nature philosophy of Schelling (*Ideas for a Philosophy of Nature*, 1797), and which has been aptly described as a surrogate deity for a scientific age.[11] Döblin was generally reticent, and sometimes evasive, about the likely influences on his thinking, and there is no firm evidence that he ever read Bergson. But if we now turn to the arguments that Bergson develops in *L'évolution créatrice*, the affinities between Döblin's concerns and Bergson's will be readily apparent.

The way Bergson conducts his argument about the nature of life in *L'évolution créatrice* is to test the tenability of views that the biologists themselves hold, and having discredited these in particular ways, to propose an alternative view as more plausible. The text contains abundant references to biological findings on such matters as embryology and aging, cell development and regeneration, protoplasma and primitive life forms, as well as on evolution and heredity.[12] Bergson has noted, for example, the 'germ plasma' theory of August Weismann which, in the 1880s, first drew a clear distinction between those features of an organism that are genetically passed on and those which are, so to speak, physiologically fixed within the life-span of the individual; and he has read the book on evolution and adaptation by Thomas Hunt Morgan, which gives a fairly comprehensive account of the key issues in evolutionary biology at the start of the twentieth century (Morgan's experimental work helped to lay the foundations for genetics as we know it).[13] The main focus in Bergson's opening chapter, however, is on two broad ways of conceiving the underlying nature of organic processes: the 'mechanistic' view, which thinks in terms of causal principles, and the 'finalist' view, which thinks in terms of purpose.

Bergson's aim is to refute both schools of thought. For him, 'mechanistic' thinking is appropriate only to the interpretation of phenomena that our intellect has separated out from the continual passage of time (*durée*), and therefore inappropriate to the interpretation of organic

---

11 Barbara Belhalfaoui-Köhn, 'Alfred Döblins Naturphilosophie – ein existentialistischer Universalismus', *Jahrbuch der deutschen Schiller-Gesellschaft*, 31 (1987), 354-82 (p.362).

12 For a very helpful exposition of the biological arguments with which Bergson engages, see Yvette Conry, *L'évolution créatrice d'Henri Bergson: Investigations critiques* (Paris: L'Harmattan, 2006).

13 Thomas Hunt Morgan, *Evolution and Adaptation* (London: Macmillan, 1903).

processes; and 'finalism' is equally unacceptable because it amounts to a kind of 'mechanism' in reverse, which assumes (with Leibniz, for example) that all ultimate purposes are given from the outset, and that there is therefore no possibility of creative development along the way (pp.39-40). Bergson is more sympathetic to vitalism, i.e. the notion that organic development is determined by an internally given principle, but he sees a difficulty with that position too: nature, he observes, knows neither a sense of purpose that is purely determined from within, nor a sense of individuality that exists in isolation, so it is futile to seek a principle of finality that is located within the individual organism.[14]

It is therefore rather by a process of elimination that Bergson establishes a basis of credibility – which he underscores with a rhetorical 'croyons-nous' (p.44) – for the proposition for which *L'évolution créatrice* is perhaps most commonly remembered, and which is commonly quoted without specific reference to its context in Bergson's argument. The notion of 'élan vital,' which he endorses towards the end of his first chapter, is a way of referring to the impetus towards development that must be *supposed* to be inherent in life as such if we are to conceive of the evolutionary process in terms which do not become trapped in the schools of thought that Bergson has rejected. Bergson would have had good grounds for claiming he had been misconstrued, for example, when Julian Huxley quipped in 1929 that explaining biological progress in terms of an *élan vital* is like explaining the movement of a train by the *élan locomotif* of the engine, because Huxley was simply reducing the implications of Bergson's phrase to mechanistic categories.[15] At a later point of *L'évolution créatrice* Bergson emphasises, indeed, that the notion of impetus is only an image, but a necessary one, to express an idea of life that eludes the concepts of physics and chemistry.

The other main thrust of Bergson's argument is that there is a fundamental difference between intellect and instinct, and that it is the latter, not the former, that enables us to grasp the true nature of evolution and the life process. For Bergson, the intellect (or intelligence) is itself a product of evolution, developed as an analytical tool which enables us to interact effectively with our environment. The intellect is good at recognising the significance of events that repeat themselves (a description which echoes the commonplace notion of scientific 'proof' as consisting in an experiment that can be reproduced), but it is inadequate, Bergson argues, for apprehending the nature of developments that are continuous and irreversible over time. Instinct, by contrast, does not analyse, but is in 'sympathy' with the world it

---

14 Henri Bergson, *L'évolution créatrice*, edition critique, ed. by Frédéric Worms and Arnaud François (Paris: Presses Universitaires de France, 2007), pp.37-39.

15 Julian Huxley, *Essays of a Biologist* (New York: Knopf, 1929), pp.32-35.

inhabits, as illustrated by the capacity of certain insects to sting their prey in such a precise way as to paralyse it without killing it. It is instinct and intuition (which Bergson describes as instinct that has become self-conscious) that provide us with an intimate awareness of life's processes and thereby point the way to a metaphysical, as opposed to scientific, understanding of the world. This sense of a sharp distinction between intellect and instinct instantly led to Bergson being classified as an anti-intellectualist; but as a French commentator observed in 1927, the mistake commonly made by his admirers and detractors alike was to overlook the dialectical dimension of his argument which resolves tensions and teases out connections between apparent antitheses.[16] In a formulation worthy of Pascal, for example, Bergson states that there are things that only the intellect is capable of seeking, but that on its own it would never find them, whereas instinct can find them, but left to itself would never be impelled to seek.

That dialectical quality kept serious thinkers interested in Bergson's writings *in spite of*, rather than because of his popular following. But there is one more dimension of *L'évolution créatrice* that understandably invited controversy. Bergson insists that no philosophy can dispense with the insights provided by evolutionary theory and the empirical inquiries associated with it, but the construction he places on the concept of evolution is that of an eternally creative process set in train by an initial impulse. He likens that process to a consciousness liberating itself, conceives it as non-spatial and immaterial, and even equates the idea of God with the continual creative life process in which we humans participate through our freedom of action. It is this aspect of *L'évolution créatrice* that was quickly recognised as pantheistic.[17] At the same time (and by contrast with Nietzsche), Bergson cleaves to the notion that the human race can be seen as a 'goal' of evolution, by virtue of its capacity for the exercise of freedom.

A number of developments towards the end of the nineteenth century can be seen to have contributed to making German readers sympathetic to the general thrust of Bergson's arguments. They include Nietzsche's critique of academic culture and his emphasis on creative renewal, which had been arousing fervent interest since the 1890s, and the vulgarisation of Nietzsche's views by Julius Langbehn, who had proclaimed in his bestseller

---

16 See in particular Julien Benda, *Le Bergsonisme ou une philosophie de la mobilité* (Paris: Mercure de France, 1912); Henri Jourdain, 'Französischer Brief', in *Probleme der Weltanschauungslehre*, ed. by Erich Rothacker (Darmstadt: Reichl, 1927), pp.459-93 (p.473).

17 For this aspect of the text and the critical responses to it in France, see Conry, pp.25-39.

Rembrandt als Erzieher (*Rembrandt as Educator*, 1891) that the age of scientific analysis was about to be superseded by an age of artistic synthesis.[18] But when it came to moulding public perceptions of the evolutionary process and of the place of mankind in nature around 1900, the most important figure is undoubtedly Ernst Haeckel. Not only had Haeckel presented himself as the great champion of Darwinian theory since the 1860s, he also fought hard to replace conventional religion with a unified world view based on science. It would be misleading to present Haeckel as straightforwardly a materialist monist because he was equally capable of speculating that all matter must in some sense be animate (*beseelt*) as he was of asserting that all the riddles of the universe could be explained on the basis of physical and chemical laws.[19] But it was primarily in response to his campaigning that public organisations promoting science-based world views on the one hand, and others seeking to interpret scientific findings in the light of traditional religious thinking on the other, became salient features of German intellectual life in the first decade of the twentieth century, and it was precisely around the time that *L'évolution créatrice* was published in France that competition between such organisations came to a head.[20] A particularly important contributor to the dissemination of Haeckel's views on evolution at that time was Wilhelm Bölsche, whose four-volume account of the 'love life' and 'descent' of natural organisms (*Das Liebesleben in der Natur*, 1898-1903) appeared with the same German publisher as the works of Bergson – Eugen Diederichs – and provided a rich source of biological imagery for literary writers to work with, including Alfred Döblin, as we noted earlier.[21]

When we consider the process by which Bergson came to be translated into German, however, we also need to note the specific roles played by two further German thinkers who were influential at the time. One was Rudolf Eucken, who had been combatting positivism and the fragmenting effect of scientific specialisation with the integrative power of philosophical idealism since the 1870s, and who was personally acquainted with Bergson. Indeed, Bergson wrote forewords for some of Eucken's works when they appeared

18 See Steven E. Aschheim, *The Nietzsche Legacy in Germany, 1890-1990* (Berkeley: University of California Press, 1992).

19 See especially Ernst Haeckel, *Die Welträthsel* (Bonn: Strauß, 1899); also Olaf Breidbach, *Visions of Nature. The Art and Science of Ernst Haeckel* (Munich: Prestel, 2006).

20 See Andreas W. Daum, *Wissenschaftspopularisierung im 19. Jahrhundert: bürgerliche Kultur, naturwissenschaftliche Bildung und die deutsche Öffentlichkeit 1848-1914* (Munich: Oldenbourg, 1998), pp.210-35.

21 For a survey of the various ways in which Darwinian motifs were treated by German literary authors in the late nineteenth and early twentieth centuries, see Peter Sprengel *Darwin in der Poesie* (Würzburg: Königshausen & Neumann, 1998).

in French translation, and it was a pupil of Eucken's who published the first German monograph on Bergson.[22] The other was Georg Simmel, an exact contemporary of Bergson's, whose studies of the cultural condition of industrialised societies drew attention to the potential for human creativity even within an apparently mechanised world. Simmel appears to have been the first German academic to offer a course of lectures on Bergson (in the winter semester of 1910-11), and he published a faithful and sympathetic account of his thinking in June 1914.[23] The key figure behind the publication of Bergson's works in German, however, was the publisher Eugen Diederichs, who undoubtedly recognised elements in them that corresponded to the cultural ideals he was already promoting.

Deeply committed to combatting philosophical materialism and the commercial spirit associated with industrialism, Diederichs sought to cultivate a spiritual aristocracy capable of sustaining a sense of higher values. He cherished the Romantic notion of the human spirit aspiring towards union with the great universal absolute, and in his publishing activities, which began in 1896, he nurtured a religiosity that showed affinities with the mystical rather than the doctrinal traditions of Christianity. The cultural elitism that Diederichs espoused meant that, in the period after the First World War, his natural clientele was to be found on the *völkisch* right rather than among the democratic sections of German society, but his outlook was never narrowly nationalistic; his programme included texts that he considered representative of the specific cultural traditions of Germany's European neighbours, as well as classic examples of Chinese, Indian, and Islamic thought.[24] The idea of including Bergson was first suggested to Diederichs by the young philosopher Max Scheler, another pupil of

---

22 See Friedrich Wilhelm Graf, 'Die Positivität des Geistigen. Rudolf Euckens Programm neoidealistischer Universalintegration', in *Kultur und Kulturwissenschaften um 1900. II: Idealismus und Positivismus*, ed. by Gangolf Hübinger et al. (Stuttgart: Steiner, 1997), pp.53-85; Cornelius Albert Steenbergen, *Henri Bergsons intuitive Philosophie* (Jena: Eugen Diederichs, 1909).

23 See *Eugen Diederichs: Selbstzeugnisse und Briefe von Zeitgenossen*, ed. by Ulf Diederichs (Düsseldorf: Diederichs, 1967), p.197; Georg Simmel, 'Henri Bergson', *Die Güldenkammer* 4 (1914), 9, 511-25; reprinted in Simmel, *Gesamtausgabe*, 24 vols (Frankfurt: Suhrkamp, 2000), XIII, pp.53-69. Cf. also Gregor Fitzi, *Soziale Erfahrung und Lebensphilosophie: Georg Simmels Beziehung zu Henri Bergson* (Konstanz: UVK Verlagsgesellschaft, 2002).

24 See George L. Mosse, *The Crisis of German Ideology: Intellectual Origins of the Third Reich* (New York: Grosset & Dunlap, 1964), pp.52-66; Meike Werner, *Moderne in der Provinz: Kulturelle Experimente im Fin de Siècle Jena* (Göttingen: Wallstein, 2003), pp.63-193; also *Versammlungsort moderner Geister: der Eugen Diederichs Verlag. Aufbruch ins Jahrhundert der Extreme*, ed. by Gangolf Hübinger (Munich: Diederichs, 1996), for further references.

Eucken's, and the translation of Bergson's works began systematically with *Matière et mémoire* (1908), the *Introduction à la métaphysique* (1909), and the *Essai sur les données immédiates de la conscience* (1911, under the amended but justifiable title *Zeit und Freiheit*).[25] Finding suitably qualified translators for these works was a persistent problem, and it was only after other philosophy graduates had been found wanting that the translation of *L'évolution créatrice* was entrusted to Gertrud Kantorowicz, a pupil of Simmel's who was also connected to the George circle. She is rightly acknowledged as the translator of the text that appeared as *Schöpferische Entwicklung* in 1912, but she had taken care to seek the help of Georg Simmel and his brother Hans with the translation of philosophical terms, and of Stefan George for some key phrases.[26] It was evidently George, for example, who suggested *Lebensschwungkraft* as the appropriate translation for *élan vital*.[27]

It is sometimes suggested that Diederichs effectively masterminded the creation of a 'German Bergson' to fit his own sense of cultural mission, so it is worth emphasising that Kantorowicz's translation is for the most part entirely faithful to Bergson's text.[28] In particular, the metaphorical imagery with which Bergson, especially in his second chapter, expresses his sense of how life develops explosively, tracing the trajectory of a sheaf or spray (*gerbe*), is rendered in ways that accurately reflect Bergson's manner of speaking. At most the German reader's understanding would have been hampered by occasional errors, which are largely attributable to a misconstruction of French syntax. At one point, for example, Kantorowicz's translation inverts the terms of Bergson's argument, giving the momentary impression that his understanding of organic phenomena is closer to the school of 'mechanism' than to that of 'finalism,' but an attentive reader would recognise here that the statement in question conflicts with the argument that Bergson has just unfolded.[29] Potentially more damaging for

---

25 'Time and free will' had already been chosen as the main title of the authorised English translation (London: Allen & Unwin, 1910).

26 Henri Bergson, *Schöpferische Entwicklung* (Jena: Eugen Diederichs, 1912). All page references for the German translation are to this edition.

27 Günther Pflug, 'Eugen Diederichs und Henri Bergson', in *Von Göschen bis Rowohlt. Beiträge zur Geschichte des deutschen Verlagswesens*, ed. by Monika Estermann and Michael Knoche (Wiesbaden: Harrassowitz, 1990), pp.158-76 (p.171).

28 See especially Friedrich Wilhelm Graf, 'Das Laboratorium der religiösen Moderne. Zur "Verlagsreligion" des Eugen Diederichs Verlags', in *Versammlungsort moderner Geister*, pp.243-98.

29 Where Bergson (p.50) writes, "la philosophie de la vie où nous nous acheminons [...] prétend dépasser à la fois le mécanisme et le finalisme; mais [...] elle se rapproche de la seconde doctrine plus que de la première", the Kantorowicz version (p.56) reads, "die Philosophie des Lebens, der wir uns zubewegen [...] erhebt den Anspruch,

the appreciation of Bergson's argument are moments where the translator introduces a sense of intention or purpose that is not there in the French. For example, where Bergson speculates on the *tendency* of early life forms to draw their energy from the sun, the German version suggests that life *aimed* at doing so: "daß das Leben zunächst danach *getrachtet* haben wird" (p.121; my emphasis). And at an earlier point, where the French text contains a series of infinitive phrases dependent upon the verb *consister à* – "pour un être conscient, exister consiste à changer, changer à se mûrir, se mûrir à se créer indéfiniment soi-même" (p.7) – the translation conveys the first of the three constructions correctly, but renders the second and third as if they were expressing purpose rather than describing a process: "Dasein [besteht] für ein bewußtes Wesen darin, sich zu wandeln; sich zu wandeln, *um zu* reifen; zu reifen, *um* sich selbst unendlich *zu* erschaffen" (14; my emphasis). Such moments when the German text erroneously introduces a teleological implication may not have been entirely without effect, because the latter passage is quoted verbatim by Ernst Robert Curtius, one of the leading German Romanists of the inter-war period, when he presents Bergson as the key figure in what he sees as the cultural and spiritual renewal of France in the period 1910-1914.[30] But such distortions are extremely infrequent, and it is to the philosophical resonances inherent in Bergson's text that we have to look in order to account for the way German readers responded to it.

Bergson's most fervent admirers not only saw him as championing the anti-mechanistic and anti-intellectual ideals they themselves professed, they also constituted a recognisable community of interests in a practical sense: in one way or another they tended to be associated with Diederichs' circle. Most prominent amongst them was Karl Joël, a professor at Basel University and author of books on the origins of nature philosophy (1903) and on free will (1908), who also published volumes with Diederichs on Nietzsche and Romanticism (1905) and the relationship between mind (*Seele*) and world (1912). In 1910 Joël published a rhapsodic essay in the most prestigious of the cultural journals of the time, *Die neue Rundschau*, in which he hailed Bergson as the 'artist' of contemporary philosophy, whose manner of arguing invited comparison with Rodin and Debussy, and whose doctrine proceeded in a truly Romantic manner from self-awareness to world vision, bringing about a 'triumph of feeling, [...] of becoming, [...] of creativity' over the constraints of causal and analytical thinking.[31] Isaac

---

Finalismus und Mechanismus gleichermaßen hinter sich zu lassen, wobei sie [...] diesem letzteren [i.e. 'Mechanismus'] nähersteht als jenem."

30 Ernst Robert Curtius, *Die literarischen Wegbereiter des neuen Frankreich* (Potsdam: Kiepenheuer, 1923), p.39.

31 Karl Joël, 'Neues Denken', *Die neue Rundschau*, 21 (1910), 549-58 (pp.554-57).

Benrubi, who proclaimed in another major cultural journal in 1911 that mechanism, scientism, rationalism and the developments in psychology that were associated with these had found their deadliest enemy in Bergson, was a pupil of Eucken's and had been involved in the early stages of translating *L'évolution créatrice*.[32] The reliable and highly readable account of Bergson's thinking by Adolf Keller, which originally appeared in the Swiss magazine *Wissen und Leben* in 1913, was subsequently also published as a book by Diederichs.[33]

Others also saw Bergson's arguments as contributing to their own purposes, whether philosophical or religious. The Berlin rabbi Julius Lewkowitz saw Bergson's critique of mechanistic science and his insistence on the existence of a creative spirit as contributing to a renewal of faith in a biblical God, while acknowledging that when Bergson speaks of God he means something more akin to Spinoza's principle of *natura naturans*.[34] There appears to have been no response from German Catholics comparable with the interest that French Catholic 'modernists' showed in Bergson, but for the Protestant pastor Emil Ott, the spiritual aspect of Bergson's interpretation of organic phenomena and his insistence on human freedom had provided important and timely stimuli for seeking God at work *within* our world, while he also saw Bergson's arguments as compatible with the idea of a transcendental God.[35] Within the scientific community, the most spectacular response to *L'évolution créatrice* came from the embryologist Hans Driesch, a leading opponent of materialist monism and the chief proponent of biological vitalism, whose views are indeed addressed in Bergson's book.[36] Driesch reviewed the French text very promptly in a specialist journal for evolutionary theory, evidently as a means of promoting his own views. Berating his fellow biologists for failing to take note of philosophical developments outside the German-speaking world, he gives a full, if somewhat hurried, summary of Bergson's arguments, notes the similarity between his ideas and those of Schelling, endorses his views on instinct and intellect, and commends Bergson's metaphysical conception of life as the best recent attempt to unify thinking about evolution. In response to

32 J. Benrubi, 'Die Renaissance des Idealismus in Frankreich', *Deutsche Rundschau*, 149 (1911-14), 388-406.

33 Adolf Keller, *Eine Philosophie des Lebens (Henri Bergson)* (Jena: Diederichs, 1914).

34 Julius Lewkowitz, *Die Philosophie Henri Bergsons* (Berlin: Rosenthal, 1913).

35 Cf. Grogin, *Bergsonian Controversy*. It is worth noting in this connection that Bergson's works were placed on the papal Index in June 1914; Emil Ott, *Henri Bergson, der Philosoph moderner Religion* (Leipzig: Teubner, 1914), especially pp.82-83.

36 See, for example, Hans Driesch, *The Science and Philosophy of the Organism: the Gifford Lectures Delivered before the University of Aberdeen in the Year 1907-08* (London: Black, 1908).

Bergson's strictures against vitalism, Driesch suggests that a first-hand knowledge of the experimental record in embryology (which had led Driesch to assume an inherent vital principle at work in organic development which, following Aristotle, he called 'entelechy') might have changed his mind.[37] Further evidence of serious, albeit more sober interest in Bergson among scientists comes from Karl Jaspers, who was working in psychopathology at the time he reviewed the German translation in 1913; he sees Bergson's emphasis on subjective experience as assisting the current momentum of psychological inquiry, but sternly warns that the problems of psychology cannot be solved by intuitive means.[38]

By 1913, however, Bergson had begun to be perceived as an author with whom it was not intellectually respectable to be associated.[39] A clear example of this backlash effect can be traced within the George circle. Friedrich Gundolf in particular had been quick to espouse Bergson's thinking, and described his book on Shakespeare in Germany (1911) as presenting history in the spirit of Bergson's conception of time; but the 'master,' as George's followers customarily called him, disapproved of this infatuation, and called upon Friedrich's brother, Ernst Gundolf, to write an essay countering it.[40] That essay, published in 1912, acknowledges Bergson as a champion of creativity and the integrity of personality, but resists his emphasis on the transience of reality, insisting indeed on the need for a traditional sense of a soul, and of an absolute that is constant and eternal.[41] The publication that lent the greatest authority to the intellectual repudiation of Bergson had, however, already appeared in 1910 in a newly launched high-profile philosophical journal entitled *Logos*.[42]

---

37 Hans Driesch, 'Bergson, der biologische Philosoph', *Zeitschrift für den Ausbau der Entwicklungslehre*, 2 (1908), 48-55 (p.51).

38 Karl Jaspers, 'Henri Bergson, *Die schöpferische Entwicklung*' (review), *Zeitschrift für die gesamte Neurologie und Psychiatrie. Referate und Ergebnisse*, 6 (1913), 885-86.

39 Samuel Saenger, 'Henri Bergson', *Die neue Rundschau*, 24 (1913), 732-33; Max Scheler, 'Versuche einer Philosophie des Lebens. Nietzsche – Dilthey – Bergson', *Die weißen Blätter* I (1913), 203-33 (p.215). Scheler's essay was reprinted, with minor refinements of style and terminology, in Scheler's *Abhandlungen und Aufsätze* (Leipzig: Weißen Blätter, 1915); alternative page references are given here to the version in *Gesammelte Werke* 3 (Bern: Francke, 1955), pp.311-39 (p.323).

40 Friedrich Gundolf, *Shakespeare und der deutsche Geist* (Berlin: Bondi 1911), pp.57f; cf. also Friedrich Gundolf, *Goethe* (Berlin: Bondi, 1916), p.264; see B. Vallentin, 'Gespräche mit Stefan George', *Castrum Peregrini* 44-45 (1961), 15-138 (pp.95-96).

41 Ernst Gundolf, 'Die Philosophie Henri Bergsons', *Jahrbuch für die geistige Bewegung* 3 (1912), 32-92.

42 For the background circumstances see Rüdiger Kramme, '"Kulturphilosophie" und "Internationalität" des »Logos« im Spiegel seiner Selbstbeschreibungen', in *Kultur und*

When the prominent Kantian philosopher Heinrich Rickert, in 1920, published a retrospective survey of the various trends that had by then come to be known as *Lebensphilosophie*, it was to the essay in *Logos* by Richard Kroner, one of his protégés, that he referred his readers for a definitive account of Bergson's arguments.[43] Kroner takes *L'évolution créatrice* as the starting point of his analysis, insisting as he does so not only that the assertion of individual freedom and personality is the driving force of Bergson's philosophy, but that his view of reality is fundamentally irrational. He objects to the subordination of philosophical reasoning to biological categories, denounces Bergson's claim that instinct can reveal the 'meaning' of evolution to us as an extreme form of Romanticism, and argues that the dependency of that insight on the intuition of the individual implies an inescapable relativism. For Kroner, a major weakness in Bergson's philosophy lies in the fact that it lacks transcendental concepts, being based on a perception of the world as eternally in flux, and he goes on to attack Bergson's belief in the power of intuition by invoking the well-known adage from the 'Transcendental Logic' of Kant's *Critique of Pure Reason* that 'intuitions without concepts are blind.'[44] What makes this jibe look crude is the fact that – as others, including Scheler, noted – Bergson explicitly speaks of the need to develop new concepts to articulate the insights that the work of intuition can bring to light.[45] Kroner's essay was, in short, a relentless application of Kantian precepts to the demolition of Bergson's arguments.

Over the following two decades, some sixteen doctoral theses on Bergson's philosophy were published in German.[46] Several of these give the impression that Bergson's texts had simply been deemed suitable material for an exercise in epistemological analysis along the lines that Kroner had established. But two of them deserve special mention here because they show clearly how, in an academic context, Bergson's thinking was assessed in relation to specific German philosophical traditions. Among those who distance themselves from the Romantic and implicitly mystical aspect of Bergson's thought, Friedrich Schäfke (Göttingen 1914) is particularly good at bringing out the senses in which Bergson's *élan vital* resembles Schopen-

---

Kulturwissenschaften um 1900. II: Idealismus und Positivismus, ed. by Gangolf Hübinger et al. (Stuttgart: Steiner, 1997), pp.122-34.

43 Heinrich Rickert, *Die Philosophie des Lebens. Darstellung und Kritik der philosophischen Modeströmungen unserer Zeit* (Tübingen: Mohr [Paul Siebeck], 1920), pp.22-24.

44 p.141. Cf. Immanuel Kant, *Kritik der reinen Vernunft*, I, ed. by Wilhelm Weischedel (Frankfurt: Suhrkamp, 1974), p.98 (B75, A51).

45 Scheler, p.220; *Gesammelte Werke*, III, pp.327-28; Bergson, *L'évolution créatrice*, pp.239-40.

46 See Gunter, *Bergson Bibliography*.

hauer's concept of the will.[47] Schäfke distinguishes more clearly than Bergson was concerned to do between the evidence that empirical science provides and the particular functions that mathematical calculation can perform as a practical and theoretical tool in the service of scientific inquiry; and he gives a lucid account of the sense in which, even as Bergson challenges the authority of the intellect, he remains inescapably dependent on it in his discussion of the role of intuition. Among those who display a greater sympathy for Romantic nature philosophy, Georg Jäger (Leipzig 1917), conducts a point-by-point comparison between Bergson and Schelling which shows both their similarity in giving priority to intuition ('intellektuelle Anschauung' in Schelling's case) and the differences between their respective conceptions of the natural world: Schelling remains at heart a rationalist, whereas Bergson emerges as a radical empiricist.[48] At the same time, Jäger notes, Bergson's pursuit of an 'absolute' leads him into a panpsychism (i.e. the sense of a 'world spirit') that is even stronger than Schelling's.

The essay that Max Scheler published in 1913, finally, is important for what it shows about the direction German philosophy took in the wake of the developments we have noted. Contrasting Bergson with Nietzsche, whose priority lay with enhancing the individual's experience of life, and Dilthey, who aimed to develop the means of grasping the historical 'world' that constituted the life of a community, Scheler sees Bergson's strength as lying in the new direction he has given to thinking about the world and the mind (*Seele*) by combining the data of immediate experience with the big questions presented by biology.[49] But he is very critical of specific elements of Bergson's thought – his view of the relation between space and time has not kept pace with mathematical thinking, his reliance on intuitive (self-)-awareness amounts to a 'highly dubious psychologistic mysticism' – and it is clear from the overall cast of his argument that Scheler is reconstructing the key steps in the development of a 'philosophy of life' with a view to superseding it.[50] More specifically, Scheler has by now absorbed Edmund Husserl's critique, in *Logical Investigations* (1900-01), of psychologism (i.e. the view that logic can be wholly accounted for in psychological terms), and at the end of his essay he intimates, without pre-empting the outcome of inquiries that are still in progress, that he has himself moved on to the kind

---

47 Friedrich Schäfke, *Bergsons 'L'évolution créatrice' in den Hauptpunkten dargestellt und beurteilt* (unpublished dissertation, University of Göttingen, 1914).

48 Georg Jäger, *Das Verhältnis Bergsons zu Schelling. Ein Beitrag zur Erörterung der Prinzipien einer organistischen Weltauffassung* (Hamburg: Lütcke & Wulff, 1917).

49 Scheler, pp.216, 229; *Gesammelte Werke*, III, pp.324, 336.

50 p.216; *Gesammelte Werke*, III, p.324; p.226; *Gesammelte Werke*, III, p.333.

of phenomenological investigation that will in due course provide the basis for his developed philosophy of ethical values. Gérard Raulet, in a recent article, also points to some senses in which the philosophical critique of Bergson may have contributed to the development of phenomenology as a movement in Germany, from Husserl to Heidegger and beyond, a process in which Roman Ingarden's early study of Bergson's thinking about intellect and intuition undoubtedly played a significant part.[51]

While the general character of Bergson's philosophy chimed in with the anti-mechanistic turn in German thought after 1900, then, there were aspects of his thinking to which German readers responded in more particular ways. He could be read as promoting subjectivity and psychological insight as well as metaphysics and spirituality, and arguments could be found both for and against him on each of these matters. In terms of the history of philosophy, the critique of Bergson appears to have played a larger part in subsequent developments than any mere assimilation of his ideas. But in terms of his impact on the German literary and intellectual culture of the time, the effect of *L'évolution créatrice* was to throw into relief the conflicting assumptions upon which educated Germans were reassessing their view of the world and of human nature in response to the implications of biological science.

---

51 Gérard Raulet, 'Histoire d'un malentendu fécond', in *Linke und rechte Kulturkritik. Interdiskursivität als Krisenbewußtsein*, ed. by Gilbert Merlio and Gérard Raulet (Frankfurt: Lang 2005), pp.23-54. Cf. also Rudolf W. Meyer, 'Bergson in Deutschland. Unter besonderer Berücksichtigung seiner Zeitauffassung', in *Studien zum Zeitproblem in der Philosophie des 20. Jahrhunderts*, ed. by Ernst Wolfgang Orth (Freiburg: Alber, 1982), pp.10-64; Roman Ingarden, 'Intuition und Intellekt bei Henri Bergson. Darstellung und Versuch einer Kritik', *Jahrbuch für Philosophie und phänomenologische Forschung*, 5 (1922), 285-461. See also Raulet, p.44.

Evolutionary Psychology as a Heuristic in Literary Studies

Katja Mellmann

Abstract: There has been a great deal of uproar about Darwinian approaches in literary scholarship. Statements range from enthusiastic prophecies of a new paradigm for literary studies to acrimonious scoldings of reductionism. Believing that the major challenge is first to find good questions to which evolutionary psychology might provide us with good answers, I outline and critically assess different veins of argumentation as revealed in recent contributions to the field. As an alternative to some simplistic mimeticism in present Literary Darwinism, I put forward the idea of evolutionary psychology as a heuristic theory that serves to resolve defined problems in interpretation and literary theory.

On the eve of Darwin's anniversary year, *Style* brought out a special double issue entitled 'An Evolutionary Paradigm for Literary Studies'. In the opening essay, Joseph Carroll describes evolutionary literary study as 'a distinct movement' which has emerged since the publication of his book *Evolution and Literary Theory* (1995) and 'is rapidly gaining in visibility and impact'.[1] The subsequent contributions by 35 respondents, however, as well as Carroll's final rejoinder, make manifest that the distinctiveness of this 'paradigm' is rather precarious and that conceptions about the goals, range of applicability, and methods of evolutionary approaches sometimes differ greatly. The emergent consensus seems quite minimalistic; in summary it consists in the belief that the human sciences do have something to tell us literary scholars; that their theories and findings are not limited to providing rich topics for humanistic meta-perspectives ('The Rhetoric of Genetics') but are co-equal contributions within the general academic pursuit of knowledge which can indeed be applied.

1 Joseph Carroll, 'An Evolutionary Paradigm for Literary Study', *Style*, 48 (2008), 103-35 (p.103). For a survey of the field see the collections: *Biopoetics. Evolutionary Explorations in the Arts*, ed. by Brett Cooke and Frederick Turner (Lexington: International Conference on the Unity of the Sciences, 1999); *Literary Biopoetics*, ed. by Brett Cooke (*Interdisciplinary Literary Studies* 2.2); *Symposium: Evolution and Literature*, ed. by Nancy Easterlin (*Philosophy and Literature* 25.2); *Anthropologie der Literatur: Poetogene Strukturen und ästhetisch-soziale Handlungsfelder*, ed. by Rüdiger Zymner and Manfred Engel (Paderborn: Mentis, 2004); *The Literary Animal. Evolution and the Nature of Narrative*, ed. by Jonathan Gottschall and David Sloan Wilson (Evanston: Northwestern University Press, 2005); *Biological Constraints on the Literary Imagination*, ed. by Katja Mellmann and Anja Müller-Wood (*Studies in the Literary Imagination* 42.2); also *Heuristiken der Literaturwissenschaft. Disziplinexterne Perspektiven auf Literatur*, ed. by Uta Klein, Katja Mellmann and Steffanie Metzger (Paderborn: Mentis, 2006).

The idea of interdisciplinarity in literary scholarship is actually not as novel as all the noise about the evolutionary 'paradigm' might suggest.[2] There have always been literary scholars who were open to ideas from other disciplines. Literary criticism has never simply had to deal 'with literature,' but also with its medium language, with what is represented in it, and with the communicative functions that literature performs within a given society. As such, literary theory has from its inception incorporated a number of context theories, including linguistics, philosophy/history of ideas, sociology/social history, iconography and history of media. While it is true that current psychology has for many years been restricted to a peripheral role in this respect, since a discipline largely dominated by behaviourism revealed few points of contact for questions relevant to literary scholars, the so-called 'cognitive turn' has crucially changed this situation. Methods of literary study have noticeably reinforced their efforts to elaborate an explicit psychological context theory, which now includes issues from the cognitive sciences and, after the emergence of evolutionary psychology, also adaptationist considerations.[3] It is this reinforcement due to enriched

---

2 To name just a few instances out of the Grand Debate: Tony Jackson, 'Questioning Interdisciplinarity: Cognitive Science, Evolutionary Psychology, and Literary Criticism', *Poetics Today*, 21 (2000), 319-47; Bradley Bankston, *Against Biopoetics: On the Use and Misuse of the Concept of Evolution in Contemporary Literary Theory* (unpublished doctoral thesis, Louisiana State University, 2004); Brian Boyd, 'Getting it All Wrong: Bioculture Critiques Cultural Critique', *American Scholar*, 75 (2006), 18-30; Jon Adams, *Interference Patterns. Literary Study, Scientific Knowledge, and Disciplinary Autonomy* (Lewisburg: Bucknell University Press, 2007), pp.77-99; Frank Kelleter, 'A Tale of Two Natures: Worried Reflections on the Study of Literature and Culture in an Age of Neuroscience and Neo-Darwinism', *Journal of Literary Theory*, 1 (2007), 153-89; Karl Eibl, 'On the Redskins of Scientism and the Aesthetes in the Circled Wagons', *Journal of Literary Theory*, 1 (2007), 421-42; Jonathan Gottschall, *Literature, Science, and a New Humanities* (New York: Palgrave McMillan, 2008); Maria E. Kronfeldner, 'Trigger Me: Evolutionspsychologie, Genzentrismus und die Idee der Kultur', *Nach Feierabend: Zürcher Jahrbuch für Wissensgeschichte*, 4 (2008), 31-45; Brian Boyd, *On the Origin of Stories. Evolution, Cognition, and Fiction* (Cambridge: Harvard University Press, 2009), pp.384-92.

3 As sort of a constitutional charter of evolutionary psychology see John Tooby and Leda Cosmides, 'The Psychological Foundations of Culture', in *The Adapted Mind. Evolutionary Psychology and the Generation of Culture*, ed. by Jerome H. Barkow, Leda Cosmides and John Tooby (New York and Oxford: Oxford University Press, 1992), pp.19-136; and, more recently, 'Evolutionary psychology: Conceptual foundations', in *The Handbook of Evolutionary Psychology*, ed. by David M. Buss (Hoboken: Wiley, 2005), pp.5-67; also, as an important discussion on discrepancies with sociobiology, John Tooby and Leda Cosmides, 'The Past Explains the Present. Emotional Adaptations and the Structure of Ancestral Environments', *Ethology and Sociobiology*, 11 (1990), 375-424.

opportunities that has occasionally been perceived as a 'new interdisciplinarity' or 'neo-naturalism,' rather than the dawn of a radically new paradigm.[4]

When in what follows I endorse evolutionary psychology as a valuable heuristic in literary studies, I do this in the sense of such a context theory, that is, as an *ancillary discipline*, the virtue of which consists in facilitating answers to questions that cannot be answered by text-immanent or cultural explanations alone. It is worth noting that the questions for which evolutionary psychology is consulted are themselves taken to arise from a genuinely *literary* epistemological interest. Accordingly, the first section of my essay is concerned to distinguish my own proposal from those Darwinian approaches that fail to identify a specifically literary *problem* in the first place, of which the adaptationist analysis is then the solution. In those approaches evolutionary psychology is accorded the status of a 'super theory' rather than that of a heuristic context theory. I shall particularly criticise the 'disguised and misplaced correspondence theory' by which those studies in practice equate literary representations with anthropological data.[5] And I shall put forward the view that Darwin's evolutionary theory does not per se apply to artifacts (such as literary characters) but, a priori, only to living things, that is, to the 'human side' within the literary communication model (producer and receiver). As an alternative, then, I outline the concept of an 'anthropological model reader' in the second part of my essay. In this concept, evolutionary psychology serves as a reception-psychological heuristic which helps to shed light on traditionally obscure domains in our understanding of the literary reading process, and thus helps to resolve correlated problems of literary text interpretation. In the third section of the essay, I give a brief discussion on the feasibility of a general 'evolutionary literary theory' or, more precisely, of writing a biological prehistory of literature (as crucial part of the history of human evolution).

Mimetistic Tendencies in 'Literary Darwinism'

In his monumental *Evolution and Literary Theory*, Carroll originally presented a very broad spectrum of potential objects of study, which included both content and formal aspects of literature, and both realistic and symbolic

4 Cf. Mary Thomas Crane and Alan Richardson, 'Literary Studies and Cognitive Science: Toward a New Interdisciplinarity', *Mosaic*, 32 (1999), 123-40; Kelleter, 'A Tale of Two Natures', p. 155.

5 Marcus Nordlund, 'Consilient Literary Interpretation', *Philosophy and Literature*, 26 (2002), 312-33 (p.327).

modes of representation.[6] The actual practice, however, has concentrated strongly on thematics and mimetic narrative. The preferred objects of examination are, for instance, male conflict behaviour in old epics and female mating behaviour in nineteenth century realism – in short: topics and texts that are especially well suited to supporting the findings of evolutionary psychology.[7] A downright caricature of this procedure can be found in the popular book *Madame Bovary's Ovaries*, in which David and Nanelle Barash explain Madame Bovary's adultery and Othello's jealousy as resulting from the adaptive value of such behaviour in evolution.[8] Of course, literary figures have neither wombs nor evolved psyches; their behaviour can thus not be explained in terms of biological evolution. But Barash and Barash are not literary scholars; their aim is an amusing illustration of evolutionary psychology for a general audience, and so there is little point in criticizing them here. A similar 'cookie-cutter' approach, however, can also be found in several studies by card-carrying literary scholars and thus give grounds for critical revision.[9]

A frequently heard argument in favour of a close comparison of human behaviour as depicted in literature with the predictions of modern psychology asserts that inaccurate interpretations can thus be avoided. By knowing about the 'Westermarck effect,' for instance, 'we can avoid erroneously importing mother/son incest into texts, such as *Hamlet*,' as Carroll puts it.[10] The so-called Westermarck effect refers to the Finnish anthropologist Edvard Westermarck's observation that closely living together during a particular period in early childhood leads to mutual sexual disinterest in adulthood (regardless of whether or not the persons in question are actually blood relatives). If we know about this innate best-bet

---

6 Joseph Carroll, *Evolution and Literary Theory* (Columbia: University of Missouri Press, 1995), pp.96-267.

7 See, for instance, Robin Fox, 'Male Bonding in the Epics and Romances', in *The Literary Animal*, pp.126-44; Jonathan Gottschall, *The Rape of Troy: Evolution, Violence, and the World of Homer* (New York: Cambridge University Press, 2008); Brian Boyd, 'Jane, Meet Charles: Literature, Evolution and Human Nature', *Philosophy and Literature*, 22 (1998), 1-30; Brett Cooke, 'Sexual Property in Pushkin's "The Snowstorm". A Darwinist Perspective', in *Biopoetics*, pp.175-204; Joseph Carroll, *Literary Darwinism. Literature and the Human Animal* (Columbia: Routledge, 2004), pp.129-45; Joseph Carroll and Jonathan Gottschall, 'Human Nature and Agonistic Structure in Canonical British Novels of the Nineteenth and Early Twentieth Centuries. A Content Analysis', in *Heuristiken der Literaturwissenschaft*, pp.473-87.

8 David P. and Nanelle R. Barash, *Madame Bovary's Ovaries: A Darwinian Look at Literature* (New York: Delacorte, 2005), pp.93-121, 14-37.

9 Jackson, 'Questioning Interdisciplinarity', p.341. Cf. also Steven Pinker, 'Toward a Consilient Study of Literature', *Philosophy and Literature*, 31 (2007), 161-77 (p.167).

10 Carroll, *Evolution and Literary Theory*, p.44.

mechanism of incest avoidance, and if we also know that Hamlet (unlike Œdipus, for example) was not separated from his mother in his infancy, then we do not get up to any mischief. Likewise, says Carroll, in Emily Brontë's novel *Wuthering Heights*: although Catherine and her adoptive brother Heathcliff seem to manifest a passionate relationship, this cannot be, according to Carroll, because they have grown up together. But, alas, 'fictional representations just do not work that way,' as Alan Richardson laconically responded on Carroll's example: 'Like dreams, fictive works can bear a number of different relations to the rules and regularities of daily experience, often giving us the inverse of the lived world. If Emily Brontë is at liberty to people the Yorkshire moors with ghosts, why not incestuous foster-siblings as well?'[11]

Richardson generally questions Carroll's "Romantic claim that literary works manifest 'intuitive' knowledge that 'often outruns scientific understanding'" (p.562). I think, though, he is not being entirely fair to Carroll here, since literary figures of course are at least very similar in certain ways to actual human beings and, as Aristotle knew, literature indeed is at some level an imitation (*mimesis*) of human behaviour. Carroll's opinion that certain 'primal forces' at work the human mind as a product of natural selection can subliminally influence the poetic imagination and the 'figurative structures' of a literary text is in some respect only an extension of the hermeneutic maxim that the author does not necessarily perfectly understand his own text, and this is not automatically ridiculous.[12] And yet Richardson hits a sensitive point in Carroll's argument, perhaps not by chiding his 'romanticism,' but certainly with the reference to the author's freedom of representative intention. Literary writers are not *compelled* to fashion verisimilar ('mimetic') representations. Hence, the argument that evolutionary psychology prevents erroneous interpretation collapses. Human nature as we understand it today can only serve as a rule of thumb for interpreting the behaviour of literary characters. Evolutionary psychology might tell us a lot about some tacit and unmarked default features of literary inventions or the reader's initial and subliminal expectations of them, and in these respects it might be useful cautiously to

---

11 Alan Richardson, 'Rethinking Romantic Incest: Human Universals, Literary Representation, and the Biology of Mind', *New Literary History*, 31 (2000), 553-72 (p.561).

12 Carroll, *Evolution and Literary Theory*, p.43; cf. also Carroll, *Literary Darwinism*, p.124, on 'the intuitive understanding of poets, novelists, and playwrights'; on the long tradition behind this maxim, see Lutz Danneberg, 'Besserverstehen. Zur Analyse und Entstehung einer hermeneutischen Maxime', in *Regeln der Bedeutung*, ed. by Fotis Jannidis et al. (Berlin: de Gruyter, 2003), pp.644-711.

apply psychology at the level of representation. But evolutionary psychology cannot provide us with hard-and-fast knock-down criteria against readings inconsistent with its predictions – unless we want to end up with an impoverished notion of literary representation according to which the author is compelled to represent 'the world as it is'.[13]

The superficial realism that Richardson criticises in Carroll is unfortunately characteristic of many other attempts in applying a Darwinian perspective to literature. In such studies, the mimetic relationship between literary representations and biology is not only generally presumed but frequently even immensely strained by the sort of questions that are posed. Take as an example a study by Jonathan Gottschall in which he examines 1,440 folktales from forty-eight different cultures in order to refute the claim 'that European tales reflect and perpetuate the arbitrary gender norms of western patriarchal societies'.[14] For what actually is the result of this study? When Gottschall discovers that the same stereotypes can be detected in all folktales he examined, he correctly refutes the 'prediction [...] that [...] an analysis of a culturally diverse sample will reveal diverse gender patterns' (pp.207-08). In testing this inferred hypothesis, however, he does not check the above-mentioned assumption *about literature* (that the tales 'reflect and perpetuate' a certain stereotype) but rather its constructivist premise (that this stereotype is an arbitrary western norm). Indeed, it seems he had not intended to do anything else – as becomes obvious when, after the job is done, he states with satisfaction that his own 'findings converge with emerging biosocial models of human behaviour and psychology' (p.219) while those of feminist literary criticism do not.

I do not however even suggest that Gottschall is wrong in preferring a biosocial to a (radically) constructivist theory of gender. My objection concerns rather the logic of his procedure. He claims to be testing a literary-critical hypothesis, but in fact he is testing an anthropological hypothesis using literary data. This strategy allows him to examine pieces of literature (folktales) just like any other piece of data, that is, without the need explicitly to justify – or at least evaluate – the material's evidentiary value and representativeness. That, of course, would be the first task in any rigorous

---

13 I want to stress that this statement holds only for *interpretative* assumptions concerning the internal coherence of the fictional world. The case is different with assumptions concerning external references (causal explanations, for example) as constructed by the literary critic, which draw on facts and knowledge beyond the textual givens in accordance with a prior 'theory of reality' (cf. Nordlund, 'Consilient').

14 Jonathan Gottschall. 'Quantitative Literary Study: A Modest Manifesto and Testing the Hypothesis of Feminist Fairy Tale Studies', in *The Literary Animal*, pp.199-224 (207).

ethnological procedure drawing on literary or other sources.[15] A detailed theory as to how folktales process (contain, manipulate, stylise, etc.) human universals (and, thereby, possibly indeed render a biological bias into a cultural norm) would not only have been indispensable here but might even have made a significant contribution to literary theory. Gottschall, however, misses this opportunity by tacitly presuming a mirroring relationship between biology and literature instead of recognising this complex and shifting relationship as the actual problem. When he proclaims that 'what ultimately drew most people to study literature was a desire to explore the behaviour and psychology of human beings,' he might well be informing us about his own motives, but surely he errs about the actual epistemological interest of the vast majority of literary scholars.[16] The sceptical aloofness which Darwinian approaches to literature are often confronted with in their home discipline is mostly due to the (not entirely wrong) impression that those mimetistic approaches of 'proving that the universals are actually present in literature' or 'simplistically mapping' them onto literary texts do not really deal with literature but rather with the world behind it, and thus indeed miss the proper object of literary study.[17]

---

15 And *is* the first thing in, for instance, Michelle Scalise Sugiyama and Lawrence Sugiyama, 'A Frugal (Re)Past. Use of Oral Tradition to Buffer Foraging Risk', *Studies in the Literary Imagination*, 42.2 (2009), 15-41. I should add here that I do not at all consider it unreasonable in principle to use literature as a source for anthropological research. For the reasons mentioned above literary texts are not the most reliable but sometimes (for instance, if one examines historic or, like in Gottschall's study on Homer, *The Rape of Troy*, even prehistoric times) still the best sources for psycho-historical, and thus also anthropological questions. You simply need to know what you are doing. As long as an epistemological interest from anthropology is disguised into one of literary study, however, one does not know. Likewise, an interdisciplinary cooperation between literary scholars and psychologists (as, for instance, John A. Johnson, Joseph Carroll, Jonathan Gottschall, and Daniel Kruger, 'Hierarchy in the Library: Egalitarian Dynamics in Victorian Novels', *Evolutionary Psychology*, 6 (2008), 715-38) can be a good venture if the literary scholars in the team administer their expert knowledge about the specificities of literature and fiction. As long as they avoid applying themselves to precisely that, however, but rather hire out as semi-skilled assistants in their neighbour disciplines, all meaning of *inter*disciplinarity gets lost.

16 Jonathan Gottschall and David Sloan Wilson, 'Introduction. Literature: A Last Frontier in Human Evolutionary Studies', in *The Literary Animal*, pp.xvii-xxvi (p.xix). Cf. also Carroll, *Literary Darwinism*, pp.124-25, who states with some enthusiasm that 'we have, for the first time, a situation in which the intuitive understanding of literary writers can converge effectively with the findings of empirical psychology.' Can it?

17 Jackson, 'Questioning Interdisciplinarity', p.341; Cf. Richardson, 'Rethinking Romantic Incest', p.569; also Nordlund, 'Consilient', p.315.

In fairness, the mimetistic shortcut was not invented by evolutionary criticism. Let us return briefly to Carroll's incest example and see what his critic makes of it. Richardson believes that Romantic writers, when representing incestuous and near-incestuous relations, were not 'blindly responding to unconscious impulses or intuitions' but rather to 'the unique and specific character of Romantic-era culture and discourse,' for which he then cites some examples that reveal period-typical associationist reasoning, and in some case even a kind of an incest avoidance theory *avant la lettre.*[18] To compare these historical discourses with our current understanding of the Westermarck effect, he says, helps to 'make sense of the two features of Romantic incest that critics have had most trouble accounting for: that incestuous desire, though idealised, nearly always ends tragically; and that this pattern holds equally for biological siblings, foster or adopted siblings, and various other co-socialized pairs' (p.570). I am quite ready to agree with this. But how, if so, can the Westermarck effect account for this without authors unconsciously responding to their intuitions about human nature, in just the manner Carroll suggested? Or, if Richardson means to replace the assumed implicit knowledge by more explicit forms of knowledge as attested in contemporary contexts, does he not stumble into the same mimetistic trap as Carroll (or at least touch it) by explaining literary inventions as broadly reflecting contemporary beliefs (a tacit assumption not uncommon in discourse-historical context studies)? To put it in his own words: if Brontë is at liberty to people the Yorkshire moors with ghosts, why not incestuous lovers living happily ever after? The fundamental freedom of representative intention in works of art applies for both implicit and explicit knowledge (if not even more strongly for the latter).

And lastly: how good is the question to which Richardson's final conclusion gives an answer? It seems to me that the Westermarck effect does not so much explain why incestuous and para-incestuous desire ends tragically in Romantic literature as *why Romantic writers take up incest*, literally or symbolically, when they want to depict an impossible love.[19] To reverse

---

18 Richardson, 'Rethinking Romantic Incest', p.563. For similar instances within the German context see Michael Titzmann, 'Literarische Strukturen und kulturelles Wissen. Das Beispiel inzestuöser Situationen in der Erzählliteratur der Goethezeit und ihrer Funktionen im Denksystem der Epoche', in *Erzählte Kriminalität*, ed. by Jörg Schönert (Tübingen: Niemeyer, 1991), pp.229-81 (pp.266-73).

19 According to Niklas Luhmann, in the eighteenth century passionate love discourse was discovered as a means to manifest and stabilise the new social phenomenon of modern individuality *Love as Passion* (*Liebe als Passion*, Frankfurt: Suhrkamp, 1994, p.167-79). In this historic idea of love the beloved is expected wholly to share and reflect the lover's singular subjective view of life. Romantic love is thus by definition impossible *in* life. It is an affair of the imagination, of literature, of hope and fantasy;

the question like that means to examine *how and by what means* an author achieves his/her goal (instead of ascertaining, in a surely redundant manner, the coherence of a literary representation with its general anthropological or culture-historical context). What cognitive programs can an author (consciously or unconsciously) count on in applying poetical techniques?[20] It is this 'relationship between the structure of literary texts and their perceived effects'[21] for which an expediently applied evolutionary approach, as a sub-discipline of cognitive poetics, can provide exciting new insights. In order to hypothesise reasonably about what innate dispositions are involved in (and how they shape) the understanding of literary texts we indeed *require* an explicit evolutionary-psychological heuristic.

Modelling the Correlation between Literary Artifact and the Human Mind: The Anthropological Model Reader

If psychology, as argued above, does not initially apply to artefacts but only to living things, our analytical focus has first to be shifted from the literary work per se to its recipient (including the author as his/her own first reader). However, since individual subjective experience is incommensurable, literary criticism normally deals with typified surrogates like the 'implied' reader of a text, an author's 'intended' reader, or an abstract or generalised 'model reader'. Discourse-historical studies in historic belief systems, habits of thought, cultural scripts and schemata, for example, implicitly construct a historical model reader in order to be able to speculate about how contemporary readers may have conceived certain issues in a

---

and literature addressed to this problem of modern love and individuality is very likely to end in tragedy on that account alone (cf. Karl Eibl, *Die Entstehung der Poesie* (Frankfurt: Insel, 1995), pp.125-33). Incestuous love thus is especially apt as a poetic symbol of this problematic, since it connotes both the total harmony of two individuals, the impossibility of fulfilment and, moreover, the sense of guilt still inhering in the attempt at radical self-realisation. The Westermarck effect readily explains incest as a symbol of the impossible, provoking strong feelings of discomfort and rejection in the reader and thus serving the author's purpose of a haunting depiction of unsolved problems; cf. also Titzmann, 'Literarische Strukturen', pp.273-77.

20 'Poetical techniques' here includes thematics (motifs, sceneries, plot schemata, etc.) as well as formal features. I think, Nordlund, 'Consilient', p.315, slightly mischaracterises positions critical of simplistic Darwinian interpretations when he imputes to them a 'fundamental primacy of poetics and formalist analysis over thematics'. Thematics is part and parcel of poetics, if literary content is not (indeed simplistically) understood as a reflex of reality but as a literary construct (based on and referring to reality, for sure, but in many different ways).

21 Reuven Tsur, *Toward a Theory of Cognitive Poetics* (Amsterdam: North-Holland Elsevier, 1992), p.1.

text.[22] In a like manner we can construct an anthropologically typified model reader in order to determine more basic regularities of text comprehension. The anthropological model reader is thus a heuristic construct representing the evolved genotype of the human mind. It consists in all plausible adaptive algorithms summarised into a skeletal surrogate of the human psyche, that is, without the phenotypic 'flesh' of individual and cultural experience that characterises the empirical reader.[23]

So what is the heuristic construct of an anthropological model reader good for? In a first approach it can provide us insight into some basic responses that a reader is likely to show toward a particular passage and thus facilitates a deeper understanding of literary meaning, including innately biased associations, emotional effects, attention rousing, hierarchies of relevance. As Nancy Easterlin has for instance demonstrated using the example of the word 'home' in a passage from Wordsworth, drawing on evolutionary thinking strongly supports our interpretive task by enabling an 'empirically grounded speculation' about even unconscious processes in literary understanding and thus helps to disclose the full spectrum of literary meaning.[24]

Second, and equally important, assumptions about likely effects also bring us back to the text itself and allow for a closer understanding of how it is composed. Since adaptationist argumentation involves the specification not only of the adaptive behaviour itself but also of the selection pressure under which it has evolved, an evolutionary analysis always entails explicit assumptions about recurring situational features in correlation with which a particular behaviour has proved successful over the long course of evolution. That is, it informs us about a specific *stimulus pattern* which should be part of the releaser mechanism of the program in question and which, then, can be traced in the literary text. Thus, rather than searching for (redundant) analogies between the behaviour of fictional persons and actual human beings, we should pay attention to the potential isomorphisms between text structure and innate releasing schemata. Literary texts provide many structural features (formal and thematic ones) that approximate the abstract search formulas built in the 'situation detectors' of our

22 To take incest as the example: Titzmann, 'Literarische Strukturen', (p.230-31), refers to his concept of 'cultural knowledge' in order to decide whether a passionate relationship as depicted in a piece of literature would have been considered incestuous by the contemporaries.

23 Tooby and Cosmides, 'The Psychological Foundations', pp.73-77, provide detailed instructions how a good and solid adaptationist argumentation has to look like if it shall not issue into 'just-so stories'. Cf. also Pinker, 'Toward', p.175.

24 Nancy Easterlin, 'Voyages in the Verbal Universe. The Role of Speculation in Darwinian Literary Criticism', *Interdisciplinary Literary Studies*, 2 (2001), 59-73, (p.65).

adaptations.[25] Literary artifacts thus can be said to act as kind of a dummy stimulus[26] on our innate dispositions.

Let me, in order to illustrate how this view can serve as a fertile heuristic for literary study in some of its most genuine concerns, refer to my own work on the development of emotive literary strategies in German eighteenth-century literature. There has long been consensus on a tendency towards 'emotionalisation' in German Enlightenment literature. However, this finding was based primarily on the poetological self-descriptions of contemporaries rather than being systematically verified in the literary texts themselves. Hypothesizing about innately biased stimuli (such as, among many others, innate fear stimuli as found in the literary Sublime or innate 'appetitive' stimuli from habitat selection which constitute the *locus amœnus* of Rococo poetry) thus turned out to be a good means to determine, and indeed even (approximately) quantify the emotional impact of a given literary text.[27] Thus, having deployed an established psychological context theory to describe fundamental effects of certain kinds of text structures, we are able to reconstruct (and reassess) particular lines of development in literary history (a general increase in emotionalising techniques, modernist attempts at 'abstraction ,' popularising tendencies).

Third, we can apply modern psychology to the historical concepts found in documents of reception or contemporary poetics in order to specify their particular psychological reference. This was my aim, for example, in examining Lessing's conception of the Aristotelian *eleos*. In comparing Lessing's descriptions of this dramatic affect with emotion programs that can be reasonably assumed from an evolutionary point of view, I found that his term of 'Mitleid' indeed refers to 'mercy,' i.e. an emotion evolved to foster pro-social activities, rather than (as it is often rendered today) to a general capability for 'empathy'. In a similar manner, Ed Tan and Nico Frijda have reconsidered the response of tears in literary sentimentalism. By observing that 'crying is a response to helplessness rather than specifically to grief ,' they can describe the sentimental response as effected by an adaptive algorithm for submission, which in a work of art can be triggered by 'any

---

25 Cf. Leda Cosmides, and John Tooby, 'Evolutionary Psychology and the Emotions', in *Handbook of Emotions*, ed. by Michael Lewis and Jeannette M. Haviland-Jones (2nd edn, New York: Guildford, 2001), pp.91-115 (p.93).

26 Cf. Katja Mellmann, *Emotionalisierung - Von der Nebenstundenpoesie zum Buch als Freund. Eine emotionspsychologische Analyse der Literatur der Aufklärungsepoche* (Paderborn: Mentis, 2006), pp.42-78, and 'Literatur als emotionale Attrappe. Eine evolutionspsychologische Lösung des 'paradox of fiction'', in *Heuristiken der Literaturwissenschaft*, pp.145-66; similarly, Clemens Schwender, *Medien und Emotionen. Evolutionspsychologische Bausteine einer Medientheorie* (2nd ed., Wiesbaden 2006 [2001]).

27 Mellmann, *Emotionalisierung*, pp.244-47, 338-44.

major resolution in a conflict where the stakes are high,' and thus debunk another myth of 'empathic' emotion.[28] An evolutionary heuristic thus enhances our understanding and analysis of historical discourse and cultural concepts.

For all these reasons I would strongly refute the frequently heard assumption that evolutionary theory, dealing with the general rather than the particular, cannot tell us much about individual periods, genres, authors, or literary works.[29] Indeed it can, if – and that is essential – our *initial* questions focus on the specific. My critique of the studies mentioned above is not a criticism of their interest in the universally human (which I share) but of their failure to state interesting questions from the point of view of literary scholarship.[30]

Given that we have such questions as a starting point, however, we would deceive ourselves were we to believe that we could treat them without any regard to the universally human as described by the human sciences. Even 'when we, as historians, explain the particular, we always draw it into the scope of more general assumptions; if we did not, we would be unable to employ any concepts at all and would be dealing with nothing but miracles.'[31] For instance, why is the open ending of a play 'open,' the supple rhythm of a poem 'supple,' the vivid and suspenseful depiction of a narrative event 'vivid,' and what actually is 'suspense'? What is it that we call 'tragic,' 'comic,' 'sentimental,' and so on? We constantly involve

---

28 Ed. S.H. Tan and Nico H. Frijda, 'Sentiment in Film Viewing', in *Passionate Views: Film, Cognition, and Emotion*, ed. by Carl Plantinga and Greg M. Smith (Baltimore: Johns Hopkins University Press 1999), pp.48-64 (pp.53-54). For general problems with the concept of empathy in the reception of literature see Mellmann, *Emotionalisierung*, pp.104-56, and Mellmann, "Objects of 'Empathy:" Characters (and Other Such Things) as Psycho-Poetic Effects', in *Characters in Fictional Worlds*, ed. by Jens Eder, Fotis Jannidis and Ralf Schneider (Berlin: de Gruyter, 2010), pp. 416-41.

29 Cf., for instance, Jackson, 'Questioning Interdisciplinarity', pp.329, 341, and Kelleter, 'Worried Reflections', pp.164-69.

30 Some 'Literary Darwinists' do not even seem to understand what this is. Joseph Carroll, for example, ('Rejoinder to the Responses', *Style*, 48 (2008), 308-70) accuses Karl Eibl and me of 'repeat[ing] the culturalist commonplace that the depiction of romantic love arose at a particular moment in the cultural history of Western Europe' (p.331), because we said that knowing about the biology of Romantic love and its occurrence in all cultures is important but not sufficient to answer the question why, 'on the ground of this general human experience, the sophisticated (predominantly literary) technique of *self articulation and world depiction* developed in the course of the eighteenth century' (Karl Eibl and Katja Mellmann, 'Misleading Alternatives', *Style*, 48 (2008), 166-71 (p.170, emphasis in the original)).

31 Eibl, 'On the Redskins', p.423; similarly, Jackson, 'Questioning Interdisciplinarity', p.423.

psychological assumptions in our discourse about literature, since many of our concepts (both in ordinary language and literary terminology) are implicitly based on intuitive commonsense knowledge about the effects of certain literary features. Although we often use these terms as if they were referring to the textual properties themselves, they are in fact referring to mental states in response to those properties.[32]

Replacing such intuitive knowledge with adaptationist reasoning seems to me to be another major benefit we can derive from an evolutionary heuristic. Explicit evolutionary-psychological hypotheses on the derivation and precise function of such basic correlations between text structure and perceived effect not only enhance our interpretative skills but also open up wider questions of poetics and aesthetics. Brian Boyd, for instance, tried to determine text structures that an evolutionary account would predict to be capable of provoking laughter, and so contributed to a theory of literary humour.[33] In a similar attempt, I described a particular strategy of fictionalisation as based on our adaptations for laughter and humour.[34] Similarly, Lisa Zunshine explored the role that our capacity for mind reading and formulating both adaptive categories (animate/inanimate) and concepts (utility, essence) might play in literary reading in two of her books.[35] And Karl Eibl, in *Animal poeta*, postulated the existence of an 'induction instinct' – an evolved cognitive tool for information gathering and experiential learning – that would account for several literary features such as the rhetorical devices of metaphor and comparison, various kinds of repetition (of motifs, sounds, metrical patterns, etc.), and narrative cohesion.[36]

---

32 See, for instance, the habit of handling suspense as a text property as criticised in Katja Mellmann, 'Vorschlag zu einer emotionspsychologischen Bestimmung von 'Spannung'', in *Im Rücken der Kulturen*, ed. by Karl Eibl, Katja Mellmann and Rüdiger Zymner (Paderborn: Mentis Verlag, 2007), pp.241-268 (pp.241-44).

33 Boyd, Brian, 'Laughter and Literature. A Play Theory of Humor', *Philosophy and Literature*, 28 (2004), 1-22.

34 Katja Mellmann, 'Das 'Spielgesicht' als poetisches Verfahren. Elemente einer verhaltensbasierten Fiktionalitätstheorie', in *Literatur als Spiel. Evolutionsbiologische, ästhetische und pädagogische Aspekte*, ed. by Thomas Anz and Heinrich Kaulen (Berlin: de Gruyter 2009), pp.65-86; also Mellmann, *Emotionalisierung*, pp.264-352, p.450-51.

35 Lisa Zunshine, *Why We Read Fiction. Theory of Mind and the Novel* (Columbus: Ohio State University Press, 2006); *Strange Concepts and the Stories They Make Possible. Cognition, Culture, Narrative* (Baltimore: Johns Hopkins University Press, 2008).

36 Karl Eibl, *Animal poeta. Bausteine der biologischen Kultur- und Literaturtheorie* (Paderborn: Mentis, 2004), pp.288-92; further elaborated in 'Eine Kuh ist eine Ziege. Zu den evolutionsbiologischen Wurzeln der Metaphorik', *Der Deutschunterricht*, 6 (2006), 44-52, 'The Induction Instinct: Evolution and Poetic Application of a Cognitive Tool', and Ralph Müller, 'Interaction in Metaphor', *Studies in the Literary Imagination*, 42.2 (2009), 43-60, 61-77.

Moreover, he suggested that a number of basic and very widespread patterns of plot structure (e.g., separation-and-reunion plots) are due to specific hard-wired mental *gestalt* schemata that once guided the behaviour of our hunter and gatherer ancestors in an adaptive manner.[37]

Investigations of that kind, establishing correlations between basic constituents of (literary) texts and adaptive algorithms, can then again be used for a deeper examination of how a given work of art is based on (exploits, perhaps violates, etc.) those evolved dispositions in a specific manner. In addition, however, they bear with great impact on our understanding of how literature works on a fundamental level. Relating cognitive adaptations like mind reading, induction, natural categories and schemata with textual features contributes to exploring the fundamental 'range of techniques that allow the simulation to run properly on the human mind,' and thus prepares the ground for literary theory.[38] As literary scholarship is by no means exclusively concerned with interpretation and literary history but also with explication of its object of study (the phenomenon 'literature'), it may well be asked to what extent an evolutionary heuristic can contribute to a general theory of literature. Does it provide new insights into what literature is, how it works, and where it comes from?

On the Origin(s) of Literature

To be sure, literature as we know it, as a vast and in itself heterogeneous component of cultural tradition in literate societies, cannot be called an 'adaptation' in the biological sense of the word. To do that one would at least have to recur to oral tradition and construct a typified surrogate conception of something especially characteristic of literature that could have already been present in ancestral times. In this spirit, there have been numerous attempts at describing the 'adaptive value' of literature or the arts in general. The problem with most of these suggestions, however, is that they are either not applicable to all forms of art/literature or not exclusive to art/literature. While it is reasonable to examine the evolutionary origins of *certain features* of art and literature, this should not be confused with determining the evolutionary origin of art or literature per se. The

37 Eibl, *Animal poeta*, pp.263-72; further elaborated in 'Epische Triaden. Über eine stammesgeschichtlich verwurzelte Gestalt des Erzählens', *Journal of Literary Theory*, 2 (2008), 197-208.

38 I owe this handy phrase to Keith Oatley and Mitra Gholamain, 'Emotions and Identification. Connections between Readers and Fiction', in *Emotions and the Arts*, ed. by Mette Hjort and Sue Laver (New York: Oxford University Press, 1997), pp.263-81 (p.273).

assumption that 'the arts fulfill a vital adaptive function' suggests a distinct psychological design that was selected as a solution to a specified adaptive problem.[39] To my knowledge, however, there have not yet been propositions suggesting how such a special cognitive design (responsible for all artistic or literary activities, and only for them) should look. And if such an assumption is to mean less than that, it is not an evolutionary explanation in the strict sense and thus not very informative; for everything we do is, in a sense, 'adaptive,' as we are members of a species that is still adapted well enough, on average, to people this planet.[40]

Narrative might be a more promising candidate for a potential cognitive adaptation in the literary field. Evidently, narrative is a human universal; also, there are indications that it develops as a distinct capacity in ontogeny; and telling stories and narrative thinking occur not only in written literature but throughout our everyday life. All these facts might hint at the possibility of a distinctly evolved cognitive feature. While most of the advocates of a biological adaptation 'narrative' content themselves with pointing to its ubiquity or the natural proclivity to a narrative organisation of information, Michelle Scalise Sugiyama was the first to attempt a clear-cut adaptationist account of narrative by determining a correlated selection pressure.[41]

---

39 Carroll, 'An Evolutionary Paradigm', p.127.

40 Obviously, Carroll does mean less, when he says, for example, that '[i]f various cultural activities are 'biologically adaptive', and if the arts are among those activities, then the arts are 'biologically adaptive'' (Carroll, 'Rejoinder', p.356), in order to refute Karl Eibl's and my demur that the arts are not a distinct biological adaptation. For rules of explanation see note 23 and Steven Pinker, *How the Mind Works* (London: Penguin, 1999), pp.525-26; cf. also Tooby and Cosmides, 'The Past', p.400, on 'two separate meanings for the word 'adaptive''.

41 This natural proclivity often is related to neurological evidence according to which the way our brain processes experiential information and produces the sense of a self resembles narrative organisation. However, this would rather be a reason not to consider narrative as a distinct adaptation. For if the way verbal narrative is 'formatted' (cf. John Tooby and Leda Cosmides, 'Does Beauty Build Adapted Minds? Towards an Evolutionary Theory of Aesthetics, Fiction, and the Arts', *SubStance*, 30 (2001), 6-27 (p.24)) can be traced back to neural working principles, there is no need to additionally explain it as a special design. And inasmuch as narrative thinking involves higher degrees of narrative order, it overlaps with other adaptations (such as induction, causality, etc.), so that narrative rather seems to be a complex cultural congregation of innate dispositions than a distinct biological adaptation itself. For a more general evolutionary view on narrative without the correlation of a distinct cognitive design see H. Porter Abbott, 'The Evolutionary Origins of the Storied Mind: Modeling the Prehistory of Narrative Consciousness and its Discontents', *Narrative*, 8 (2000), 247-56, and Katja Mellmann, 'Voice and Perception. An Evolutionary Approach to the Basic Functions of Narrative', in

According to her, narrative is an especially effective 'information storage and transmission system' that expedites inculturation and social learning in a species adapted to the 'cognitive niche'.[42] This is a valid suggestion for an explanation at the 'ultimate' level, i.e. for an answer to the question how the assumed special design could have come into being, and as such distinguishes itself from the countless suggestions exposing only a general 'adaptiveness' of narrative. Whether narrative really relies on a distinct cognitive program (or if it is rather a contingent universal resulting from a plurality of innate algorithms that are combined similarly over and over again in all cultures) is anything but clear at the moment; but Scalise Sugiyama's pioneering considerations demonstrate in an exemplary manner how such questions can be pursued in future.[43]

The search for specifically literary adaptations, however, is not the only way that evolutionary psychology can contribute to literary theory. As may already have become evident, literature involves a large number of adaptations from many areas of human behaviour (environmental experience, emotions, social cognition, and so on) which are not especially related to literature, yet indeed seem to constitute its very core.[44] John Tooby and Leda Cosmides, in their seminal essay 'Does Beauty Build Adapted Minds?,' have taken this as a starting point for explaining why at all we spend a significant amount of time engaged with fictions, which from a Darwinian standpoint at first seems to be a puzzling anomaly (p.7). In their view the pleasure we derive from reading may be due to the fact that certain textual features enable the execution of our adaptations in an 'organizational mode' and thus facilitate the ontogenetic development and adjustment of our phylogenetic adaptations.[45] They 'think that the task of organizing the brain both physically and informationally over the course of the lifespan is

---

Toward a Cognitive Theory of Narrative Acts, ed. by Frederick Aldama (Austin: University of Texas Press, 2010), pp. 119-140.

42 Scalise Sugiyama, Michelle, 'Reverse-Engineering Narrative: Evidence of Special Design', in *The Literary Animal*, pp.177-96 (p.190). The essay is only one in a series of articles of hers exploring the incorporation of adaptive information in hunter-and-gatherer folklore; see her own bibliography in Scalise Sugiyama and Sugiyama, 'A Frugal (Re)Past'.

43 For the most recent study on this topic (too recent to be included here) see Boyd, *On the Origin.*

44 Cf. also Pinker, *How the Mind Works*, pp.538-43.

45 Tooby and Cosmides, p.16. - The idea of an 'organizational mode' of performance was adopted in literary studies by Eibl, *Animal poeta*, pp.277-352; Karl Eibl, 'Adaptationen im Lustmodus. Ein übersehener Evolutionsfaktor', in *Anthropologie der Literatur*, pp.30-48; Mellmann, *Emotionalisierung*, pp.69-74; Karl Eibl, 'Zwei Kulturen? Zwei Denkweisen und ihre biologischen Ursprünge' in *Im Rücken der Kulturen*, pp.31-48; Mellmann, 'Das Spielgesicht'.

the most demanding adaptive problem posed by human development. Building the brain, and readying each of its adaptations to perform its function as well as possible is [...] a vastly underrated adaptive problem' (p.14) which, in their opinion, was resolved by evolving specialised pleasure circuits that function as 'motivational guidance systems' (p.16) to provoke such seemingly pointless behaviours as play and fictional engagement. In short, literary works are so appealing because they provide occasions for an intrinsically rewarded organisation of our neurocognitive apparatus (pp.13-18).[46]

This is not so far away from Steven Pinker's suggestion that the arts are kind of cheesecake for the mind in that they are simply 'pressing our pleasure buttons'.[47] But Tooby's and Cosmides' additional assumption of 'an

---

[46] For a similar consideration see Boyd's conception of art as a form of 'cognitive play' (*On the Origin*, pp.80-98, 381). Boyd's critique of Tooby's and Cosmides' account ('Evolutionary Theories of Art', in *The Literary Animal*, pp.147-76 [pp.169-70]) relies on several misconceptions: (1) His objection that according to their theory 'interest in art should taper off beyond childhood' (Boyd, p.169) disregards that they explicitly speak of '[b]uilding, readying, *and maintaining* the brain' (Tooby and Cosmides, p.14, my emphasis). (2) Further, he criticises that they 'concentrate strongly on representation, which is only one component of art', and that 'their hypothesis would not account for music', for instance (Boyd, p.169), while they repeatedly mention music among their examples (cf. Tooby and Cosmides, pp.17-18, 25). (3) A really absurd objection is that they 'ignore actual experience' (Boyd, p.170) by overemphasising learning processes mediated through fiction. Yet their question is not how learning works (with the wrong answer: exclusively by fiction) but why we like fiction (answer: because it facilitates pleasure-rewarded learning effects). Learning through fiction is thus by no means considered to compete with or displace other learning processes but is in fact reasonably put in conjunction with them (cf. Tooby and Cosmides, p.12). (4) Finally, Boyd thinks that their account for the 'proximate mechanism, our immediate motivation for fiction, is lost in fog', for 'untruth per se does not make us attend to stories' (Boyd, p.170). But this was not their argument. Fictions need to match *specified preference systems* (cf. Tooby and Cosmides, pp.13-14) – *that* is the crucial criterion according to their theory, rather than 'untruth' (which is only one of several possible conditions that allow for a riskless execution of immature programs). And the proximate motivation for our involvement with fiction is *pleasure*. For a critical discussion of Boyd's own suggestion see Pinker, 'Toward', pp.173-77.

[47] Pinker, *How the Mind Works*, p.525. It might seem odd to some that I see Tooby's and Cosmides' analysis as so similar to Pinker's, since theirs is often taken as representing the 'functional' and his the 'byproduct' position. However, Tooby and Cosmides themselves 'still consider the byproduct hypothesis to be the default hypothesis, with a great body of logic and evidence in favor of it', and Pinker's 'argument [...] a powerful one,' that 'successfully explains many features of the arts,' but also they think 'that there is much that it leaves unexplained' (p.11), namely those motivations for art and fiction that cannot be explained immediately by particular preferences

entire suite of developmental adaptations' 'that were selected to involve humans in aesthetic experiences and imagined worlds' makes an important amendment to the picture drawn by Pinker, as their hypothesis accounts for the outstanding significance of play, imagination, fiction, literature, and so on, within human culture.[48] Phenomena like these then appear to be more than mere by-products of human evolution: they are paradigmatic, albeit interchangeable, expressions of a fundamental characteristic of our species. That said, though, Tooby and Cosmides resist the temptation to propose a fundamental *impulse* to produce fiction (or 'the arts,' narrative, creativity, imagination, etc.) in the sense of a distinct cognitive program.[49] Rather, the 'additional layer of adaptations' supposed by them consists in a heterogeneous number of discrete adaptations all across human behaviour, including, for example, motoric activities like dancing as well as cognitive ones.[50]

In conclusion: even if evolutionary contributions to literary theory will presumably rarely achieve the form of straightforward explanations of literary phenomena, the evolutionary perspective may indeed, in time, greatly enhance our general understanding of literature by specifying its particular place within human behaviour as a whole.[51] The fact that evolution can account for literature only from multiple angles need not to be considered a deficiency but can also be seen as an advantage, because it draws attention to the multi-factorial complexity – that is, the very nature – of literature. Having once acknowledged that there is no such story as 'The Emergence of Literature in Human Evolution' but rather an entire range of evolutionary histories of various proto-forms of literature,[52] we can set out

---

pertaining to fully developed algorithms (but necessitate an indirect explanation by additional preference systems guiding the development and organisation of those algorithms). Conversely, Pinker only 'proposed that *many* of the arts may have no adaptive function at all' and even contemplates that, '[u]nlike other forms of art, fiction lends itself to at least a prima facie case that it is also an adaptation' (Pinker, 'Toward', p.171, my emphasis).

48 Tooby and Cosmides, 'Does Beauty', p.14, p.11.

49 I wonder why Boyd, *On the Origin*, maintains that '[a]rt in this broad sense [of cognitive play] is a human *adaptation*' (p.381, emphasis in the original) and that there are 'features common to all forms of art' (p.69), although he is actually well aware of the multiple origins of art behaviours and in fact sketches out in his book a multi-layered and rather complex evolutionary history of art, fiction and storytelling.

50 Tooby and Cosmides, 'Does Beauty', p.11.

51 Cf. also Boyd, *On the Origin*, pp.384-85.

52 Cf. Lorenz Welker, 'Kategorien musikalischen Verhaltens in evolutionärer Perspektive. Geschlechtsdifferenzen, Universalien und ein Schichtenmodell musikalischer Wahrnehmung', in *Im Rücken der Kulturen*, pp.271-90, who distinguishes several proto-forms of music which he takes to have emerged from different

to write those many stories. That, surely, is a worthwhile interdisciplinary research programme.[53]

behavioural domains but can all be said to be among the ancestors of what today we call 'music'.

53 Thanks to Karl Eibl and Nancy Easterlin for their comments on a draft version of this essay.

Contributors

JON ADAMS is a researcher, writer, and filmmaker based at the London School of Economics. He is interested in the epistemology and methodology of literary study. In 2007 he published a book on the history of attempts to make a science of literary criticism called *Interference Patterns.* He has also written on plot taxonomies, and on the popularisation and social diffusion of scientific information through fictions.

DAVID AMIGONI is Professor of Victorian Literature at Keele University, UK. He has published widely on Victorian English literature, and is co-editor, with Jeff Wallace, of *Charles Darwin's Origin of Species: New Interdisciplinary Essays* (1995), and author of *Colonies, Cults and Evolution: Literature, Science and Culture in Nineteenth-Century Writing* (2007).

DAVID BAGULEY is Emeritus Professor at the University of Durham and at the University of Western Ontario. He has published extensively on nineteenth-century French literature, notably on the Naturalist movement and the works of Émile Zola. His books include *Naturalist Fiction: The Entropic Vision* (1990, new edition 2005) and *Napoleon III and His Regime. An Extravaganza* (2000). His latest book is a critical edition of Edmond de Goncourt's novel *La Fille Élisa* (2010).

ANNA BARTON is a lecturer in English at Sheffield University, specialising in nineteenth-century poetry. Her first monograph, *Tennyson's Name: Identity and Responsibility in the Poetry of Alfred Lord Tennyson*, was published in 2008.

ALISTAIR BROWN is a postdoctoral teaching assistant in the Department of English at Durham University, and an Associate Lecturer at the Open University. He has published journal articles on digital literary studies, science fiction, and the two cultures debate. He is currently seeking a publisher for his PhD thesis on *Demonic Fictions: Cybernetics and Postmodernism*, whilst he is also planning a monograph on *Reading Games: Computer Games and the Limits of Literature.*

JOSEPH EMONDS has published four books on syntactic and morphological analysis: *Transformational Approach to English Syntax*, *Unified Theory of Syntactic Categories*, *Lexicon and Grammar: the English Syntacticon*, and *Discovering Syntax*. He is American but moved to England in 1992. He has also taught at universities in France, Holland, Japan, Austria, Spain and the Czech Republic.

JOHN HOLMES is a Senior Lecturer in English at the University of Reading. He is the author of *Dante Gabriel Rossetti and the Late Victorian Sonnet Sequence: Sexuality, Belief and the Self* (2005) and *Darwin's Bards: British and American Poetry in the Age of Evolution* (2009), and the co-editor of *Horae Amoris: The Collected Poems of Rosa Newmarch* (2010). He is a founder member of the British Society for Literature and Science.

SIMON J. JAMES is Senior Lecturer in Victorian Literature at Durham University. He is the author of *Unsettled Accounts: Money and Narrative in the Novels of George Gissing* (2003) and of articles on Gissing, Charles Dickens and H. G. Wells. He has edited *The Wellsian*, five Wells novels for Penguin Classics, and Gissing's *Charles Dickens: A Critical Study*, and is writing a study of Wells and high culture.

DAVID KNIGHT is Professor emeritus of History and Philosophy of Science in Durham University. His most recent books are *Science and Spirituality: the Volatile Connection* (2004), *Public Understanding of Science; a History of Communicating Scientific Ideas* (2006), and *The Making of Modern Science: Science, Technology, Medicine and Modernity, 1789-1914* (2009).

CHRISTOPHER LLOYD is Professor and Head of French at Durham University. He has published nine books on modern French culture, including *J.-K. Huysmans and the fin-de-siècle novel* (1990), *Collaboration and Resistance in Occupied France* (2003) and *Henri-Georges Clouzot* (2007). He is currently co-editing and co-authoring two books on French Occupation narratives with Professor Margaret Atack of Leeds University.

LOUISE LYLE is Lecturer in French and Comparative Studies at the University of London Institute in Paris. Her research focuses on the influence of the life sciences on French literature of the nineteenth and twentieth centuries, and she has recently published articles on social Darwinism, Lamarckism, Lysenkoism and eugenics. She is currently working on a monograph entitled *Social Darwinism in the French Ideological Novel, 1880-Present.*

JOHN A. McCARTHY is Professor of German and Comparative Literature and Professor of European Studies at Vanderbilt University. His teaching and research interests range from the eighteenth to the twentieth century and includes philosophy and literature, the nature and representation of evil, and science and literature. His publications include eleven books, including recently *Remapping Reality: On Chaos and Creativity in Science & Literature* (2006) and over ninety articles and book chapters.

KATJA MELLMANN is a Research Fellow in the German Department of the University of Göttingen. Her research interests include an evolution-inspired psychology of art reception, poetic metre, eighteenth- and nineteenth-century German literature, narratology and literary theory. She is the author of *Emotionalisierung* (2006) and co-editor of *Heuristiken der Literaturwissenschaft* (2006), *Im Rücken der Kulturen* (2007) and special issues of both the *Journal of Literary Theory* (2008) and *Studies in the Literary Imagination* (2009).

DAVID MIDGLEY is Reader in German Literature and Culture at Cambridge Unversity, and Fellow of St John's College, Cambridge. He is the author of *Arnold Zweig. Zu Werk und Wandlung 1927-1948* (1980) and *Writing Weimar. Critical Realism in German Literature, 1918-1933* (2000), and has published many essays on German modernist writers.

DOUGLAS MORREY is Senior Lecturer in French at the University of Warwick. He has published widely on modern and contemporary French film and literature. He is currently completing a book on Michel Houellebecq.

NICHOLAS SAUL is Professor of German at the University of Durham. He has published widely on literature and philosophy, literature and science, literature and homiletics. His latest books are *Gypsies and Orientalism in German Literature and Anthropology of the Long Nineteenth Century* (2007) and *The Cambridge Companion to German Romanticism* (2009). He is writing a monograph on German literary Darwinisms, 1859-2008.

PATRICIA WAUGH is Professor of English Literature at Durham University. She has published many essays and books on modern literature and intellectual history, literary and cultural theory, literary modernism, postmodernism and feminism. She is currently completing two books: a history of modern British fiction and its intellectual contexts and a book on modern literary studies and the biologisation of culture.

WENDY WHEELER is Professor of English Literature and Cultural Inquiry at London Metropolitan University where she supports research in semiotic and biosemiotic systems from a humanities, arts, and cultural studies perspective. She is the author of *A New Modernity: Change in Science, Literature and Politics* (1999); *The Political Subject: Essays on the Self from Art, Politics and Science* (2001); and *The Whole Creature: Complexity, Biosemiotics and the Evolution of Culture* (2006), and has published widely in journals and book collections.

Index

www.ingramcontent.com/pod-product-compliance
Lightning Source LLC
LaVergne TN
LVHW020651300926
849161LV00004B/893